S0-AVW-286

Chemistry Library

ACS SYMPOSIUM SERIES 722

Service Life Prediction of Organic Coatings

A Systems Approach

David R. Bauer, EDITOR
Ford Motor Company

Jonathan W. Martin, EDITOR
National Institute of Standards and Technology

American Chemical Society
Oxford University Press

Chemistry Library

SEP/AE
Chem

Library of Congress Cataloging-in-Publication Data

Service life prediction of organic coatings: a systems approach /David R. Bauer, editor, Jonathan W. Martin, editor.

pp.   cm. — (ACS symposium series, ISSN 0097-6166 ; 722)

"Developed from a symposium sponsored by the Division of Polymeric Materials: Science and Engineering, at a special meeting held in Breckenridge, Colorado, September 9-14, 1997."

Includes bibliographical references and index.

ISBN 0-8412-3597-X (alk. paper)

1. Plastic coatings--Congresses. 2. Service life (Engineering)-Congresses.

I. Bauer, David R. (David Robert), 1949-
II. Martin, Jonathan W. III. American Chemical Society. Division of Polymeric Materials: Science and Engineering. IV. Series.

TP175.S6S955 1999   98-32089
667'.9--dc21   CIP

The paper used in this publication meets the minimum requirements of American National Standard for Information Sciences—Permanence of Paper for Printer Library Materials, ANSI Z39.48-94 1984.

Copyright © 1999 American Chemical Society

Distributed by Oxford University Press

All Rights Reserved. Reprographic copying beyond that permitted by Sections 107 or 108 of the U.S. Copyright Act is allowed for internal use only, provided that a per-chapter fee of $20.00 plus $0.50 per page is paid to the Copyright Clearance Center, Inc., 222 Rosewood Drive, Danvers, MA 01923, USA. Republication or reproduction for sale of pages in this book is permitted only under license from ACS. Direct these and other permissions requests to ACS Copyright Office, Publications Division, 1155 16th Street, N.W., Washington, DC 20036.

The citation of trade names and/or names of manufacturers in this publication is not to be construed as an endorsement or as approval by ACS of the commercial products or services referenced herein; nor should the mere reference herein to any drawing, specification, chemical process, or other data be regarded as a license or as a conveyance of any right or permission to the holder, reader, or any other person or corporation, to manufacture, reproduce, use, or sell any patented invention or copyrighted work that may in any way be related thereto. Registered names, trademarks, etc., used in this publication, even without specific indication thereof, are not to be considered unprotected by law.

PRINTED IN THE UNITED STATES OF AMERICA

Advisory Board

ACS Symposium Series

Mary E. Castellion
ChemEdit Company

Arthur B. Ellis
University of Wisconsin at Madison

Jeffrey S. Gaffney
Argonne National Laboratory

Gunda I. Georg
University of Kansas

Lawrence P. Klemann
Nabisco Foods Group

Richard N. Loeppky
University of Missouri

Cynthia A. Maryanoff
R. W. Johnson Pharmaceutical
    Research Institute

Roger A. Minear
University of Illinois
    at Urbana–Champaign

Omkaram Nalamasu
AT&T Bell Laboratories

Kinam Park
Purdue University

Katherine R. Porter
Duke University

Douglas A. Smith
The DAS Group, Inc.

Martin R. Tant
Eastman Chemical Co.

Michael D. Taylor
Parke-Davis Pharmaceutical
    Research

Leroy B. Townsend
University of Michigan

William C. Walker
DuPont Company

TP1175
S6 S47
1999
CHEM

Foreword

THE ACS SYMPOSIUM SERIES was first published in 1974 to provide a mechanism for publishing symposia quickly in book form. The purpose of the series is to publish timely, comprehensive books developed from ACS sponsored symposia based on current scientific research. Occasionally, books are developed from symposia sponsored by other organizations when the topic is of keen interest to the chemistry audience.

Before agreeing to publish a book, the proposed table of contents is reviewed for appropriate and comprehensive coverage and for interest to the audience. Some papers may be excluded in order to better focus the book; others may be added to provide comprehensiveness. When appropriate, overview or introductory chapters are added. Drafts of chapters are peer-reviewed prior to final acceptance or rejection, and manuscripts are prepared in camera-ready format.

As a rule, only original research papers and original review papers are included in the volumes. Verbatim reproductions of previously published papers are not accepted.

ACS BOOKS DEPARTMENT

Contents

vi

Preface

What exposure should I use? How long do I need to expose my paint to meet specification?

These questions are commonly asked by coating suppliers who seek to sell coatings to industry and government. In turn, coating users or manufacturers who are responsible for setting specifications often ask the following question of people in the business of performing durability tests: How many hours of exposure in your box is equal to a year in Florida?

The answers to these questions determine how coatings are selected, tested, and how they are ultimately used. Coatings have been evaluated using a variety of outdoor and laboratory equipment for more than 80 years. Although specific testing chambers have varied, the basic process has remained virtually unchanged. Outdoor exposures are considered the most reliable. Laboratory exposures are evaluated in terms of their ability to "correlate" with outdoor exposure results and their ability to "fail" coatings rapidly. Because outdoor exposures often take excessive periods of time, decisions (both product development and material selection) are often made using these accelerated laboratory tests. Despite the maturity of this testing protocol, premature coating failures can be estimated to cost manufacturers, government, and consumers easily in excess of a billion dollars a year. In addition, development cycle times are still too long adding to coating costs, and fear of failure often prevents the implementation of new, improved coating systems.

Clearly, service life prediction is critical to the coatings industry. We believe that a fundamentally new approach is needed. The new approach is based on trying to answer a different question from those posed above: How do I learn enough about my coating and the environment in which it is placed to assess the risk of failure in that environment?

To answer this question it is necessary to take a systematic approach to characterizing the environment, determining how the exposure variables affect failure processes, and determining how material and processing variables interact. The new systems approach is multidisciplinary and draws on expertise not traditionally found either in coating manufacturers or in coating users. The purpose of the Symposium on which this book is based was to bring together experts in a wide variety of fields to discuss the state of the art in service life prediction. The symposium was held in Breckenridge, Colorado, in September 1997.

After first giving an overview of the current state of the art in testing, this book describes the key aspects of service life prediction that we feel are necessary for successful prediction. The first key aspect is the characterization of the environment. This includes everything from global measurements of UV radiation to characterization of the microclimates inside a particular building or joint of an automobile. It is also necessary to understand how to measure exposure conditions.

For example how do coating temperatures vary in outdoor and laboratory exposures? Characterization of conditions in accelerated, laboratory chambers must be compatible with characterization of outdoor conditions. It is important to characterize exposures in terms of dose or cumulative damage events rather than in terms of time.

Laboratory exposure testing is often expensive and we must maximize the information gained from all exposures. Another critical point is that a given coating system never fails at a single time. Rather failure has to be described in terms of a distribution. Reliability methodologies widely used in other industries provide robust ways to handle such information and to relate laboratory testing to field performance. Predicting performance ultimately involves understanding the fundamental mechanisms of failure. There have been many recent advances both in the chemical characterization of degradation and in the physical manifestation of those changes that ultimately leads to failure. Understanding failure mechanisms not only will lead to improved service life prediction but also to improved products.

After discussing these issues, the book concludes with chapters that address questions of data and data management. Every panel ever exposed, and every coating on every product ever sold is a potential source of information concerning reliability. Managing and interpreting that wealth of data is a daunting task. Of particular interest is the issue of "meta-data", that is the data that is necessary to describe the experiment. Without the correct meta-data, experimental results are often meaningless and cannot be compared to other experiments. Managing extensive service life databases will be a challenge for the success of the systematic approach that is advocated.

The Breckenridge Symposium concluded with a group discussion of what was learned and more importantly what still needs to be done. From an industry perspective, it is clear that we need to both improve the durability of our products and to shorten development times. In To support those goals, we need to develop further the tools that are described in this book. In particular we need to develop a better characterization of the environment that is easily accessible to coating scientists. We need to continue to develop the fundamental understanding of failure mechanisms. We need to agree on a common protocol for handling service life data. Finally, we need to develop concrete examples that demonstrate how this methodology can work to meet the needs of the coating industry. While we believe that this book represents the current state of the art in service life prediction in coatings, it should be clear that this is very much a work in progress. Further discussions will be necessary and a future meeting is in the planning stage. Those interested in the current status and detailed conclusions of the meeting are welcome to visit SLP web site at **http://ciks.cbt.nist.gov.80/slp/.**

Finally we thank the Breckenridge organizing committee and all the authors and participants for making the symposium a truly memorable event. We also thank the National Institute of Standards and Technology, the National Renewable Energy Laboratory, the Federal Highway Administration, Wright Patterson Air Force Base, and the Forest Products Laboratory for their financial support of the Symposium.

DAVID R. BAUER
Research Laboratory MD 3135
Ford Motor Company
P. O. Box 2053
Dearborn, MI 48121

JONATHAN W. MARTIN
National Institute of Standards and Technology
Quince Orchard Boulevard
Route 270, Building 226, Room B350
Gaithersburg, MD 20899

Chapter 1

A Systems Approach to the Service Life Prediction Problem for Coating Systems

Jonathan W. Martin

National Institute of Standards and Technology, Quince Orchard Boulevard, Route 270, Building 226, Room B350, Gaithersburg, MD 20899

ABSTRACT

The conventional and reliability-based service life methodologies for coating systems are compared with respect to their predictive abilities. It was concluded from this comparison that the scientific merit of several of the underlying premises of the conventional methodology are suspect. Specifically, the premise which draws into question the merit of the conventional methodology is the belief that the weather repeats itself over some time scale. Unlike the conventional methodology, the reliability-based methodology has a strong scientific basis and has had an outstanding record in predicting the service life of a wide variety of materials, components, and systems. The application of this methodology to coating systems, however, will require dramatic changes in the way that the industry views its service life prediction problem. Specifically, major changes will be required in 1) the missions and objectives assigned to the primary sources of service life data; 2) the characterization of the unaged coating system and coating constituents, 3) the characterization of the exposure environment and coating system degradation; and 4) the collection, analysis, storage, and retrieval of experimental data. Application and implementation of the reliability-based methodology to the coatings service life prediction problem is discussed.

INTRODUCTION

Over the last two decades, the organic coatings industry has undergone rapid technological and structural changes. These changes have been largely induced by federal and state legislative actions such as restrictions pertaining to hazardous chemicals, toxic effluents, waste disposal, and volatile organic compounds; and have led to increased competitive pressures to produce environmentally and user friendly coatings without sacrificing ease of application, initial appearance, or, most

U.S. government work. Published 1999 American Chemical Society

importantly, significantly reducing the expected **service life** (SL) of a coating system. Other consequences of this legislation include the gradual displacement of almost all commercially-important, well-established coatings (largely high-solvent coatings) by newer systems (water-borne, high solids, and powder coatings); the formulation and application of which are often based on different chemistries and technologies.

Unlike the displaced coatings, however, performance histories for these new coatings are neither available nor has there been time to generate them, since, at present, the generation of a reliable performance history for a new coating system requires an extensive in-service or outdoor exposure program, often taking between five and ten years to complete. Moreover, attempts at avoiding this task have had limited success and have, in a few cases, led to expensive litigation, loss in customer good will, and product substitution.

The coatings industry, therefore, is faced with the problem of generating service life data in a timely manner. This dilemma is not shared by all industries, however. For example, the electronics, medical, aeronautical, and nuclear industries make quantitative service life estimates for their products and have long since made the transition from an overwhelming dependence on long-term in-service tests to a heavy reliance on laboratory results. This transition has been accomplished through the implementation of a service life prediction methodology called **reliability theory and life testing analyses** (hereinafter, called the reliability-based methodology) or, equivalently in the medical industry, survival analysis. The feasibility of applying this methodology to coating systems has already been demonstrated by Tait [1993], Tait et al. [1993], Schutyser and Perera [1993], and Martin et al.[1985, 1989, 1990].

Implementation of a reliability-based methodology will require dramatic changes in the way the coatings industry views its service life prediction problem. Specifically, major changes will be required in 1) the missions and objectives assigned to the primary sources of service life data; 2) the characterization of the unaged coating system and coating constituents, 3) the characterization of the exposure environment and coating system degradation; and 4) the collection, analysis, storage, and retrieval of experimental data. The reward for these efforts will be a greatly reduced time-to-market for new coatings and better communication of service life results within the coatings industry and with coating consumers. This paper describes the reliability-based methodology and its implementation.

SOURCES OF SERVICE LIFE DATA

Regardless of the material, product or system, quantitative service life data are only available from three sources: 1) **accelerated laboratory,** 2) **outdoor,** and 3) **fundamental mechanistic experiments**. One common feature of all three data sources is that the generation of experimental data is expensive both in terms of time and money and the quantity of data generated from individual experiments is almost always small. Other features of greater practicality include the following:

1) In both accelerated laboratory and fundamental mechanistic studies, exposure variables can be monitored and controlled; whereas outdoors, exposure variables cannot be controlled, but can only be monitored.

2) Well-designed accelerated laboratory exposures provide an effective means for sorting through a large number of independent variables (material, environmental, processing, application, and design) affecting the service life of a coating system. This effort is necessary to identify influential and non-influential variables affecting the service life of a coated product.

3) Well-designed outdoor exposure experiments provide valuable information on the dominant failure mode and the expected failure times for a product exposed at a specific location. Such information is valuable in designing accelerated laboratory experiments.

4) Once the number of variables has been pared down, fundamental mechanistic studies provide a powerful means for isolating underlying failure mechanisms causing degradation.

In the following sections, the conventional and proposed reliability-based service life methodologies are compared with respect to their missions, objectives, and how the above stated issues are satisfied. It should be remembered, however, that in making this comparison, that the conventional service life methodology was implemented a long time before much thought was given to the service life prediction problem, service life prediction methodologies, and before sophisticated theory or tools were available for making service life estimates.

CONVENTIONAL SERVICE LIFE PREDICTION METHODOLOGY

In the United States, the paradigm for the conventional coatings service life methodology had its genesis in a meeting held under the auspices of ASTM Committee E (the forerunner of ASTM Committee D1) in 1902. The purpose of this meeting was to propose improved standards for assessing the durability of maintenance coatings [Pearce, 1954]. Outcomes from this meeting included 1) the designation of a bridge in 1905 in Havre de Grace, Maryland as the first test bridge for exposing new maintenance coatings; 2) the construction of several outdoor sites for exposing coated panels in Virginia, North Dakota, and Pennsylvania from 1905 to 1907 [Gardner, 1911]; 3) the establishment of a task group within ASTM D1 on accelerated laboratory experiments in 1910 [Pearce et al., 1954], and 4) the introduction of crude, by today's standards, weathering devices between 1915 and 1920 [Muckenfuss, 1913; Capp, 1914; Nelson, 1922]. Thus, by 1920, all of the ingredients for the conventional methodology were in place.

A schematic of this methodology is shown in Fig. 1. In it, accelerated laboratory experiments are designed to capture "the balance of exposure conditions" occurring outdoors; that is, they are designed to simulate outdoor environments. Once this balance has been captured, the accelerated laboratory experiment should consistently generate results which are highly correlated with those obtained from outdoor experiments.

The validity of the conventional methodology, therefore, depends on three implicit premises. They are as follows:

Premise 1:   The performance of nominally identical coated panels exposed in the same environment at the same time exhibits little or no variability;

Premise 2:   The results from outdoor exposure experiments are the de facto standard to which accelerated laboratory exposure results must duplicate (correlate); and

Premise 3:   The results from a successful accelerated laboratory experiment should correlate with exposure results generated anywhere outdoors.

As discussed in the following sections, none of these premises appears to have any scientific validity. The premise which questions the scientific merit and usefulness of the conventional methodology, however, is premise 2; that is, are outdoor exposure results a good standard for which to judge the adequacy of laboratory experiments [Reinhart, 1948].

RELIABILITY-BASED SERVICE LIFE PREDICTION METHODOLOGY

Unlike the conventional methodology, the reliability-based methodology has had a short, but highly successful history of predicting the service life of a wide variety of products [Nalos, 1965; Nelson, 1990]. This methodology differs from the conventional technology in that 1) reliability theory and life testing analyses were specifically designed to address the service life prediction problem; 2) the reliability-based methodology has a strong scientific and theoretical bases; 3) the methodology is constantly evolving from inputs from many disparate branches of science and technology; and 4) the output from this methodology is a quantitative estimate of the service life of a product exposed in its intended service environment.

A schematic of the reliability-based methodology is depicted in Fig. 2. The reliability-based methodology attempts to integrate the data generated from each of the primary sources of service life data into estimating a coating system service life. Thus, all three data sources are viewed as generating complementary and comparable data. Successful implementation of this methodology, therefore, requires that the data collected from each source has a scientific basis; are quantitative and comparable; and are of known precision and accuracy.

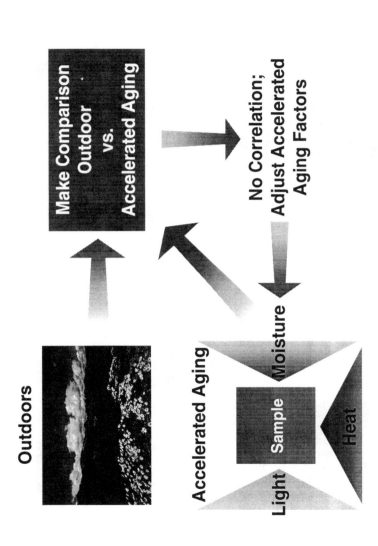

Figure 1—Schematic depicting the test strategy for the conventional service life prediction methodology.

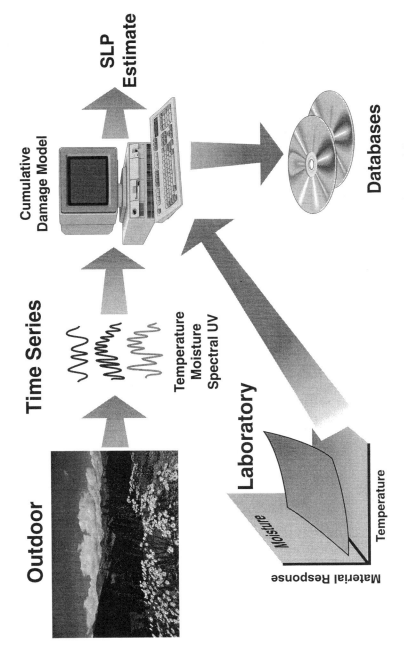

Figure 2—Schematic depicting the test strategy for the reliability-based service life prediction methodology.

The major differences between the conventional and reliability-based methodologies are in the missions assigned to accelerated laboratory and to outdoor exposure experiments. Specifically,

1. Outdoor exposure experiments are viewed as just another laboratory experiment; albeit one in which individual weathering variables cannot be controlled, but can be monitored.
2. Outdoor weathering variables must be monitored and characterized in the same manner that they are in laboratory experiments.
3. Accelerated laboratory experiments are statistically designed to systematically cover the range of each weathering variable to which the coated product is expected to be exposed in-service. **No attempt is made to design an accelerated laboratory experiment which simulate or captures "the balance of exposure conditions" occurring outdoors.**
4. The major difference between accelerated and fundamental mechanistic laboratory experiments are in the number of independent variables investigated. Accelerated laboratory experiments are designed to sort through the effects of a large number of variables; whereas fundamental mechanistic experiments are designed to thoroughly investigate the effects of a few variables.
5. Laboratory and outdoor exposure results are mathematically related through a cumulative damage mode. Cumulative damage models describe the irreversible accumulation of damage occurring throughout the life of a coating system exposed in its intended service environment.

Detailed descriptions of the reliability-based methodology and theory are presented in Nelson [1990] and in journals like Technometrics and the IEEE Transactions on Reliability. The application of these techniques to coatings has been reviewed by Martin et al. [1996].
In the next section, various aspects of the reliability-based service life prediction methodology are briefly discussed.

Reliability Theory and Life Testing Analysis

A coating system functions to protect and enhance the appearance of a coated object. Thus, it has failed whenever it no longer performs its intended function or, more specifically, whenever at least one of its critical performance properties has been exceeded; this is commonly called a **failure mode**. Examples of failure modes for loss of appearance or loss of protection include corrosion, cracking, chalking, and color change (see Fig. 3).

Each failure mode can be related to one or more **root faults**. Examples of root faults include the exposure environment, coating composition, material processing, application variables, and the design of the coated product. Under each

8

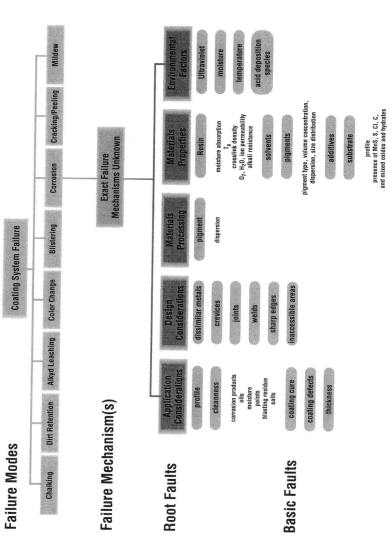

Figure 3—Fault tree for loss of protection due to corrosion where the underlying failure mechanisms are poorly understood. This lack of understanding is graphically portrayed by the black box between the observed failure mode and the root faults.

root fault are a number of **basic faults** (see Fig 3) which actually cause a coating system to fail. A major objective of the reliability-based methodology is to isolate the basic fault(s) initiating the failure. This can be accomplished through proper experimental design of the accelerated laboratory and fundamental mechanistic experiments.

At a higher level of investigation, the objective of the reliability-based methodology is to establish the connection between a failure mode and its root and basic faults. This is seldom an easy task since it requires the elucidation of the intermediate degradation steps (physical, chemical, or physical and chemical) causing a coating system to fail. If the degradation steps have are well-elucidated, then fundamental mechanistic experiments can be employed and the results from these experiments used in making estimates of the service life of a coating system on the chemical **degradation kinetics** of the study coating. Bauer et al. [1991, 1993a, 1993b] and Gerlock et al.[1985], for example, have made great strides in elucidating the photodegradation kinetics of several clear coatings used in automotive applications. Unfortunately, most commercially viable products are chemically too complex to isolate the underlying failure mechanisms. In these cases, the linkage between a failure mode and its root and basic faults is more tenuous and can only be empirically made through **cause-and-effect** or **dose-response relationships.** In the case of loss-of-protection, for example, the connection between the observed failure mode and basic faults can be best described by a gray box (see Fig. 3).

Laboratory experiments designed to isolate basic faults and to elucidate the linkage between a failure mode and its underlying faults are called **life tests**. In a life test, a number of performance properties of a coated panel are monitored over time. Associated with each performance characteristic is a user-defined maximum or minimum **critical value, h_{crit},** above or below which the coated panel is said to have failed (see fig. 4) [Tait, 1993a, 1993b; Martin et al., 1985, 1989, 1990; Gertsbakh et al., 1966]. The **time-to-failure**, t, of a coating system, therefore, is the time after a coating is applied at which a critical performance value is first exceeded.

When a number of nominally identical coated panels are exposed at the same time and in the same exposure environment, the times-to-failure for these panels almost always exhibit wide temporal variation [Tait, 1993; Tait et al., 1993; Martin et al., 1985, 1989; Schutyser et al., 1992, 1993; Crewdson, 1993]. (A violation of premise 1 of the conventional service life prediction methodology). In Fig. 5 for example, 24 nominally identical specimens were immersed in a 5% salt solution for 6000 h and the degradation state of each panel tracked [Martin et al., 1990]. The weakest or first panel exceeded the critical performance value after approximately 1000 h of immersion, while 6 of the 30 panels displayed no sign of degradation after 6000 h of immersion. Thus, the performance of these nominally identical specimens ranged from poor (times-to-failure less than 1000 h) to excellent (times-to-failure greater than 6000 h). A key decision in estimating the service life of a product is the fraction of failures of nominally identical coated panels before which an end user

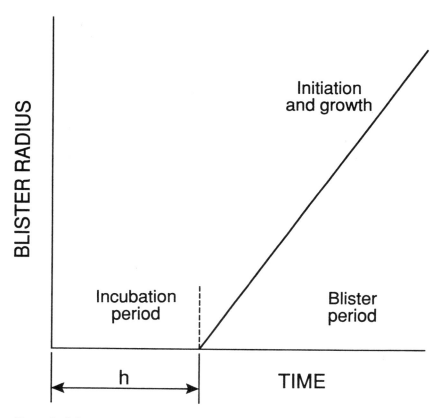

Figure 4—Schematic representation of changes in a critical coating system performance characteristic over time.

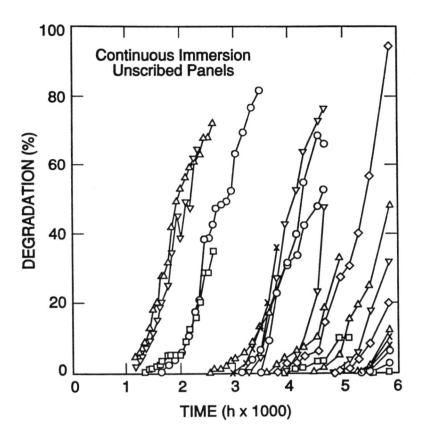

Figure 5—Percent area blisters versus immersion time for 24 nominally identical and simultaneously immersed coated steel panels containing no intentionally induced defects. The panels were continuously immersed in 5% NaCl solution. Six of the 24 panels displayed no degradation after 6000 h of immersion (taken from Martin et al. 1990).

deems that a coating system has failed. For most applications, this fraction or percentage will be much less than 1%. The mean or median time-to-failure seldom has any practical significance.

From an experimental viewpoint, it is fortunate that the times-to-failure of nominally identical panels are always ordered from the weakest to the strongest. This ordering allows one to predict the service life of a product without observing the failure times for all of the coated panels on test; that is, it is possible to estimate the times-to-failure for the specimens on exposure after observing the first few ordered times-to-failure. The coated panels whose failure times are not observed are said to be **censored.** Censoring also arises from other situations including damage to a panel during handling, loss in shipment, and removal of a panel for destructive analyze (see discussion in Nelson, 1990). In life testing, censoring is ubiquitous and is of great practical value (censoring can easily reduce exposure times by a decade or more). It is not be surprising, therefore, that estimating the service life of a product from censored samples receives a lot of attention in reliability analysis.

Finally, when a coating system is exposed in the field or laboratory, a number of performance characteristics begin to change simultaneously [Walker, 1974]. Each performance characteristic is in effect competing with the other performance characteristics in causing a coating system to fail (often termed **competing risks** [David et al., 1978]). The failure mode which "wins out" is the **dominant failure mode** for a given exposure environment over a specified period of time. The dominant failure mode often changes for nominally identical coated panels exposed at the same location over different exposure periods or at different locations [Rychtera, 1979; Degussa, 1985]. For example, the dominant failure mode for a coating system exposed in a semi-desert environment like Arizona is often associated with a loss of appearance due to the high spectral ultraviolet irradiance at this site; whereas, the dominant failure mode for the same coating system exposed in Florida may be associated with a loss of protection, which is attributable to the long time of wetness common to semi-tropical environments.

Characterization of Outdoor Exposure Environments

A major implicit assumption of the conventional methodology (premise 2) is that the weather repeats itself over some time interval. If this does not occur, then the use of outdoor weathering results as of standard of performance to which laboratory experiments must correlate is unjustified and the scientific merit and predictive abilities of the conventional methodology must be questioned.

Although a myriad of weathering variables may affect the service life of coatings systems, three variables (ultraviolet radiation, moisture, and temperature) are commonly viewed as being primarily responsible for the weathering of coating systems. It is the repeatability of these variables which will be discussed, although the discussion which follows also applies to all other weathering variables.

The non-repeatability of the weather and of weathering results is supported by three sources of information: 1) time series analyses for individual weathering variables, 2) testimonials from coating researchers, and 3) field exposure results.

The meteorological community has long since concluded that the weather and individual weathering variables do not repeat over any time interval. Recent reviews include those by Burroughs [1992] and the Climate Research Council [1995]. Based on trend analysis (determining temporal changes in the mean and variance values for a weathering factor) and spectrum analysis (determining if a weathering variables exhibits any cyclic behavior), these authors have concluded that there is no scientific evidence that the weather repeats itself over any time scale.

The proposition that the weather does not repeat itself over any time scale is also supported by testimonials from coating researchers. For example, it has long been recognized that dominant failure mode for nominally identical specimens exposed for the same duration and at the same time often changes from one environment to another [Scott, 1983] and that the rankings of outdoor exposure results do not agree for coated specimens exposed 1) at the same site and at the same time of year, but in different years [Grinsfelder, 1967] 2) at the same site, but at different times [Stieg, 1975; Ellinger, 1977; Lindberg, 1982; Stieg, 1966; Rosendahl, 1976; Grossman, 1993; Greathouse and Wessel, 1954; Morse, 1964; Singleton et al, 1965; Grinsfelder, 1967; Rosato, 1968; Mitton et al., 1971; Gaines et al., 1977; Scott, 1977] 3) at the same site, same year, and the same time of year, but for different durations [Reinhart, 1958], and 4) at different sites, but at the same time of the same year [Stieg, 1975; Kamal, 1966; Hoffman and Saracz, 1969; Morse, 1964; Singleton et al., 1965]. **In fact, no study was found claiming that outdoor exposure results are reproducible.** This is a violation of premises 2 and 3 of the conventional methodology.

Finally, more quantitative outdoor exposure studies indicating the lack of reproducibility include those published by Ashman, G.W. [1936], Wirshing, R.J. [1941], and Epple [1968]. These researchers have conducted experiments in which nominally identical coated panels were exposed at the same location and for the same duration, but the exposure experiments were started at different months of the same year. Exposure results differed by a factor of two or more.

Although individual weathering variables cannot be controlled in outdoor experiments, they can be monitored and, in order to relate laboratory and field exposure results, individual weathering variables must be monitored and characterized in the same manner that they are monitored and characterized in laboratory experiments. Efforts have been initiated to make such characterizations; examples include the following: spectral ultraviolet solar radiation [Thompson et al.,1997; Lechner and Martin,1993; Martin, 1993], panel temperature [Saunders et al.,1990], and moisture content [Burch and Martin, 1998].

Quantification of Coating System Degradation

Over the last two decades, significant advances have been made in quantifying both appearance and corrosion degradation. This is particularly true for laboratory measurements. Examples of advances in appearance measurements at the microscopic and molecular level include infrared spectroscopy [Bauer, 1993; van der Ven and Hofman, 1993], x-ray photoelectron spectroscopy [Wilson and Skerry, 1993], and electron spin resonance [Gerlock et al., 1985]. Improvements in macroscopic appearance measurements have largely revolved around the computerization of existing optical appearance measurements [Schläpfer, 1989]

Examples of advances in corrosion protection measurements at the microscopic level include chemical property measurements of coating system degradation using Fourier transform infrared spectroscopy [Nguyen et al., 1987, 1991], changes in the electrochemical properties using AC impedance spectroscopy [Tait et al., 1993; Kendig et al., 1987; Leidheiser, 1992], and changes in the internal mechanical stress properties in a coating system as it ages [Croll, 1979; Perera, 1990; Perera et al., 1987]. Improvement in macroscopic corrosion protection measurements include computer image processing of corrosion and blistered areas using visible or thermographic [McKnight et al., 1984, 1989; Bentz et al., 1987; Duncan et al., 1993; Pourdeyhimi et al., 1994].

Although significant advances have been made in quantifying the degradation of a coating system at the sub-macroscopic level, the most common method for characterizing loss of protection degradation is still via visual standards. Such characterization is known to be subjective and the continued use of visual standards is a major hindrance to the implementation of any quantitative service life prediction methodology. At the time that visual standards were introduced, they were a significant advance over even more qualitative characterization metrologies which they supplanted. The usefulness of visual standards, however, has long since passed and they should be replaced with more quantitative, cost-effective, accurate, and precise degradation measurements achievable through computer image processing.

Data Bases and Integrated Knowledge Systems

Probably the greatest change from the conventional methodology to the reliability-based methodology is the quality and quantity of data collected. The reliability-based methodology is very data intensive and the collected data are viewed as having great intrinsic economic and technical value.

The worldwide effort in establishing databases is extensive. The most advanced efforts are meteorological variables [World Climate Programme, 1986a, 1986b], chemicals [Buchanan et al., 1978; Langley et al., 1987], superconductors

[Munro et al., 1995], medicine [Wiederhold, 1981; Blum, 1982; Kissman et al., 1969], electronics [Munro and Chen, 1997], aerospace [Whittaker et al., 1969]. For construction materials, the most advanced databases are for metals, metal alloys [Westbrook, 1993] and metal corrosion [Rumble and Smith, 1990]. Efforts in creating standardized databases for polymers, coatings, and composites are still in their infancy [Moniz, 1993].

The general steps in creating a technical database have be described by Rumble and Smith [1990]. Efforts in establishing a database include 1) selection of the raw data to be collected, 2) evaluation of the collected data, 3) formation of an electronic database and in advanced applications, 4) creating an expert system, and 5) developing algorithms to query the data.

Selection of the raw data is most difficult and the most crucial part of the process. The difficulty lies in that users of the data have different views as to what data should be collected, how it should be collected, and how it should be reported. Careful and extensive efforts are required. Guidance for the selection of raw data are provided in ASTM E1484, Rumble and Smith [1990] and Moniz [1993].

Data evaluation is the process of ensuring the reliability and usefulness of the collected data. It is the process by which one enhances the confidence in a database. Extensive national and international efforts to standardize the data evaluation process are on-going. Excellent descriptions of the steps involved in data evaluation are described by Barrett [1993] and Munro and Chen [1997].

Finally, steps in establishing an electronic database are described by Rumble and Smith [1990]; while expert systems and data mining techniques are described in Piatetsky-Shapiro [1991].

SUMMARY

The conventional and reliability-based service life prediction methodologies are compared with respect to their ability to predict the service life of coated objected. It was concluded from this comparison that the scientific merit of several of the underlying premises of the conventional methodology were suspect; specifically, the premise that the weather repeats itself over some time scale.

The reliability-based methodology, on the other hand, has had an outstanding record in predicting the service life of numerous materials, components, and systems. Implementation of a reliability-based methodology, however, requires substantial changes in the way that coating service life prediction problem is viewed. The greatest changes will be in the missions assigned to accelerated laboratory and outdoor experiments; the mission of fundamental mechanistic studies will remain essentially unchanged. In a reliability-based methodology, outdoor experiments are viewed just like a laboratory-based experiment, albeit one in which individual

weathering variables cannot be controlled. Individual weathering variables can
characterized, however, in the same manner as they are characterized in the
laboratory. Such a characterization would greatly facilitate the comparison of
outdoor and laboratory results via cumulative damage models. Accelerated
laboratory experiments, on the other hand, are systematically designed to determine a
coating system's degradation response over the range of exposure conditions that the
coating system is expected to encounter in-service and to isolate influential and non-
influential variables affecting the service life of the coating system. This is
accomplished through appropriate experimental designs. Results from laboratory
and outdoor exposure experiments are stored in a computerized database for future
retrieval and analysis. This is made possible since all the collected data are
quantitative and are comparable from one data source to another. The power of
computerized databases is that it allows the researcher to query the database for
relationships which were not previously recognized without conducting the
experiment.

ACKNOWLEDGMENTS

Funding for this paper was provided in part by the Federal Highway Administration
(FHWA) and by a government/industry/university consortium on Coating Service
Life Prediction Methodologies at NIST.

LITERATURE CITED

Ashman, G.W. Journal of Industrial and Chemical Engineering **1936**, 28, 934.

Barrett, A.J. "Manual on the Building of Materials Databases", C.H. Newton [Ed.],
ASTM Manual Series MNL 19, American Society for Testing and Materials,
Philadelphia, PA, **1993**, 53.

Bauer, D.R. et al. Ford Research Center Technical Report No SR-91-101, **1991.**

Bauer, D.R. Proceedings of the American Chemical Society Division of Polymeric
Materials: Science and Engineering, **1993a**, 68, 62.

Bauer, D.R. Progress in Organic Coatings, **1993b**, 23, 105.

Bentz, D.P.; Martin, J.W. Journal of Protective Coatings and Linings **1987**, 4, 38.

Blum, R.L. "Lecture Notes in Medical Informatics", Springer-Verlag, New York,
1982.

Buchanan, B.; Mitchell, T. "Pattern-directed Inference Systems", D.A. Waterman
and F. Hayes-Roth [Eds.] Academic Press, NY, **1978**, 297.

Burch, D.M.; Martin, J.W. This publication, **1998**.

Burroughs, W.J. "Weather Cycles: Real or Imaginary", Cambridge University Press, New York, **1992**.

Capp, J.A. "Proceedings of the Annual Meeting of the American Society for Testing Materials", **1914**, 14, 474.

Climate Research Committee "Natural Climate Variability on Decade-to-Century Time Scales", National Academy Press, Washington, D.C., **1995**.

Crewdson, M.J. "Proceedings of the American Chemical Society Division of Polymeric Materials: Science and Engineering", **1993**, 69, 143.

Croll, S.G. Journal of Applied Polymer Science, **1979**, 23, 847.

David, H.A.; Moeschberger, M.L. "The Theory of Competing Risks", Griffin's Statistical Monographs and Courses No. 39, London, **1978**.

Degussa, Farbe und Lacke, **1985**, 91: 906 (reprinted as Degussa Technical Bulletin No. 22).

Duncan, D.J.; Whetton, A.R. Proceedings of the American Chemical Society, Division of Polymeric Materials: Science and Technology, **1993**, 68: 157.

Ellinger, M.L. Journal of Coatings Technology, **1977**, 49, 44.

Epple, R. Journal of the Oil and Colour Chemists' Association, **1968**, 51, 213.

Gaines, G.B. et al. Energy Research and Development Administration, **1977**, ERDA/JPL-954328-77/1.

Gardner, H.A. "Paint Technology and Tests", McGraw Hill, New York, **1911**.

Gerlock, J.L. et al. Journal of Coatings Technology, **1985**, 57, 37.

Gertsbakh, I.B.; Kordonskiy, Kh. B. "Models of Failure", Springer-Verlag, New York, **1966**.

Greathouse, G.A.; Wessel, C.J. "Deterioration of Materials: Causes and Prevention Techniques", Reinhold, New York, **1954**.

Grinsfelder, H. Applied Polymer Symposium, **1967,** 4, 245.

Grossman, D.M. "Accelerated and Outdoor Durability Testing of Organic Materials", Warren D. Ketola and Douglas Grossman [Eds.], American Society for Testing and Materials, Philadelphia, PA, ASTM STP 1202, **1993**.

Hoffmann, E.; Saracz, A Journal of the Oil and Colour Chemists' Association, **1969a**, 52:113.

Hoffmann, E.; Saracz, A. Journal of the Oil and Colour Chemists' Association, **1969b,** 52, 1130.

Kamal, M.R. Polymer Engineering and Science, **1966**, 6, 333.

Kendig, M.W. et al. Corrosion Protection by Organic Coatings, M.W. Kendig and H. Leidheiser [eds.], The Electrochemical Society, Pennington, NJ, **1987**, 253.

Kissman, H.M.; Wexler, P. Journal of Chemical Information and Computer Sciences, **1985**, 25, 212.

Langley, P. et al. "Scientific Discovery: An Account of the Creative Processes", MIT Press, Cambridge, MA, **1987**.

Lechner, J.A.; Martin, J.W. (1993) Proceedings of the American Chemical Society Division of Polymeric Materials: Science and Engineering, **1993**, 69, 230.
Leidheiser, H. Jr. Journal of Coatings Technology, **1992**, 63, 21.

Lindberg, B. Proceedings of the XVI FATIPEC Congress, meeting held in Liege, Belgium, May 14, 1982, **1982**, 1272.

Martin, J.W.; McKnight, M.E. Journal of Coatings Technology, **1985,** 57, 39.

Martin, J.W. et al. Journal of Coatings Technology, **1985**, 61, 39.

Martin, J.W. et al. , Embree, E., and Tsao, W. Journal of Coatings Technology, **1990**, 62, 25.

Martin, J.W. Progress in Organic Coatings, **1993**, 23: 49.

Martin, J.W. et al. "Methodologies for Predicting the Service Life of Coating Systems, Federation of Societies for Coatings Technology Monograph Series, Blue Bell, PA, **1996**.

McKnight, M.E.; Martin, J.W. "New Concepts for Coating Protection of Steel Structures", ASTM STP 841, D.M. Berger and R.F. Wint [Eds.], American Society for Testing and Materials, **1984**, 13.

McKnight, M.E.; Martin, J.W. (1989) Journal of Coatings Technology, **1989**, 61, 57.

Mitton, P.B.; Richards, D.P Journal of Paint Technology, **1971**, 43, 107.

Moniz, B. "Manual on the Building of Materials Databases", C.H. Newton [Ed.], ASTM Manual Series MNL 19, American Society for Testing and Materials, Philadelphia, PA, **1993**, 34.

Morse, M.P. Official Digest, **1964**, 36, 695.

Muckenfuss, A.M, The Journal of Industrial and Engineering Chemistry, **1913**, 5, 535.

Munro, R.G.; Begley, E.F. "NIST Standard Reference Database No. 62: High Temperature Superconductors", Standard Reference Data Program, National Institute of Standards and Technology, **1995**.

Munro, R.G.; Chen, H, "Computerization and Networking of Materials Databases: Fifth Volume", ASTM STP 1311, S. Nishijima and S. Iwata [Eds.], American Society for Testing and Materials, **1997,** 198.

Nalos, E.J.; Schultz, R.B. IEEE Transactions on Reliability, **1965**, R-14, 120.

Nelson, H.A. Proceedings of ASTM, **1922**, 22, 485.
Nelson, W. "Accelerated Testing: Statistical Models, Test Plans, and Data Analysis", Wiley, New York, **1990.**

Nguyen, T.; Byrd, E., Journal of Coatings Technology, **1987**, 59, 39.

Nguyen, T. et al., Journal of Adhesion Science, **1991**, 5, 697.

Pearce, W.T. et al. American Society for Testing Materials, ASTM STP 147, **1954**, 1.

Perera, D.Y.; Van den Eynde, D. Journal of Coatings Technology, **1981**, 53, 41.

Perera, D.Y.; Van den Eynde, D. Journal of Coatings Technology, **1987**, 59, 55.

Perera, D.Y. Proceedings of the 16th International Conference on Organic Coating Science, Athens, Greece, **1990**.

Piatetsky-Shapiro, G. and Frawley, W.J. Knowledge Discovery and Databases, AAAI Press, Menlo Park, CA, **1991**.

Pittock, A.B. Review of Geophysical Space Physics, **1978,** 16, 400.

Pourdeyhimi, B.; Nayernouri, A. Journal of Coatings Technology, **1994,** 66, 51.

Reinhart, F.W. SPE News, September, **1948,** 4:3.

Reinhart, F.W American Society for Testing Materials, ASTM STP 236, **1958,** 57.

Rosato, D.V. "Environmental Effects on Polymeric Materials", Vol. I, D.V. Rosato and R.T. Schwartz [eds.], Interscience, New York, **1968.**

Rosendahl, F. Proceedings of the XIII FATIPEC Congress, Juan les Pins, **1976,** 563.

Rumble, J.R. and Smith, F.J. "Database Systems in Science and Engineering", Adam Hilger, New York, **1990.**

Rychtera, M. "Deterioration of Electrical Equipment in Adverse Environments", Daniel Davey & Company, Inc., Hartford, CT, **1970.**

Saunders, S.C. et al. National Institute of Standards and Technology Technical Publication, NIST-TN 1275, **1990.**

Schläpfer, K. European Coatings Journal, May, **1989,** 388.

Scott, J.L. Journal of the Oil Colour Chemists' Association, **1983,** 66, 129.

Schutyser, P.; Perera, D.Y. Proceeding of the FATIPEC Congress, Amsterdam, **1992.**

Schutyser, P.; Perera, D.Y. Proceedings of the 1993 ACS Polymeric Materials Science and Engineering Symposium on the Durability of Coatings, Denver, Colorado, **1993,** 68, 141.

Scott J.L. Journal of Coatings Technology, **1977,** 49, 27.

Singleton, R.W. et al. Textile Research Journal, **1965,** 35, 228.

Stieg, F.B. Journal of Paint Technology, **1966,** 38, 29.

Stieg, F.B. Journal of Paint Technology, **1975,** 47, 54.

Tait, W.S. Proceedings of the 1993 ACS Polymeric Materials Science and Engineering Symposium on the Durability of Coatings, **1993,** 68, 101.

Chapter 2

Risk Management: The Real Reason for Long Product Development Time Cycles

F. Louis Floyd

Duron, Inc., 10414 Tucker Street, Beltsville, MD 20705

Abstract

We all know that developing a new product is difficult, time-consuming, and risky (low probability of success). The total manpower cost is directly related to the sum of the technical and the marketing degrees-of-difficulty. While it is certainly true that technically-difficult assignments can absorb unexpected time in finding elusive solutions to persistent technical problems, the majority of time expended during a typical development program is usually spent on reducing the total risk to the corporation represented by the new product. By carefully analyzing the risk management portion of the product development process, alternate routes can be found which take less time to implement, and deliver more useful product performance-in-use information.

Introduction

I remember the first business course I took as a lad. In it, our professor surprised us by pointing out that the prime directive for any company is not to make a profit, or to make a given product, or to improve our community, or to provide jobs. The real prime directive is "to survive!" After all, if one doesn't survive, none of the rest of the dialog matters. So it is not at all surprising that conservatism and risk aversion rein in corporations world-wide. It is also not surprising that individuals adopt similar conservatism, since few want to risk their own jobs.

The most recent form of this view, in somewhat loftier form, can be found in the work of Arie de Geus in his book "The Living Company."[1] De Geus's work is quite interesting in that he focuses on companies that have survived over very long times (200+ years), against a backdrop of companies that rarely survive beyond 50 years after founding. Surviving companies are viewed by de Geus as living organisms, which are capable of learning and growing. But more to the point of the present discussion, these companies appear to have a better view of risk management than the rest of us.

The field of reliability theory has arisen to deal with the issue of product liability within the broader topic of risk management. The aerospace, electronics, nuclear, and medical industries have all utilized reliability theory to substantially reduce the time needed to develop new products, while simultaneously reducing failures and product costs.[2] The thesis of this paper is that the coatings industry could achieve similar advances by adopting the tenants of reliability theory.

The Dimensions of Risk

Risk Perception

Studies[3] have shown that a person's perception of risk can differ significantly from an analytical assessment of risk. *The level of perceived risk seems to be inversely related to the level of involvement one has in the decision process for taking the risk, and directly related to the degree of surprise*. Thus, people who have a decision imposed on them view the risk associated with that decision as being higher than people who participated in the decision making process. And people who are surprised to learn that they are at risk see that risk as being even higher. In addition, *risks with a catastrophic potential are seen as more severe than those with much greater actual damage spread over a longer period of time*. To complicate matters further, these opinions appear to be impervious to any scientific evidence which is contrary to their personal belief. For example, nuclear power is seen as far more risky by the general public than cigarette smoking, even though substantial evidence exists to the contrary.

Considerable corporate energy goes into discovering the risks inherent in a given decision or activity and developing contingency plans to deal with the problems when they arise. Unknown risks deny this opportunity, so when such a failure occurs, the attending staff are surprised and have no prepared responses available to them. People responding to surprise situations commonly make mistakes and often compound the problem with even more mistakes in their efforts to correct the problem. This can escalate a small problem into a serious situation and challenge a corporation. Therefore, *an unknown (undefined) risk is perceived to be far worse than a known risk.*

Conservative Response to Risk

Business risk deals with the probability that a given decision (or product) will result in financial harm to the company. [Moral and ethical issues are not considered here.] In both frequency of occurrence and size of actual impact, the most common form of risk is that a business decision (strategy) will fail to produce the desired profitability. Product liability risks are less of a problem for corporations, but they consume considerable attention due to their potentially catastrophic nature.

A good example legitimizing the concern for risk is the recent problem which auto manufacturers were having with base-coat / clear-coat technology being used for automotive coatings. Tens of millions of cars and trucks manufactured in the U.S. between 1985 and 1990 were finished using this new technology, which was subsequently found to experience peeling failures. The price tag for correcting the problem could easily exceed a billion dollars. The problem: an unexpected mode of failure with the new system.

While entrepreneurial companies tend to take big risks (and have commensurate failure rates), major corporations are distinctly risk-averse. In particular, major corporations abhor the catastrophic potential of unknown (undefined) risk. Managers in such organizations learn early in their professional lives that putting their company at unknown risk will jeopardize their careers.

Separate from specific financial claims, there is also the risk that product failures will erode a corporation's credibility to the point that customers prefer to deal with their competitors. Whole markets can be lost without necessarily paying out a lot of claims. Which is all the more reason for conservatism.

Over the years, corporations have evolved a wide variety of ways to minimize the inherent risks of doing business. It starts with taking only moderate risks, and follows up with a whole plethora of tests, checks, and balances designed to insure that no major risk goes forward unrecognized, unchallenged or unquantified. Decisions are frequently delayed by efforts to develop more information to reduce risk.

Actual Sources of Product Failures

Reliability theory has shown, by collecting and plotting large amounts of failure data collected over many years, that most materials, components, and systems fail by one of three failure modes, depending on the age of the product (see figure 1):[4]

Failure Mode A. Early in a material's life, the failure frequency is high, but declines rapidly with age. These kinds of failures are frequently referred to as infant mortality (biological), or burn-in (electronics) failures. These early failures can be caused by design flaws or manufacturing errors, but are more frequently associated with flaws or defects in the way a material is installed or applied. For example, failure to properly prepare the substrate prior to painting is by far the most common cause of exterior architectural paint failures. *Most product liability claims occur during this interval.*[5]

Failure Mode B. After a brief time, the hazard rate drops to a low level, and remains there for a long period of time. Failures during this interval are usually associated with accidental damage, rather than material deficiencies. Coatings examples include storm damage, construction accidents, and normal household wear-and-tear. [This is why touch-up is an important property in architectural coatings.] *Very few product claims originate in this region.*

Failure Mode C. All materials have the common characteristic that they *fail due to accumulated damage.* This accumulated damage is the summation of small incremental damage events, each one of which is insufficiently small to cause failure, and usually not readily detectable. After long times in service, sufficient damage is accumulated that materials start to fail by a wearing-out process. This kind of failure can be truly described as characteristic of the life expectancy of the material. This manifests itself as a steady increase in hazard rate. However, these failures should cause no concern, provided the time scale is consistent with customer expectations.

A universal model, containing these three failure modes, takes on the form of a bathtub (figure 1), when hazard rate is plotted vs. time. Hazard rate is defined as the conditional probability of failure in the next time interval, given survival up till time *t.*

The hazard rate plot illustrates three definitions relating to service life issues:

1. The *risk* of a given situation is directly related to the quantity and nature of early failures occurring during period A. Reflecting this point, a majority of product failure claims occur during period A.

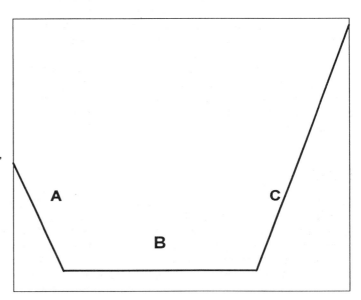

Hazard Rate

(Conditional probability of failure at time t, given survival up to time t)

A

B

C

Units of Time

Caption:

Region A: early failures can be due to design deficiencies or manufacturing errors, but are more frequently related to improper installation, deficient training or capabilities of operators, or using for inappropriate purposes. This can be seen in some cases as a lack of *robustness* on the part of the product.

Region B: failures are random, and caused by "freak" events such as storms or accidents.

Region C: failures represent the normal "wearing-out" process; indicative of end-of-life. These can be termed as related to the *capability* of the product.

<u>Figure 1</u>. Bathtub-Shaped Curve of Hazard Rate vs. Total Lifetime of Population

2. A product is considered to be *robust* if it does not exhibit any early failures. The rationale is that a product should be designed with sufficient robustness to accommodate the varying installation conditions which might be reasonably encountered. And if the customer is a professional painter, that expectation of robustness is absolute, even if not reasonable.

3. The *normal lifetime* for the product, absent improper installation or usage, or freak accidents, can be defined as the end of period B. This is commonly referred to as the *capability* of a given product in service in a given environment. One implication: the product should be taken out of service (or renewed) before the end of period B.

Managing Risk

The present author believes that the magnitude of effort which goes on in a company to manage and control risk is one of the larger consumer of resources of any activity undertaken in a company. Virtually every decision is influenced by perceptions of risk, and is frequently delayed by efforts to develop more information to reduce the uncertainty associated with that risk. If true, then how we manage risk is the best place to look for economies, if one desires to significantly reduce the costs and time inherent in the product development cycle.

The **role of risk manager** (minimizer) can be difficult to fix with regards to business decisions and credibility. For product issues, this role naturally falls on the shoulders of the R&D function, not only because they usually originate the product, but also because they are usually the only function in an organization with the resources and skills necessary to carry out detailed testing and quantification of the elements of risk (a long-time-frame process). Unfortunately, R&D personnel are often led to believe that this risk-reduction part of their jobs is an intrusion on their "true" mission of designing new products. This misconception can come from both general and R&D management, if they have a limited understanding of the overall risk management process.

For R&D, the components in risk reduction involve: verifying that the design goals are appropriate (i.e., that commercial success can reasonably be expected to follow technical success); testing to prove that the design goals have been met; and testing for problems which might be encountered after product introduction. The clear message here is that *it is far more important to be effective than efficient*, since all the efficiency in the world cannot correct for a faulty goal.

Example of
Current Product Development Protocol

Prior to engaging in product development, R&D must first have an on-going program of formulating and testing alternatives from the wide variety of possible raw materials and technologies available. This development of options is a key step in providing a competitively healthy future for the corporation, but predates the start of the actual development process. It also operates on a long time scale of years-to-decades.

As an example of the our current product development process, let us consider the development of a new architectural coating intended for exterior use on commercial and residential construction. Our typical approach involves a number of sequential steps, which can cycle back in an iterative fashion until success is achieved:

1. Marketing specifies a target cost and performance balance for a new product. This is usually a fast response to a specific market-place opportunity. [time scale: hours to days]
2. R&D develops candidate coatings selected from the options developed in step 1 plus the totality of prior formulating experience in the corporation. Only enough testing (*ca.* 10 to 15 properties) is conducted to determine if key properties are obtained within the specified cost constraint. These results are compared to the performance of relevant competitors in the marketplace, to check the validity of the stated goal. This stage is frequently referred to as demonstrating feasibility. [time scale: weeks]
3. Extensive formulation studies follow, to guarantee that over 200 properties are simultaneously achieved, within the specified cost parameters. This usually involves numerous iterations, not only technically, but in revising the actual goal and cost constraints. [time scale: months to quarters]
4. Extensive panel exposure studies are initiated on test fences, involving different orientations, locations, and substrates, to determine if the candidate is capable of withstanding the natural weathering process. It is common to encounter unexpected results during this process, usually in proportion to the degree of departure from known systems. Accelerated testing may be conducted simultaneously, to build a database of comparative information, but it is rare to make decisions based on accelerated results alone, due to lack of confidence in those results. [time scale: years]

5. Formulation studies continue to refine the application (liquid state) and appearance properties of the candidate, and to correct any solid state property deficiencies uncovered in step 5. [time scale: months to quarters]

6. After the completion of a year or two of fence testing, and a lot of lab testing, test houses are painted to check real-world application and performance of candidate products. This inevitably uncovers some deficiencies in robustness, which requires further formulation revisions, followed by further testing. [time scale: years]

7. At the conclusion of house testing, the product is scaled up, production batches are made, and the product is test marketed in a narrow geographical region (field testing). This allows one to move forward in the commercialization process with controlled risk. It is here that most remaining problems of insufficient robustness to varying application and lifetime conditions are uncovered. [time scale: quarters to years]

8. Once the test marketing is successfully concluded, the product is introduced across the entire market with full fanfare. [time scale: months]

The total cycle time for this process is on the order of 5 to 7 years, and can be depicted graphically in figure 3. Here we are plotting the relative amount of risk reduction achieved as a result of the information generated (Y-axis) vs. calendar time (x-axis). This is plotted as a continuous function, so that each stage adds on to the results of the preceding stage. Each stage has a spurt of learning, followed by a more gradual increase. This tendency to plateau after an initial spurt is frequently referred to as reaching diminishing returns. The judgments which go into the decision to move onto the next stage are governed by the preceding discussion on risk.

Caveat: It should be noted that while all this *testing does lower risk*, it can do so *only within the context of current knowledge and experience*. Unknown modes of failure with new technologies can remain undetected throughout this entire process. Therefore, the greatest risk will always lie with new technologies. And competition and regulations inexorably push our industry into new technologies.

WHERE CAN WE SAVE TIME ?

There are many issues inherent in our current testing protocol which contribute to long time cycles. From figure 2, one can surmise that if the various plateau regions can be shortened, considerable time could be saved without significant sacrifice in learning or risk. There are also some hidden opportunities for time savings, as the following discussion will highlight.

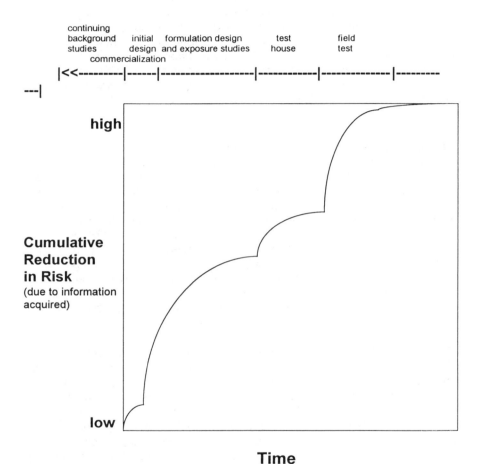

Figure 2. Cumulative Value of Information
Generated at Various Stages of Product Development

Maintaining a clear goal. The problem of changing goals is endemic in R&D. We should certainly put considerable effort into defining the goal of the program, and vigorously resist mission-creep. However, we should also be realistic by clearly acknowledging our limited ability to accurately forecast precise needs. Thus, there is a discovery process which occurs during product development where feasibilities are constantly compared to costs, competitive activity, and customer needs to iteratively arrive at both the definition and the accomplishment of the goal at about the same time. By subjecting our on-going activities to frequent reality checks with our commercial colleagues, we can avoid the considerable time wastage which occurs when R&D develops a product which is inappropriate for the market need.

Quantification of failure. In the coatings industry, failure is typically a defined phenomenon, rather than an abrupt loss of performance. The definitions tend to be ambiguous, and the measurements of degradation tend to be qualitative rather than quantitative. The main reason for low precision is that end-points are frequently judgment calls, based on some qualitative illustration of failure (e.g. photographic standards). Thus we live with the dreaded ASTM type scale of 1 (complete failure) to 10 (no failure), which is qualitative, situation-specific, and very open to interpretation. To compensate for this, experimenters typically repeat their work a number of times to demonstrate "reproducibility." As an alternative, Martin and McKnight[6] have shown that by using image analysis techniques, one can substantially reduce the error in the evaluation process, and with that reduce the time needed to arrive at a reliable conclusion.

Replication deficit. No matter how much is written on the subject, the coatings industry still persists in employing inadequate replication of results. Inadequate replication leads to low precision, which leads to additional formulation iterations, which continues the spiral. Worse yet, what little replication is done is focused on defining average lifetime, instead of quantifying early failures, where the risks truly lie. Given enough time, the formulator can overwhelm the uncertainty with enough data, to convince himself and others of the merit of moving forward. Proper replication up front will always save time by increasing the credibility of the observed result.

Experimental design and data treatment. Because failure data are not normally-distributed, conventional experimental designs and data analysis techniques are inappropriate.[7] Maximum-likelihood techniques deal with skewed failure data much better, and should be employed for more rigorous data treatment.

We also need to shift our attention away from the differential distribution (bell shaped curve) to cumulative distributions (S shaped curve), in order to bring into focus the issue of early failures as opposed to expected lifetime. We have been taught to look at experimental data in the form of differential distributions ("bell" shaped curve), which focuses attention on the overall average performance, and de-emphasizes the extremes as "outliers." Unfortunately, this attitude causes us to overlook the major source of risk with new products: early failures. Thus, we spend a lot of time, and still fail to quantify the risk we set out to. A switch in mindset to cumulative distributions would materially improve this process, but only with proper replication, as mentioned above.

Role of field testing. All testing for robustness up to the point where it is put into the hands of the customer for regular use is inherently biased! The simple reason for this is that up to that point, the people conducting the tests are controlling variables, and hence risks. It is only when real people use our products in the real world that we find out about robustness issues for certain. We would all be better served by proceeding rapidly to this stage, bypassing the fence test stage.

Role of exposure (panel, fence) testing. To this author, exterior fence testing has a role, but not in product development. Fence testing is what suppliers should be doing to establish the capability of their products under controlled conditions. It should also be conducted by paint companies to establish a record of their own capabilities against which failure complaints can be judged. But fence testing can say very little about robustness-in-use. It is simply too controlled a process, and takes too long to produce results. Properly conducted lab testing is actually better suited to testing for robustness than exterior exposure testing.

New activity: life testing analysis. LTA refers to testing to determine what agents in nature cause a given mode of degradation in a paint film, and then developing dose-response information. Done properly, this can save more testing time than all other items covered in this paper combined. Up till now, this has not been done in any consistent or comprehensive way in the coatings industry. This is ultimately the key to our doing things differently, and therefore having a meaningful positive impact on our product development time cycle.

For example, experiments have already shown that paints degrade in appearance properties due to photochemical degradation,[8] and in protective properties by a fatigue process.[9] While we know that UV light causes photodegradation of polymers (binders), with resultant loss in gloss, or chalking, we don't know the exact wavelength-dependence of this process, nor do we know anything about synergistic effects of other ingredients in the paint. If we did, we could develop an "action spectrum" which quantifies the effective dosage vs. wavelength for a given paint system. And with an action spectrum, we could construct and run more meaningful lab tests.

As another example, it has been shown that at least some cracking behavior is related to cyclic environmental conditions (hot/cold, freeze/thaw, wet/dry), and that this may be aggravated by chemical changes occurring during these physical cycles. If we could quantify this, and learn about the deficiencies in film formation or cure which contribute to premature failure via this mode, considerable progress could be made in achieving better protective qualities from paint.

Role of lab tests. Only after life testing analysis is done, can meaningful lab tests be constructed and run. Once we learn from LTA what we should be testing for and how, it is then a relatively simple matter to construct and run lab tests which will give meaningful information about the capability of our particular product.

For example, one significant problem which we have today is in the area of UV testing. Carbon and xenon arc weathering devices use light sources which contain far more IR than UV radiation, with the result that panels are baking (annealing) during testing, which is not characteristic of the natural environment. As a result, it is virtually impossible to study cracking tendencies of paints in such devices. Other cold light sources of UV contain non-natural bands of UV or distort the natural distribution among bands, relative to nature. This introduces both false positives and false negatives as a result.

In order to make this process work, we will have to have access to cold UV light sources which can either be either tuned, or filtered, in narrow bands in a highly reproducible manner. Such a device could then be used to study the dose-response behavior of a candidate paint, knowing what wavelengths are important, without interfering with other durability issues, such as crack resistance.[10]

Conclusion

A large amount of the time spent during product development is actually devoted to issues of risk management. As a result, product development cycle times will not likely be reduced by working harder and faster. They will only be reduced by taking a different approach to the way risk is managed in a company. Reliability theory provides alternate protocols which can be used in R&D to both shorten the product development cycle, and to simultaneously reduce the total risk implicit in the process. If the various functions of a company can learn to collaborate on risk issues, and share the resulting risk of moving forward, product development cycle times can be reduced even further.

Acknowledgments

I wish acknowledge the contributions of Prof. Sam Saunders of the Department of Applied Mathematics at Washington State University, and Dr. Jonathan W. Martin of the National Institute of Standards and Technology. They taught me about reliability theory and how it can be used to enhance the product development process. I also turn to them frequently for reality checks as I translate my learning into management language and practice.

REFERENCES

[1] de Geus, Arie, The Living Company, Harvard Bsns School Press, Boston, MA, 1997.

[2] Shooman, M.L., Probabilistic Reliability: An Engineering Approach, McGraw-Hill, New York, 1968.

[3] Slovic, P., Fischhoff, B., and Lichtenstein, S., "Rating the Risks," in Readings in Risk, T. S. Glickman and M. Gough, editors, Resources for the Future, 1990. [distributed by Johns Hopkins Univ. Press]

[4] Carhart, R.R., A Survey of the Current Status of the Reliability Problem, Rand Corporation Research Memorandum, RM-1131, August 14, 1953.

[5] Jensen, F., and Petersen, N.E., Burn-in: An Engineering Approach to the Design and Analysis of Burn-in Procedures, Wiley, New York, 1982.

[6] Martin, J. S., and McKnight, M. E., "Prediction of the Service Life of Coatings on Steel II: Quantitative Prediction of the Service Life of a Coating System," *J. Coatings Technology,* 57 (724), May 1985, p. 39.

[7] Nelson, W., Accelerated Testing, Wiley, New York, 1990.

[8] Martin and McKnight, *ibid.*

[9] Floyd, F. L., "Predictive Model for Cracking of Latex Paints Applied to Exterior Wood Surfaces," *J. Coatings Technology,* 55 (696), January, 1983, pp. 73-80.

[10] Martin, J. A., "Quantitative Characterization of Spectral Ultraviolet Radiation-Induced Photodegradation in Coating Systems Exposed in the Laboratory and the Field," *Progress in Organic Coatings,* 23, 1993, p. 49.

Chapter 3

Weather Cycles: Real or Imaginary?

W. J. Burroughs

Squirrels Oak, Clandon Road, West Clandon, Surrey GU4 7UW, United Kingdom

The evidence of cycles with periodicities in the range 2 to 100 years is examined in terms of their being the product of either the natural variability of the global climate, or the result of solar activity and/or tidal forces.   Conclusions are then drawn on whether either the evidence or the physical explanations of the observed changes are sufficient to influence investment decisions and planning the maintenance of plant and equipment.

The ability to predict changes in the climate on the timescale of a few years to several decades would have a profound effect on how we manage our lives.   Leaving aside predictions of global warming, there is no possibility of standard numerical weather forecasts addressing such lengthy changes.   If, however, we can develop a better understanding of how the atmosphere interacts with more slowly varying components of the global climate (e.g. sea surface temperatures, snow cover, pack ice and soil moisture) then it might be possible to predict the probabilities of periods of extreme weather occurring months and years ahead.   Evidence of weather cycles is a useful measure of whether such progress is a realistic prospect.

The subject of weather cycles has been of peculiar fascination to many meteorologists. Whether the product of the natural variability of the climate system or the result of interactions between the Earth's atmosphere and oceans, and external agencies including periodic variations of solar activity and the astronomical motions of the Moon and the planets, there has been a long and largely fruitless search for clear evidence for predictable cycles.   Even more intriguing is how this search has waxed and waned as if to parallel the elusive periodicities that it sought.

As far as solar activity is concerned, a good starting point is the comprehensive review (*1*) which concluded that "despite a massive literature on the subject, there is at present little or no convincing evidence of significant or practically useful correlation between sunspot cycles and the weather or climate".   This conclusion, together with a similar general scepticism about other periodic variations in the climate, meant that interest in weather cycles had reached a low ebb.   Apart from the well established diurnal and annual cycles in the weather, it is probably fair to say that only the Quasi-Biennial Oscillation (QBO) in the stratosphere was accepted by the meteorological community as being real (*2*).   In addition, a similar QBO was widely acknowledged as being a feature of many tropospheric weather records, but it was regarded as a "statistical will o'the wisp" (*3*).   But for the rest, the majority verdict was at best 'not proven'.   Since the late 1970s there has, however, been a number of interesting developments.

© 1999 American Chemical Society

Which cycles?

Because of the breadth of work done on cycles, it is not possible to cover the entire field. In terms of making investment decisions and planning the maintenance of plant and equipment, it is best to concentrate on periodicities in the range 2 to 100 years, with particular emphasis on the QBO, 3 to 5-year quasi-cycles, 11, 18 to 22, and 80 to 90-year cycles, and in the case of the last three their possible links with solar activity, and also the alternative lunar explanation for the 18 to 22-year cycle. This omits the 30 to 60-day cycles in tropical cloudiness (4) and their possible links with some of the cycles mentioned above. It also excludes how the evidence for cycles is extracted from time-series which are reviewed elsewhere (5). Also omitted is the major area of work associated with the astronomical theory of the origin of the Pleistocene ice ages (6–8) which became widely accepted in the 1980s.

QBO

As noted above, the evidence for the QBO in the stratosphere has become well established (2). This regular reversal of the winds in the stratosphere over the equator had been closely studied since the early 1950s (Figure 1). The interesting developments in recent years have been associated with whether there was a link between this stratospheric periodicity and the weather feature that appeared in so many tropospheric weather records. In particular, it has been linked with the fluctuations in the El Niño–Southern Oscillation (ENSO) (see next section) and identified as a factor in the extent to which the weather is influenced by the 11-year variation in solar activity associated with the incidence of sunspots. The latter provides a particularly good example of how a new set of observations can breathe life into the dying embers of a hypothesis about weather cycles, only to extinguish the new hopes when predicted events fail to materialise.

It has been known since 1980 that the north polar stratosphere during winter tended to be colder during the west phase of the QBO than during the east phase. It was then observed that at the maximum in solar activity the polar stratosphere was unusually warm in the QBO was in its west phase (9). By sorting out the east and west phase winters it was possible to show a strong correlation. For the period 1956 to 1988, west phase winters showed a marked positive correlation with warmer winters when the Sun was active and with colder winters when the Sun was quiescent, and vice versa for east phase winters. There is a less than 4 in 1000 chance that this combination of correlations could be the product of chance.

In terms of tropospheric weather, an examination of the 19 west phase winters during the period 1956 to 1988 showed an interesting correlation between surface pressure and solar activity. Over northern Canada pressure was abnormally high when solar activity was high, while at a point in the western Atlantic (25°N, 55°W) there was an equal and opposite negative correlation. These results suggested that during west phase winters at times of high solar activity, pressure over North America should be higher than normal while over the western Atlantic it would be lower than normal. This pattern brings cold northerly winds to the east coast of the USA more frequently than usual (10). So it was predicted that, with the QBO in the westerly phase and solar activity at high levels, the winters of both 1988/89 and 1990/91 would have temperatures well below average. When the first cold winter failed to materialise, it was argued that abnormally low sea surface temperatures (SSTs) in the equatorial Pacific (see next Section) might have disrupted the weather patterns over North America. In 1990/91 this explanation did not wash and, when a mild winter in the east was capped by the warmest February the USA had experienced this century, the reputation of seasonal forecasts based on the QBO–solar activity connection was in tatters.

ENSO

The interannual variability of the atmosphere and ocean of the tropical Pacific basin – generally referred to as the ENSO – has been the subject of widespread discussion (11, 12). Instead of recapitulating the broad features of this globally important climatological phenomenon, we will concentrate on the evidence of quasi-cycles in the behaviour of the tropical Pacific and how computer models have been developed to simulate this interaction between the ocean and the atmosphere. The basic feature of the temperature of the central equatorial Pacific is a periodicity of between 3 and 5 years (Figure 2). In

36

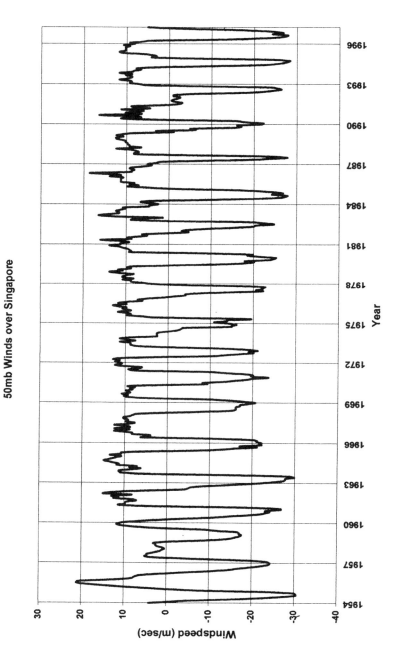

Figure 1. The monthly windspeed at the 50 millibar-level (21 km) above Singapore showing the regular (quasi-biennial) reversal of the wind direction every 27 months or so.

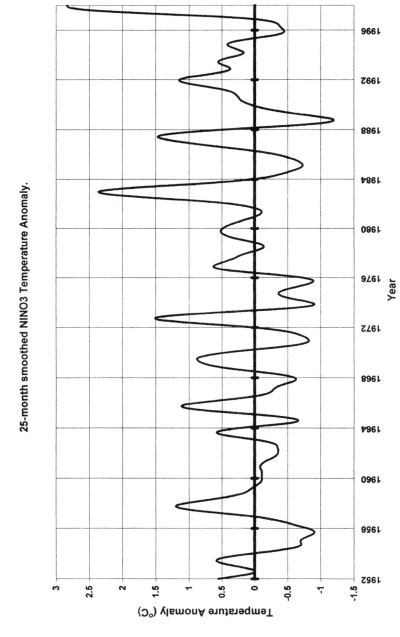

Figure 2. The temperature anomaly in the central equatorial Pacific (150 °W–90 °W, 5 °N– 5 °S), between 1952 and 1997, smoothed with a 25-month binomial filter to show quasi-periodic fluctuations greater than about a year in duration.

addition there is some evidence of a biennial oscillation (*13*), which may be linked to the stratospheric QBO, but it displays variable amplitude and appears to miss a beat from time to time.   The longer periodicity appears to be a natural product of the way the atmosphere and the ocean interact across the Pacific basin.

Out of the many efforts to model the Pacific basin (*12*) one particular model has addressed the possibility of quasi-cyclic behaviour.   This approach (*14*) seeks to handle the basic problem of what causes the oscillation between the El Niño and non-El Niño conditions.   During normal conditions (non-El Niño) the winds blow from the east towards the area of rising air over the warmer water in the western Pacific.   As these winds pile up warm water in the west they draw deeper colder water to the surface in the east, accentuating the temperature contrast which drives the wind, so strengthening the contrast.   During the El Niño the reverse applies.   As unusually warm water extends eastwards, it is accompanied by wind from the west into this rising air over the warm water.   As this cuts off the cold water that normally upwells in the east, it strengthens the westerly winds enhancing the anomalous conditions.   So, in principle, either the El Niño or non-El Niño could last indefinitely.

The way out of this impasse is that when the sea-level rises in, say, the east it transmits a 'signal' to the west to lower the sea-level in the west.   This 'signal' is in effect a travelling displacement of the thermocline.   The important feature of this process is that hydrodynamical models show that the ideal case of a symmetrical bell-shaped depression of the thermocline will disperse into two waves – an eastward-travelling Kelvin wave and a westward-travelling Rossby wave.   The former is a gravity-inertia wave which shows no meridional velocity variations and on which the restorative forces are due to the stratification of the ocean and the rotation of the Earth.   The latter, by contrast, is governed by the restorative force of the latitudinal variation of the Coriolis parameter, so its speed varies with distance from the equator.

Within 2 to 3 degrees of the equator – often termed the equatorial waveguide – processes are dominated by the zonal component of the wind stress which causes changes in the slope and the thickness of the warm upper layer.   The balance of these stresses and the weak Coriolis forces results in downwelling and upwelling Kelvin waves crossing the Pacific basin in 2 to 3 months.   Rossby waves move westward at speeds which depend on latitude.   Near the equator they are trapped in the equatorial waveguide and cross the Pacific basin in about 9 months, but at 12°N and 12°S they take around 2 years.

Changes in the depth of the thermocline, and hence the thickness of the warm layer, can be separated into these two types of waves.   During the El Niño the anomalous pattern is sustained by eastward-travelling equatorial Kelvin waves.   Simultaneously, Rossby waves are generated in the east Pacific.   Close to the equator they are initiated by the wind anomalies.   At higher latitudes they are generated by presently unknown interactions of coastal Kelvin waves (generated by the equatorial Kelvin waves) and the bathmetry along the North and South American coasts.   The equatorial Rossby waves can cancel out the positive feedback sustaining the El Niño.

When the warm layer off South America is abnormally thick (i.e. during an El Niño) the effect is to initiate upwelling Rossby waves.   These are reflected at the westerly rim of the Pacific basin and return in the equatorial wave-guide as upwelling Kelvin waves which tend to reverse the growth in the warm upper layer.   This delayed negative feedback serves to switch off the El Niño conditions.

By contrast, the much slower off-equatorial Rossby waves reach the western rim a year or two later, returning as downwelling Kelvin waves which tend to increase the thickness of the warm layer in the east.   This much delayed feedback serves to set up an oscillation between the El Niño and non-El Niño conditions.

The simple computer model of the Pacific basin (*14*) has been able to produce a quasi-periodic oscillation of equatorial SSTs.   These are irregular, occurring at intervals of about 3 to 5 years and resemble the observed fluctuations in the Pacific.   In spite of its limitations, this simple linear model can produce a physically realistic mechanism which combines a coupled oscillator and delayed feedback.   It also suggests that when the oscillations are small the system has little predictability, being dominated by random fluctuations in the global climate.   When the fluctuations are large the delayed feedback mechanism and the behaviour are more predictable.   This may be the key to many transient examples of apparently cyclic behaviour in the weather.   A major chance disturbance in a slowly varying component of the climate may trigger a delayed feedback process which results in a few oscillations before the 'cycle' fades away.

North Atlantic Oscillation

A similar process to the ENSO may be at work in the North Atlantic during the winter. Known as the North Atlantic Oscillation (NAO), this see-saw behaviour shifts between a deep depression near Iceland and high pressure around the Azores, which produces strong westerly winds, and the reverse pattern with much weaker circulation (15). The strong westerly pattern pushes mild air across Europe and into Russia, while pulling cold air southwards over western Greenland. The strong westerly flow also tends to bring mild winters to much of North America. One significant climatic effect is the reduction of snowcover, not only during the winter, but well into the spring. The reverse meandering pattern often features a blocking anticyclone over Iceland or Scandinavia which pulls arctic air down into Europe, with mild air being funnelled up towards Greenland. This produces much more extensive continental snowcover, which reinforces the cold weather in Scandinavia and eastern Europe, and often means that it extends well into spring as long as the abnormal snow remains in place.

Since 1870, the NAO has fluctuated appreciably on timescales from several years to a few decades (Figure 3). It assumed a strong westerly form between 1900 and 1915, in the 1920s and, most notably, from 1988 to 1995. Conversely, it took on a sluggish meandering form in the 1940s and during the 1960s bringing frequent severe winters to Europe but exceptionally mild weather in Greenland. How these variations might play a part of an organised atmospheric-ocean interaction is not yet clear. If it can be demonstrated that a period of one phase of the oscillation produces the right combination of sea surface temperatures to switch it into the opposite phase, then this may provide insight into more dramatic changes in ocean circulation which can occur in the North Atlantic. The fact that this oscillation shows some evidence of a 20-year periodicity suggests that there may be some more predictable mechanism at work.

Whatever the impact of the NAO on longer term developments, it remains central to understanding recent climatic events, because of the influence it exerts on average temperatures in the northern hemisphere. Of all seasons, winters show the greatest variance, and so annual temperatures tend to be heavily influenced by whether the winter was very mild or very cold. When the NAO is in its strong westerly phase, its benign impact over much of northern Eurasia and North America outweighs the cooling around Greenland, and this shows up in the annual figures. So, a significant part of the global warming since the mid-1980s has been associated with the very mild winters in the northern hemisphere. Indeed, since 1935 the NAO on its own can explain nearly a third of the variance in winter temperatures for the latitudes 20° to 90° N (16). So understanding more about this major natural factor in climatic change is central to explaining the causes of global warming in the twentieth century.

Solar and lunar cycles

There are many possible cycles of solar activity and astronomical effects, of which the lunar tides are the most important, which could influence the weather (5, 17). The most significant of these in terms of appearing in many meteorological records and surviving critical scrutiny is the periodicity of around 20 years. This has been widely attributed to both the 22-year double sunspot (Hale) cycle and to the 18.6-year lunar tidal cycle. The most significant evidence of this cycle is to be found in the analysis of records of drought in the western and central USA (18, 19), Central England Temperature records (20), global marine temperature records (21), and Greenland ice-core isotope records for the periods 1244 to 1971 (22).

The debate about the origin of this ubiquitous signal in climatic records has yet to be resolved. Perhaps the most interesting feature of this work is the apparent discrepancy between differing interpretations of the US drought indices. The work of Mitchell et al (18) clearly pointed to a 22-year cycle of solar origin, whereas Currie's (19) work appeared to support the lunar explanation. Subsequent work by Mitchell (23) showed that the 22-year cycle was the dominant feature in the US drought series between 1600 and 1962, and there was virtually no sign of the lunar cycle. But if the series was cut in half and the two halves analysed separately then the lunar cycle appeared as strongly as the solar one. The reason for this behaviour was that around 1780 the phase of the lunar cycle shifted by 180 degrees. As yet there is no adequate explanation as to why any lunar influence on the weather might shift in this way. So at the moment the jury is still out on the climatic significance of the 20-year cycle and, to the extent it is real, whether it is due to solar and/or lunar effects on the weather.

40

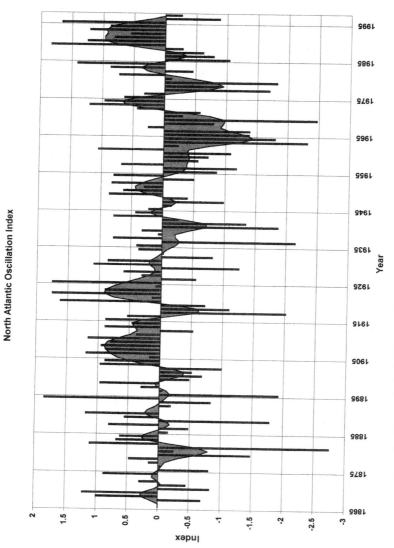

Figure 3. The standardised difference of December to February atmospheric pressure between Ponta Delgada, Azores, and Stykkisholmur, Iceland, 1867 to 1996. The smoothed curve is a 21-year binomial filter to show fluctuations longer than about 10 years.

One other aspect of possible solar influence of the weather deserves a mention. This is the fact that the 80 to 90-year variation both in the peak in solar activity and in the period between successive maxima shows a marked parallelism with global temperature trends (Figure 4). Satellite measurements have confirmed that solar output does vary with solar activity (24), but they are an order of magnitude too small to explain the observed global trends in terms of direct solar input (25). So unless some amplification mechanism is at work, possibly linked to the fact that a disproportionate amount of this variation occurs in the ultraviolet (UV), it is not realistic to attribute observed trends to solar forcing. Estimates of past variations in solar UV output do, however, show a remarkably high correlation with observed temperature trends in the northern hemisphere (26), while modelling work suggests that these UV variations could be amplified by circulation responses in the stratosphere (27). So the jury is still out on whether observed varaitions in solar activity can exert an appreciable impact on the climate.

Autovariance and chaos

Recent work in a number of areas has shed new light on various aspects of how the natural variability of the climate may hold the key to understanding many aspects of quasi-cycles in the weather. Work at Reading University (28) has shown that a relatively crude model of the climate, but one which handles the non-linear dynamics of the atmosphere in a reasonably sophisticated manner, produces the circulation of the atmosphere which varies strongly for periods longer than a year with the strongest response being at around 10 years. The general result that large-scale features of the atmosphere exhibit significant periodicities in the range 10 to 40 years was obtained from a number of different experiments.

This is a fascinating result. Because earlier linear simulations have not produced similar results, it emphasises the importance of the non-linear behaviour of the atmosphere. Furthermore, the fact that the atmosphere alone can exhibit such long-term fluctuations shows that it is not necessary to involve either the slowly varying components of the climate or extraterrestrial agencies to explain long-term quasi-periodic fluctuations. So it is possible that, say, the 20-year cycle could be largely the product of atmospheric autovariance rather than due to solar and/or lunar influences.

The slowly varying components of the climate remain strong candidates for much of the observed quasi-cyclic behaviour including the 20-year cycle. Modelling work of the thermohaline circulation of the North Atlantic (29), in which warm surface water flows polewards and colder water returns to lower latitudes at depth (30), has shown the possibility of substantial and often chaotic variations. The controlling factor is how temperature and salinity vary in the region where the surface water sinks to form bottom water. Salinity is the dominant factor in the rate at which polar water sinks and, unlike temperature, this is not subject to local feedback. This is because salinity is controlled by the balance between precipitation and evaporation, which is not affected by local salinity but by wider atmospheric and oceanic circulation. When the model uses a distribution of precipitation and evaporation which resembles the climatology of the North Atlantic it produces quasi-periodic but chaotic fluctuations on a decadal time-scale. The NAO could be related to this phenomenon as it clearly has a significant impact on deep water formation (31).

This modelling work has also received indirect support from the latest set of Greenland ice-core data (Greenland Ice Sheet Project 2, GISP-2). Results from these data have shown that the climate can change dramatically in just a few years. Observations of both the amount of dust in the ice, which is correlated with average temperature (32), and the amount of snow that fell at the end of the last Ice Age (33) show abrupt and substantial changes. In particular, the dust levels show frequent rapid and chaotic fluctuations between mild and cold conditions in as little as 3 to 5 years, and often lasting only for a decade or so. This led to the phrase a 'flickering switch' being used to describe this behaviour, which bears a marked resemblance to the models of thermohaline circulation in the North Atlantic. While there remains a considerable challenge to explain how changes of this scale could happen so rapidly, this recent combination of modelling and observations shows how swiftly the subject of climatic change is developing.

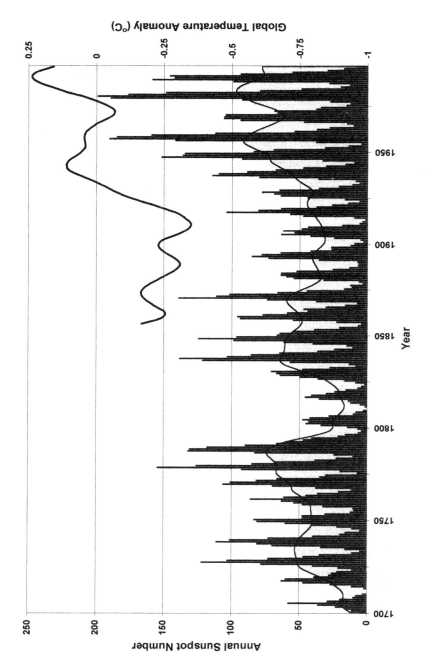

Figure 4. Curves showing (a) the annual sunspot number since 1700, (b) the annual sunspot number smoothed using a 21-year binomial filter, and (c) the annual global temperature anomaly since 1856 smoothed with a 21-year binomial filter.

Conclusions

This limited review of the search for weather cycles and the explanation of quasi-periodic and chaotic variations in the climate suggests that the prospects for long term predictions are not rosy. With the exception of the 20-year cycle, which can be found in many records, the classic search for cycles has made little progress in recent decades. In contrast, the analysis of the feedback processes in the global climate has grown apace. These advances have, however, highlighted the complexity of the interactions and raised real doubts about whether useful forecasts can be made over periods longer than a year or so. The best that can be said is that when there is a major disturbance in the climate system it can produce a decaying quasi-cyclic response which behaves in a more predictable manner for an oscillation or two before disappearing back into the noise. This conclusion will be put to the test as events unfold in the equatorial Pacific following the sudden onset of a strong El Niño during 1997.

Parallel investigations of non-linear systems (chaos theory) have combined with increasing evidence of the capacity of the climate to switch between different states abruptly and frequently to show that even more erratic behaviour is possible. These changes are by definition wholly unpredictable: we can only hold our breath and hope that neither natural fluctuations nor human activities trigger any such dramatic shifts. Against this uncertain background the safest course for planners is to work on the basis of climatological statistics, taking account only of global warming should it continue as predicted and its implications for extremes become apparent.

References

1. Pittock, A. B. *Rev. Geophys. Space Phys.*. **1978**, *16*, 311–314.
2. Naujokat, B. *J. Atmos. Sci.* **1986**, *43*, 1873–1877.
3. Landsberg, H. E.; Mitchell, J. M., Jr; Crutcher, H. L.; Quinlan, F. T. *Mon. Wea. Rev.* **1963**, 549–556.
4. Madden, R. A.; Julian, P. R. *J. Atmos. Sci.* **1972**, *29*, 1109–1123.
5. Burroughs, W. J. *Weather Cycles: Real or Imaginary?;* Cambridge University Press: Cambridge, UK, 1994.
6. Imbrie, J.; Imbrie, J. Z. *Science,* **1980**, *207*, 943–953.
7. Martinson, D. G.; Pisias, N. G.; Hays, J. D.; Imbrie J.; Moore, T. C., Jr; Shackleton, N. J. *Quat. Res.* **1987**, 1–30.
8. Imbrie, J.; et al. *Paleoceanography,* **1992**, *7*, 701–738.
9. Labitzke, K.; van Loon, H. *J. Atmos. Terres. Phys.* **1988**, *50*, 197–206.
10. Barnston, A. G.; Livesey, R. E. *J. Climate.* **1989**, *2*, 1295–1313.
11. Bigg, G. R. *Weather,* **1990**, *45*, 2–8.
12. Philander, S. G. *El Niño, La Niña and the Southern Oscillation;* Academic Press: London, UK, 1990.
13. Ropelewski, C. F.; Halpert, M. S.; Wang, X. *J. Climate,* **1992**, *5*, 594–614.
14. Graham, N. E.; White, W. B. *Science,* **1988**, *240*, 1293–1302.
15. van Loon, H.; Rogers, J. C. *Mon. Wea. Rev.* **1978**, *106*, 295–310.
16. Hurrell, J. W. *Geophys. Res. Lett.* **1996**, *23*, 665–668.
17. Lamb, H. H. *Climate: present, past and future;* Methuen. London, UK, 1990.
18. Mitchell, J. M.; Stockton, C. W.; Meko, D. M. In *Solar-terrestrial influences on weather & climate;* Editors McCormac, B. M.; Seliga, T. A.; D. Reidel Publishing Co., 1979; pp 125–143.
19. Currie, R. G. *J. Geophys. Res.* **1981**, *86*, 11055–11064.
20. Mason, B. J. *Q. J. R. Meteorol. Soc.* **1976**, *102*, 478–498.
21. Newell, N. E.; Newell, R. E.; Hsuing, J.; Wu, Z. *Geophys. Res. Lett.* **1989**, *16*, 311–314.
22. Hibler, W. D., III; Johnson, P. M. *Nature,* **1979**, *280*, 481–483.
23. Mitchell, J. M. *Climatic Change,* **1990**, *16*, 231–246.
24. Willson, R. C.; Hudson, H. S. *Nature,* **1991**, *351*, 42–44.
25. Kelly, P. M.; Wigley, T. M. L. *Nature,* **1992**, *360*, 328–330.

26. Lean, J.: Beer, J.: Bradley, R. *Geophys. Res. Lett.* **1995,** *22,* 3195–3198.
27. Haigh, J. D. *Science,* **1996,** *272,* 981–984.
28. James, I. N.; James, P. M. *Nature,* **1989,** *342,* 53–55.
29. Weaver, A. J.; Sarashik, E. S.; Marotze, J. *Nature,* **1992,** *353,* 836–838.
30. Taylor, N. K. *Weather,* **1992,** *47,* 146–151.
31. Sy, A.; Rhein, M.; Lazier, J. R. N.; Koltermann, K. P.; Meincke, J.; Putzka, A.; Bersch, M. *Nature,* **1997,** *386,* 675–679.
32. Taylor, K. C.; Lamorey, G. W.; Doyle, G. A.; Alley, R. B.; Grootes, P. M.; Mayewski, P. A.; White, J. W. C.; Barlow, I. K. *Nature,* **1993,** *361,*432–434.
33. Alley, R. B.; Meese, D. A.; Shuman, C. A.; Gow, A. J.; Taylor, K. C.; Grootes, P. M.; White, J. W. C.; Ram, M.; Waddington, E. D.; Mayewski, P. A.; Zielinski, G. A. *Nature,* **1993,** *362,* 527–529.

Chapter 4

Application of Spectral Weighting Functions in Assessing the Effects of Environmental UV Radiation

P. J. Neale

Smithsonian Environmental Research Center,
P.O. Box 28, Edgewater MD 21037

Abstract

The energy distribution of solar ultraviolet (UV) radiation reaching the earth's surface is a strong function of wavelength, and the spectral distribution is a function of several climatic parameters, including ozone concentration. Similarly, most chemical and biological effects of UV also show wavelength dependence with generally greater effectiveness shown by shorter wavelengths. Prediction of the effects of UV exposure requires specification of effectiveness as a function of wavelength, i.e. a spectral weighting function, which is applied to measured or modeled irradiance spectra. Two basic types of spectral weighting functions are action spectra and biological weighting functions. Action spectra are based on responses to monochromatic exposure while biological weighting functions are based on polychromatic (differential broad band) techniques. Experimental and analytical techniques for determining spectral weighting functions are briefly reviewed. Though polychromatic approaches have mainly been applied to predicting the biological effects of UV, possible application to chemical effects, including aging of surface coatings, is also discussed.

The photochemical activity of the ultraviolet radiation (UV, 280-400 nm) in solar irradiance reaching the earth's surface has a number consequences for materials and organisms in the environment. Some of these consequences directly affect man and society. Degradation of coatings on environmentally exposed surfaces such as that of buildings or vehicles is the focus of this symposium, but there are also significant human health effects of UV exposure, such as sunburn and cancer. Environmental UV has other effects which do not directly impact man, but are nevertheless significant from a global perspective, such as UV effects on photosynthesis and growth of micro-organisms. Over the last couple of decades, concern over UV effects has intensified due to observations of stratospheric ozone depletion, which selectively increases solar radiation in the short wavelength UV-B (280-320 nm). Although attention frequently focuses on the relationship between UV and the ozone content of the atmosphere, UV intensity and spectral distribution are also strongly affected by other climatological factors such as sun angle, cloud cover, and aerosols. The spectral variation in natural UV radiation is important because most effects of UV are a strong function of wavelength. Usually, the effectiveness of UV increases as wavelength decreases. Thus, the wavelengths most enhanced by ozone depletion also have the highest potential (per unit energy or photon) for inducing effects.

The strong spectral dependence in both environmental UV and responses to UV necessitates a spectral dimension to any predictive approach to assessing UV effects. The primary requirements are UV spectral irradiance with high resolution (1-2 nm) and estimation of spectral weighting functions for the effect(s) of interest. Spectral UV irradiance is measured by a spectroradiometer which resolves the spectral structure of solar irradiance with a diffraction grating or interference filters. The latter approach is used by the Smithsonian SR-18 spectroradiometer(1). Contributions elsewhere in this volume present further discussion of radiometer design and use [refs as appropriate]. If spectral measurements of sufficient resolution are lacking, another possibility is to use a spectral irradiance model (2).

This contribution discusses spectral weighting functions, how they are determined and how they are applied. A spectral weighting function (SWF) is simply any continuous measure of UV effects as a function of wavelength. Two specific types of SWFs are action spectra and biological weighting functions. Action spectra are based on responses to monochromatic irradiance and are defined for both biological and chemical effects. Biological weighting functions are determined under broad band (polychromatic) irradiance. The implications of this difference will be discussed in detail.

Spectral weighting functions scale an irradiance spectrum for its effectiveness at producing an outcome of UV exposure. The concept is analogous to computing a weighted average in that the shorter (more damaging) wavelengths of the spectrum are given greater emphasis:

$$E^* = \sum_{\lambda=280\,nm}^{700\,nm} \varepsilon(\lambda) \cdot E(\lambda) \cdot \Delta\lambda \qquad (1).$$

where E^* is effective irradiance (dimensionless), $E(\lambda)$ is irradiance at each wavelength (mW m^{-2} nm^{-1}) and the $\varepsilon(\lambda)$ are the scaling coefficients ((mW m^{-2})$^{-1}$). The coefficients, $\varepsilon(\lambda)$, are the spectral weighting function; generally they increase in magnitude (effect per unit irradiance) as wavelength decreases (Fig. 1). An alternative approach is to define weighted irradiance, E_{eff}, with units (mW m^{-2})$_{eff}$, in which case the coefficients, $\varepsilon(\lambda)$, are dimensionless. The weighted exposure (E^* or E_{eff}) is the quantitative measure of the effectiveness of the original $E(\lambda)$ spectrum.

Action spectra are one type of function that can be used to define $\varepsilon(\lambda)$ in Eq 1. An action spectrum is defined by measuring the response to narrow-band (monochromatic) radiation at many wavelengths and plotting the results as a function of wavelength. However, the precise determination of an action spectrum is a demanding exercise for which reason there are few action spectra for UV responses. Coohill (3) discussed six criteria that are required for analytically useful action spectra. The criteria include transparency of the sample at the wavelengths of interest, minimization of scattering, and reciprocity - the equivalence of intensity and time in determining total exposure. The criteria are all directed towards being able to quantitate exactly the energy incident to and absorbed by a chromophore. When this is possible, and the chromophore absorption is known, the quantum yield of UV effects on the process can be calculated as

$$\psi(\lambda) = \frac{\varepsilon(\lambda)}{A(\lambda)} \qquad (2)$$

where $\psi(\lambda)$ is quantum yield (moles photoproduct per mole of absorbed photons), and $\varepsilon(\lambda)$ is the action spectrum (moles photoproduct per unit exposure) and $A(\lambda)$ (moles photons absorbed per unit exposure) is the absorption spectrum. The advantage of estimating a quantum yield is that it is a fundamental constant that will be unaffected by other exposure conditions.

The requirements for precise action spectrum determination can usually be satisfied for investigations of chemical processes in transparent aqueous media, but it can be more difficult to define action spectra for biological processes and chemical processes in complex materials. Light scattering and absorption by structures and molecules other than the target chromophore is usually high in these latter contexts. Thus, even if we know that a certain effect is based, say, on UV damage to DNA, it is difficult determining how much UV is absorbed by the DNA within the organism because of absorption and scattering by surrounding tissue. Moreover, for UV effects on many physiological processes, the candidate chromophore(s) are either uncertain or unknown. Finally, using action spectra as measured by monochromatic radiation implicitly assumes that each waveband ($\Delta\lambda$) contributes independently to the overall effect, i.e. that there is no interaction between different wavebands. For some photochemical effects, this is probably not a bad assumption.

In contrast to chemical processes, UV effects on *in vivo* biological processes usually involve interactions between spectral regions. A simple example of how interactions can occur is for *in vivo* damage to DNA. Illumination with UV-A and blue light induces photorepair of DNA (4,5), so net DNA damage by UV-B in intact organisms depends on whether the UV-B is accompanied by longer wavelength background irradiance (5). Thus to obtain experimental results that are more representative of biological responses to environmental UV, measurements are made using treatments composed of a range of wavelengths, i.e. with a polychromatic source. One approach is to add monochromatic UV to a broadband background, which shows how much each wavelength enhances a baseline response. This approach retains the precision of an action spectrum, but is still quite unlike natural illumination (3). A more realistic polychromatic approach is to generate a set of spectra that employs cutoff filters, i.e. a series of filters that pass longer wavelength light starting at successively shorter cutoff wavelengths. Here, short wavelength light (e.g. UV-B) is always added to a background of higher intensity, long wavelength irradiance (UV-A and visible), as is the case with solar irradiance. The tradeoff for greater realism is that effects can no longer be precisely attributed to specific wavelengths. Instead, the weights are composites of the effects at that wavelength and the interactive effects of other wavelengths (3)[2]. These empirical weights are referred to here as 'biological weighting functions' (BWFs). BWFs have also been referred to as polychromatic action spectra (3) or sometimes no specific distinction was made between an action spectrum and a BWF [i.e. both are 'action spectra' (6)].

Biological weighting functions were identified some time ago as a primary research objective if the effects of ozone depletion were to be understood. Considerable effort has therefore been directed towards developing experimental and analytical approaches to defining BWFs. The next sections discuss approaches to BWFs, followed by consideration of how those techniques may be applied in non-biological (i.e. chemical) contexts.

BWF Methodology

The basic approach for obtaining environmentally relevant BWFs is to measure responses to polychromatic irradiance treatments as defined by cutoff filters (Fig. 2). The particular method used to measure UV effects will depend on the process being studied. However, common elements to all experimental protocols are the choice of UV source and the number of filters used. Sources that have been used include filtered incident solar irradiance alone, solar irradiance supplemented by UV lamps, and solar simulator (xenon arc) lamps. Solar simulators can approximate the spectral distribution of solar irradiance, but no artificial source is completely accurate. Spectral features of solar irradiance at the earth's surface that are difficult to simulate are: 1) the sharp drop in energy with wavelength in the UV-B (7), and 2) the high ratio of UV-A and visible to UV-B. Fluorescent UV lamps, which have wide commercial availability due to their use as tanning lamps, have much higher amounts of short wavelength UV-B than found in solar irradiance even after filtering through cellulose acetate sheets to remove the UV-C component (8). Since the BWF weight at any wavelength implicitly includes interactions with other wavelengths, there can be some question about the capability of a BWF that was estimated using artificial sources to predict responses to solar irradiance. On the other hand, use of solar irradiance alone provides little reliable variation in the shortest UV-B

---

[2] In theory, interactive effects could be directly, and precisely, accounted for by adding coefficients for cross-products of irradiance at different wavelengths to Eq. 1. But in most cases the specification of the relevant cross product coefficients would be an arduous, if not impossible, experimental objective.

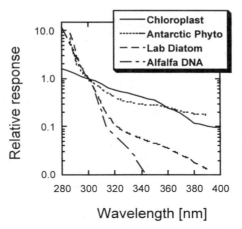

Figure 1. Examples of spectral weighting functions for biological processes. Shown are the SWFs for UV inhibition of chloroplast activity (27), inhibition of photosynthesis in Antarctic phytoplankton (17), inhibition of photosynthesis in a laboratory culture of phytoplankton (diatoms) (22), photodamage of DNA in alfalfa seedlings (28). Functions are compared by setting weight at 300 nm equal to one.

Figure 2. Examples of polychromatic treatment exposures produced by passing light emitted by a xenon arc lamp through Schott long-pass cutoff filters with nominal 50% transmission wavelengths of (from right to left): 280, 295, 305, 320, 335, 345, 365 and 400 nm. Spectra have been normalized to 1 at 395 nm. Adapted from (23).

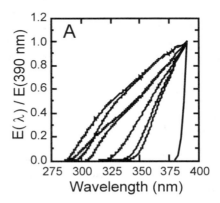

Figure 2. Examples of polychromatic treatment exposures produced by passing light emitted by a xenon arc lamp through Schott long-pass cutoff filters with nominal 50% transmission wavelengths of (from right to left): 280, 295, 305, 320, 335, 345, 365 and 400 nm. Spectra have been normalized to 1 at 395 nm. Adapted from (23).

wavelengths (< 305 nm), though the rotation of the ozone hole provides a natural source of spectral variation in Antarctica (9). The best test of the predictive power of a BWF is to compare predictions to observations in an independent set of response - irradiance observations (10).

Materials used for long-pass cutoff filters include Schott filter glass and various types of other glasses and plastics. Schott filter glass is the best choice from the standpoint of durability and optics, however it is expensive and some filter types are no longer manufactured. Other materials used include acrylic sheet ('Plexiglas'), mylar, and polycarbonate. It is well to keep in mind that cutoff properties vary between manufacturers (11) and have to be monitored through any experimental period. Choice of the number of filters used is a tradeoff: resolution will improve as the number of filters is increased but there are practical limits to the number of spectral treatments. Typically, filter cutoffs are spaced at 10 to 20 nm intervals. Finally, determining a BWF, and its subsequent application in any other context, depends directly on defining spectral irradiance at adequate resolution. Because of the typical steep slopes of spectral weighting functions in the UV-B (e.g. Fig. 1), resolution should be at least 2 nm.

Analytical considerations

Calculation of an action spectrum is unambiguous: effect is related to exposure at each wavelength. Calculation of a BWF is more complex. As for an action spectra, one starts with a set of responses and their associated treatment spectra (e.g. Fig. 2). There are three strategies that can be used to derive a spectrum: 1) simple difference, 2) adjusting an assumed function, and 3) component analysis. In general, the data show an increasingly greater overall effect as successively shorter wavebands of UV are transmitted to the treatment. A simple analysis is to order the results by irradiance of UV exposure (mW m^{-2}) and to compute the difference in effect between successive treatments. Then the biological weight for each waveband is estimated as the differential effect divided by the difference in energy between the treatments that include and exclude that waveband. However, as Rundel points out (12), the estimated weight may be inaccurate if the actual response changes rapidly over the wavelengths for which the treatments differ, this could be the case if the differential bandwidth exceeds 10 nm in the UV-B or 15-20 nm in the UV-A. Moreover, there is no objective method to determine the effective center wavelength for a weighting without already knowing a high resolution spectrum.

The advantage of the difference method is its simplicity, and it was used to estimate useful, albeit crude, BWFs for the effect of UV on photosynthesis by suspended microalgae (phytoplankton) in the oceans around Antarctica. Smith et al (16) measured in situ photosynthesis under a visible, visible + UV-A and visible+UV-A+UV-B for phytoplankton from the Bellingshausen Sea (9). The simple difference method was used to calculate what is, in effect, a two point BWF. A greater number of spectral treatments (four-five) was used in exposures of coastal Antarctic assemblages of phytoplankton (13-15). Despite the limitations of the difference approach, these initial results were useful in obtaining the first estimates of the relative increase of biologically damaging irradiance associated with Antarctic ozone depletion.

The only way to improve upon the coarse resolution of the difference method is to make some assumption about the shape of the BWF. The overall concept is to start with a spectral 'template' and adjust it so as to best mimic the observed responses to the spectral treatments. There are several variations on this approach, which mainly differ in how modification of the template is constrained. An approach that tightly constrains the BWF is to assume that the BWF is similar in shape to an already known action spectrum. A good example is the pioneering study of Smith *et al.* (16) on the spectral sensitivity of marine photosynthesis to UV. The relative change in photosynthesis in each treatment was compared to irradiance weighted by spectral weighting functions for DNA damage and inhibition of chloroplast electron transport (Fig. 1). Inhibition was more closely predicted by the latter weighting function. The authors did formulate a dose-response relationship based on chloroplast spectrum but were cautious in assigning general significance to this result: they did not conclude that the chloroplast spectrum predicts effects of UV on marine photosynthesis, rather that the DNA spectrum was not an acceptable predictor. Now that more detailed BWFs have been estimated for UV inhibition of phytoplankton photosynthesis, it does appear that constraining the shape of the BWF may have lead to a substantial underestimation of the weights in the UV-B (17).

Constraining the BWF to the same shape as a pre-existing action spectrum is probably only necessary if a limited number of spectral treatments are used (three in the case of Smith *et al.,* (16)). If a larger number of treatments is available, the constraints can be relaxed. A general property of most action spectra is an approximate log-linear decrease in weight as wavelength increases (12,18). Thus, a simple, but general, template for a BWF is an exponential function, i.e. $\varepsilon(\lambda) \propto e^{-a\lambda}$, with the characteristic log-linear slope (a) to be determined for each particular BWF. The slope giving the best fit to observed response can be readily estimated using nonlinear regression (19). Behrenfeld et al. (18) used this approach to estimate a generalized BWF for UV-B effects on photosynthesis by phytoplankton. However, a single log-linear slope is not apparent in other BWFs for inhibition of photosynthesis, even when just considering weights for wavelengths < 340 nm (Fig. 1). If a single log-linear slope is assumed, even though multiple slopes are present, then the fitted slope will correspond to the spectral region that contributed most to the differences between experimental treatments (8,17). For example, the Behrenfeld *et al.* analysis could have been relatively insensitive to variation in weights in the longer UV-B, since it emphasized explaining the relative decrease between treatments in which spectral irradiance differed primarily at wavelengths shorter than 310 nm (18).

Rundel recognized that BWFs could have a more complex shape than a single log-linear slope and advocated using a more generalized exponential function (12), e.g., that the natural log of the BWF is a polynomial:

$$\varepsilon(\lambda) = e^{-(a_0 + a_1\lambda + a_2\lambda^2 + \ldots)} \qquad (3.3)$$

Many BWFs can be approximated by such a general exponential function (20). Again, the parameters for the BWF polynomial (a_i) can be estimated by non-linear regression. A complete analytical protocol is given by Cullen and Neale (8). The constraints on the template can be even further relaxed by introducing thresholds, offsets (which allow negative weights), etc (12), but these complexities place additional statistical demands on the base data set. The principle of Occam's razor always applies: the model should be chosen to explain the most variation possible with the least number of parameters. The regression results can also be used to estimate the uncertainty in the BWF, based on the technique of propagation of errors (21).

A third method of estimating BWFs places a minimum of constraints on the template. In this technique (8,22,23), UV spectral irradiance in each treatment is analyzed by Principal Component Analysis (PCA) to generate up to four principal components (essentially, statistically independent shapes defined by weights for each wavelength) which account for nearly 100% of the variance of the treatment spectra relative to the mean spectrum. Component scores (the relative contribution, c_i, of each principal component, i, to a given UV spectrum, normalized to total visible irradiance, 400 to 700 nm) are derived for each of the treatment spectra. The estimation of the BWF then proceeds by non-linear regression as for the Rundel method, except that the parameters estimate the contribution of each spectral component to weighted irradiance. For the case of photosynthesis, coefficients m_i are estimated such that $E^* = E_{VIS}(m_0 + \sum_{i=1}^{4} m_i c_i)$, where E_{VIS} is visible irradiance in W m^{-2}, m_0 is the coefficient for any visible irradiance effects that are not part of the spectral treatments, plus the weight of the mean spectrum, and m_i are the contributions of the spectral components. Again, only as many components are incorporated into the final estimate as can be justified based on variance explained. Finally, once the m_i are estimated by regression, the $\varepsilon(\lambda)$ estimates and their respective statistical uncertainty can be calculated via the original spectral components. Details of the method are described elsewhere (22,23), and a step-by-step protocol is provided by Cullen and Neale (8). An example of spectra calculated using the PCA method are shown in Fig. 1 (Lab Diatom). Typically, the BWFs explain > 90% of the variance in photosynthesis of 72 spectral treatments (8 cutoff filters and 9 intensities), and the weighting functions show more complex features than simple exponential slopes (24).

Application of polychromatic techniques to material photodegradation

The preceding discussion has emphasized using polychromatic exposures to obtain weighting functions for biological responses. However, these experimental and analytical approaches may also be appropriate for chemical effects in complex systems where different types of UV effects could be interacting. This could occur, for example, if some of the photoreactants and photoproducts are involved in several reactions with different wavelength dependencies. Depending on the photochemistry involved, surface coatings may be an example of such a system. Other examples may be in some types of photochemistry in seawater. For example, UV induces production of hydroxyl radicals in seawater through photoreactions occurring in the heterogeneous mixture of dissolved compounds that absorb UV (25). Therefore, measurements of hydroxyl radical production under polychromatic exposure were recently made at Palmer Station, Antarctica using the same incubator that previously has been used to determine UV effects on phytoplankton (Miller, Kieber, Mopper and Neale, unpublished data). The results will then be compared with action spectra determined using monochromatic exposures.

The purpose of this contribution is not to show whether or not wavelength interactions are occurring in chemical settings, but to recommend a course of action if such interactions are known, or suspected to be, occurring. The experimental and analytical techniques used in estimation of biological weighting functions could easily be adapted to studying materials photodegradation, as our preliminary work measuring seawater photochemistry has shown. The spectral weighting functions so obtained could still be termed polychromatic action spectra to differentiate them from analytical action spectra measured with monochromatic irradiance (as suggested by (3)), or the term "photodegradation weighting function" may be appropriate.

In summary, measurements of UV spectral irradiance and spectral weighting functions are the two main requirements for determining effectiveness of environmental UV in inducing a particular effect, whether it is reduction in service life of a surface coating or interference in some cellular function. Another question, not treated by this report, is how a measurement of effectiveness (E^* or E_{eff}) is translated into a prediction of effect on some process. The conceptual framework for this is the exposure response curve (ERC; (26)) which is based on the kinetics of UV induced effects. For simple chemical processes, in which availability of substrate is not limiting and there are no back reactions, the response will be directly proportional to properly weighted UV dose and reciprocity (equivalence of time and intensity in determining dose) will apply. In contrast, biological processes can have highly non-linear kinetics (17). Under these circumstances, reciprocity fails and predicted effects will depend on the intensity as well as the duration of exposure. Examples of different types of exposure response functions are described by Neale (17).

Finally, there has been more than a decade of efforts at developing models to predict the effect of ozone depletion on biological processes and this has resulted in steady advances in the comprehensiveness and resolution of the models. It is hoped that some of these principles and approaches can be adapted to solve problems in prediction of UV effects on materials.

Literature Cited

(1)   Thompson, A.; Early, E. A.; Deluisi, J.; Disterhoft, P.; Wardle, D.; Kerr, J.; Rives, J.; Sun, Y.; Lucas, T.; Duhig, M.; Neale, P. J. *National Institutes of Standards and Technology Journal of Research* **1997**, *102*, 270-322.

(2)   Arrigo, K. R. *Mar. Ecol. Prog. Ser.* **1994**, *114*, 1-12.

(3)   Coohill, T. P. *Photochem. Photobiol.* **1991**, *54*, 859-870.

(4)   Friedberg, E. C. *DNA repair*; W. H. Freeman: New York, 1985, pp 614.

(5)   Hanawalt, P. C.; Setlow, R. B. *Molecular mechanisms for repair of DNA*; Plenum: New York, 1975; Vol. 5, Parts A and B, pp 418.

(6)   Caldwell, M. M.; Camp, L. B.; Warner, C. W.; Flint, S. D. In *Stratospheric Ozone Reduction, Solar Ultraviolet Radiation and Plant Life*; R. C. Worrest and M. M. Caldwell, Eds.; Springer: 1986; pp 87-111.

(7)   Iqbal, M. *An introduction to solar radiation*; Academic Press: 1983.

(8)   Cullen, J. J.; Neale, P. J. In *Effects of ozone depletion on aquatic ecosystems*; D.-P. Häder, Ed.; R. G. Landes: Austin, 1997; pp 97-118.

(9)   Smith, R. C.; Prézelin, B. B.; Baker, K. S.; Bidigare, R. R.; Boucher, N. P.; Coley, T.; Karentz, D.; MacIntyre, S.; Matlick, H. A.; Menzies, D.; Ondrusek, M.; Wan, Z.; Waters, K. J. *Science* **1992**, *255*, 952-959.

(10)   Lesser, M. P.; Neale, P. J.; Cullen, J. J. *Molecular Mar. Biol. Biotech.* **1996**, *5*, 314-325.

(11)   Karentz, D.; Bothwell, M.L.; Coffin, R.B.; Hanson, A.; Herndl, G.J.; Kilham, S.S.; Lesser, M.P.; Lindell, M.; Moeller, R.E.; Morris, D.P.; Neale, P.J.; Sanders, R.W.; Weiler, C.S.; Wetzel, R.G. *Arch. Hydrobiol. Beih Ergebn. Limnol.* **1994**, *43*, 31-69.

(12)   Rundel, R. D. *Physiol. Plant.* **1983**, *58*, 360-366.

(13)   Mitchell, B. G. In *Response of Marine Phytoplankton to Natural Variations in UV-B Flux*; B. G. Mitchell, O. Holm-Hansen and I. Sobolev, Eds.; Chemical Manufacturers Association: Washington, D.C., 1990; Appendix H.

(14)   Lubin, D.; Mitchell, B. G.; Frederick, J. E.; Alberts, A. D.; Booth, C. R.; Lucas, T.; Neuschuler, D. *Journal of Geophysical Research* **1992**, *97*, 7817-7828.

(15)   Helbling, E. W.; Villafañe, V.; Ferrario, M.; Holm-Hansen, O. *Mar. Ecol. Prog. Ser.* **1992**, *80*, 89-100.

(16)   Smith, R. C.; Baker, K. S.; Holm-Hansen, O.; Olson, R. S. *Photochem. Photobiol.* **1980**, *31*, 585-592.

(17)   Neale, P. J.; Cullen, J. J.; Davis, R. F. *Limnol. Oceanogr.* **1998**, *in press*,

(18)   Behrenfeld, M. J.; Chapman, J. W.; Hardy, J. T.; Lee, H. I. *Mar. Ecol. Prog. Ser.* **1993**, *102*, 59-68.

(19)   Marquardt, D. W. *J. Soc. Ind. Appl. Math.* **1963**, *11*, 431-441.

(20)   Boucher, N. P.; Prézelin, B. B. *Mar. Ecol. Prog. Ser.* **1996**, *144*, 223-236.

(21)   Bevington, P. R. *Data Reduction and Error Analysis for the Physical Sciences*; McGraw Hill: New York, 1969, pp 336.

(22)   Cullen, J. J.; Neale, P. J.; Lesser, M. P. *Science* **1992**, *258*, 646-650.

(23)   Neale, P. J.; Lesser, M. P.; Cullen, J. J. In *Ultraviolet radiation in Antarctica: Measurements and biological Effects*; Weiler, C. S.; Penhale, P. A., Ed.; Am. Geophysical Union: Washington, D.C., 1994; Vol. 62; pp 125-142.

(24)   Neale, P. J. In *The effects of UV radiation on marine ecosystems*; de Mora, S. J.; Demers, S.; Vernet, M. Eds.; Cambridge Univ. Press: Cambridge, 1998; pp in press.

(25)   Mopper, K.; Zhou, X. *Science* **1990**, *250*, 661-664.

(26)   Coohill, T. P. In *Stratospheric ozone depletion/ UV-B radiation in the biosphere*; Biggs, R. H.; Joyner, M. E. B. Eds.; Springer-Verlag: Berlin, 1994; Vol. 118; pp 57-62.

(27)   Jones, L. W.; Kok, B. *Plant Physiol.* **1966**, *41*, 1037-1043.

(28)   Quaite, F. E.; Sutherland, B. M.; Sutherland, J. C. *Nature* **1992**, *358*, 576-578.

Chapter 5

Statistical Analysis of Spectral UV Irradiance Data Collected at Ocean City, New Jersey

Nairanjana Dasgupta[1], Jonathan W. Martin[2], and Xiaomei Guan[1]

[1]Program in Statistics, Washington State University, Pullman, WA 99164
[2]National Institute of Standards and Technology, Gaithersburg, MD 20899

The goal of this study is to find a statistical model to study the dependencies among wavelengths. 2 nm half band pass solar spectral ultraviolet measurements were made at twelve-minute intervals for each of 18 wavelengths from 290.2 nm to 323.6 nm from April to September 1996 in Ocean City, NJ. Two data sets, the daily maximum and the hourly mean irradiance were created from the raw data. Auto-regressive models and transfer function models were used for both data sets. For the analysis 305.9 nm was chosen as the explanatory variable to predict 291.5 nm which was the response variable. For the final auto-regressive models R^2 values of .89 and .96 were obtained for the daily maximum and hourly mean data respectively. The forecasts from both models were reasonably good. This study will provide valuable insight into design of future spectral radiometers for characterizing solar ultraviolet radiation.

There is a growing concern in the scientific community and the public about the harmful effects of solar ultraviolet (UV) radiation. The effects of biologically damaging ultraviolet-B, which may be a consequence of stratospheric ozone depletion, are of major concern. As a consequence, several agencies in the world are collecting spectral UV radiation data.

The present analysis examines data collected by a spectro-radiometer designed and constructed by the Smithsonian Environmental Research Center which measures spectral UV-B data continuously in 18 wavelengths centered at nominal 2 nm increments from 290 to 324 nm. The 18 wavelengths are determined by interference filters that, for the instrument used, had actual center wavelengths of

© 1999 American Chemical Society

290.2, 291.5, 294.0, 296.1, 297.9, 300.0, 301.4, 303.9, 305.9, 307.7, 309.8, 312.4, 314.3, 315.7, 318.4, 320.1, 322.1, and 323.6 nm. Detailed information on the design and performance of the instrument are given in (*1*). Measurements were made from April 30, 1996 to September 14, 1996 in Ocean City, New Jersey. The data were initially acquired at intervals of 4 s, and averaged to 1 min. Archived data were corrected for drift between initial and final calibrations (generally less than 5% over the operation period), and used to calculate 12 minute averages. Irradiance data were measured in $(J/m^2/nm/min)/mVolt$.

It is of interest to study the dependencies among the various wavelengths. The purpose of this research is to understand the "statistical" properties of UV irradiance data and use the relationships to predict and forecast UV irradiance based on the time of day and irradiance at another wavelength. This may offer valuable insight in future design of spectral radiometers for characterizing solar ultraviolet radiation.

The irradiance data are in two dimensions, wavelength and time. Figure 1 is a graph of the data of the selected wavelengths of 305.9 nm and 322.1 nm, over 15 days. Similar graphs can be obtained for the other wavelengths from 290.2 nm to 323.6 nm. It is evident from Figure 1 that a high correlation exists between irradiance at different wavelengths and time of day. Table I below is a table of Pearson correlation coefficients calculated over 129 days of data over selected wavelengths. The correlation ranges from .89- .99 indicating a strong positive relationship among the selected wavelengths.

From the raw data that were collected every twelve minutes, two summary data sets were created for statistical analysis. The first one is of maximum daily irradiance. This is a relatively small data set that is easily handled. This can be thought of as the first step in the analysis. The second data set was compiled averaging the data for each hour. Though considerably smaller than the raw data, it provides detailed information. The irradiance for the non-daylight hours of the day were assumed to be zero. Both the described data sets effectively summarize the raw data while dealing with problems like missing values.

TABLE I. Pearson correlation coefficients calculated for selected wavelengths

wavelength (nm)	290.2	291.5	305.9	307.7	322.1	323.6
290.2	1.000	0.965	0.988	0.990	0.978	0.977
291.5		1.000	0.965	0.955	0.895	0.892
305.9			1.000	0.999	0.967	0.965
307.7				1.000	0.977	0.976
322.1					1.000	0.999
323.6						1.000

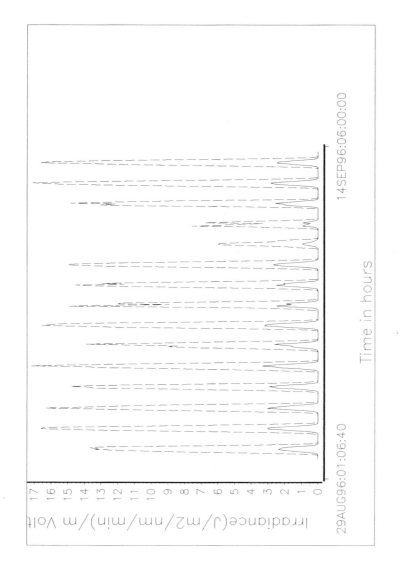

Figure1: Irradiance in (J/m^2/nm/min)/m Volts for wavelengths 305.9 nm and 320.1 nm channel.

Statistical Methodology

From Table I and Figure 1 it is evident that there is a strong correlation between different wavelengths. This suggests that one may use a simple regression model to predict the irradiance of one wavelength based on another. Multiple regression is not a good practice since any two wavelengths used as explanatory variables will be highly correlated (as seen in Table I). This will pose difficulties in building confidence limits for the estimated coefficients and predicted irradiance values. Predictions for wavelength 315 nm based on knowledge of 305 nm and 325 nm have been investigated (2). Though predictions from these models are good, the errors are auto-correlated and multicollinear.

The strategy of this analysis is to combine time series analysis done in the past (3-5) and the idea of regression analysis (2). Irradiance at 305.9 nm channel, will be used as an explanatory variable to model and predict irradiance at 291.5 nm channel, while taking into account the time series nature of the data. The choice of these wavelengths was arbitrary. 305.9 nm was chosen as one of the middle wavelengths and 291.5 nm was chosen since it was one of the shorter wavelengths. Results from this choice of wavelengths will be discussed in detail. R^2 values obtained from different choices of the explanatory variable will also be mentioned. Two different approaches will be used, auto-regressive method and transfer function method. Details of both these methods are provided in (6-8). A brief description of the methods is provided below.

Auto-regressive method. In this model, at time t, the response Y_t, is written as a linear function of the explanatory variable, X_t, and a dependent error, ε_t. The current value of the error, ε_t, is expressed as a finite linear aggregate of k previous values of the process and a random white noise, v_t. The kth order auto-regressive process can be written as

$$y_t = \beta_0 + \beta_1 x_t + \varepsilon_t \tag{2.1}$$
$$\varepsilon_t = \rho_1 \varepsilon_{t-1} + \rho_2 \varepsilon_{t-2} + \ldots + \rho_k \varepsilon_{t-k} + v_t$$

The white noise, v_t, is assumed to have 0 mean and constant variance and be uncorrelated over time. To fit this model one needs to find the order, k and estimate the k parameters in (2.1). To determine the order k, auto-correlation functions (ACF) and partial autocorrelation functions (PACF) (7-8) and plots are used. The parameters, $\rho_1, \ldots, \rho_k$, are estimated using generalized least squares methods (9).

To measure the adequacy of the model (2.1) for the given data, various diagnostics checks can be performed. These include the Q-statistics, which follow the chi-square distribution approximately (7-8). For a desirable model, the calculated Q-statistics should be less than the chi-square critical point. When more than one model satisfies the model adequacy conditions, the model with the highest R^2 is the "most desirable" model. Based on the model (2.1), y_t can be predicted and confidence limits can be obtained for the in-sample and out-of-sample forecasts.

Transfer function model. A single-input, single-output transfer function model is also called the ARMAX model. In this model, it is assumed that the spectral irradiance at wavelength X is linked to the spectral irradiance at wavelength Y. It is further assumed that a change in X affects the change in Y after a time delay; that is the change in Y lags the change in X. The form of the model is follows:

$$y_t = \nu(B) x_t + n_t \tag{2.2}$$

$$\nu(B) = \sum_{j=-\infty}^{\infty} \nu_j B^j = \frac{\omega(B)}{\delta(B)} B^b$$

where y_t is the output series and x_t is the input series and n_t is the noise, $\omega(B)$ is a polynomial in B for moving average terms and $\delta(B)$ is a polynomial in B, for auto-regressive terms. B is back shift operator and b is a delay parameter representing the actual time lag that elapses before the input variable produces an effect on the output variable. The first step in the analysis is 'pre-whitening' the input x_t. This involves time-series analysis of the input x_t using the ACF and PACF to identify its order k, and estimate its stochastic process. Prewhitening, xt, transforms the error term, n_t to the new process, ε_t. The next step is to identify and estimate the transfer function $\nu(B)$ using the cross-correlation function, CCF, (8) along with auto-correlation function (ACF) and partial auto-correlation functions (PACF).

For a desirable model, the sample cross-correlation function between the pre-whitened input series, x_t, and the final error terms, ε_t, should show no patterns. As a diagnostic measure, the modified Q_0 statistic (8) is calculated. This is compared to the corresponding critical value of the chi-square distribution and a calculated value less than the critical value indicates an appropriate model. When more than one model satisfies the model adequacy conditions, the model with the smallest AIC is the "most desirable" model.

The main difference between transfer function method and the regression method is that transfer function models use forecasted values of the explanatory variables to forecast values of the dependent variable. The variability in the forecasts of the explanatory variables is incorporated into forecasts of the dependent variable (8). Hence, the confidence intervals for the out-of-sample forecasts will be wider in this situation.

Analysis of the Daily Peak Values

Due to the short duration of the Ocean City, NJ, data, a statistical analysis without any yearly or seasonal adjustment, of daily peak values is undertaken. For the analysis the explanatory variable is the wavelength 305.9 nm (MAX306), and the response variable is the wavelength 291.5 nm (MAX292).

Auto-Regressive Method. Serial correlation is expected since the regression is performed with time series data. It is reasonable that the irradiance of consecutive days will be related. So the auto-regressive procedure with a lag of 1 is used in our analysis to correct the auto-correlation. The estimated model is

$$MAX\,292 = -0.000573 + 0.000673(MAX\,306) + \varepsilon_t \qquad (3.1)$$
$$\varepsilon_t = -0.436877\,\varepsilon_{t-1} + \nu_t$$
$$R^2 = 0.89$$

The parameter values are all statistically significant at (p-value=.05). The calculated Q-statistics are all less than the corresponding chi-square critical values. This indicates that the model (3.1) adequately fit the data.

The forecasts for MAX292 are plotted in Figure 2. In order to check the forecasting ability, the last 3 days of our available data are not used in our model building stage, so the forecasts for the last 3 days are out-of-sample forecasts. Actual peak values for wavelength 292 nm can be compared with the predicted and forecasted values of wavelength 292 nm. From Figure 2, the within sample predictions captured the original trend of the observed variable (MAX292), but the out-of-sample forecasts did not give good results compared with the real observations.

Transfer Function Model. Given the short time period, the solar spectral irradiance is assumed stationary both in the mean and variance. The ACF and PACF for the explanatory variable, MAX306, are used to pre-whiten it.

Based on the auto-correlation and partial auto-correlation, the fitted model can be written with back shift notation as

$$\varepsilon_t = (1 - 0.2548\,B - 0.1989\,B^{11})(MAX\,306_t - 3.3247) \qquad (3.2)$$

The parameters for this model are statistically significant (p-value=.05) and the Q-statistics were less than the chi-square critical value (p-value=.05). A "contemporaneous relationship" (delay parameter, b=0) is detected from the cross-correlation between input MAX306 and output MAX292. This indicates that a change in the variable, X, causes a change in the variable, Y, in the same time period. The final transfer model with the smallest AIC is estimated as

$$MAX\,292 = -9.6 \times 10^{-4} + 6.9 \times 10^{-4}/(1 - 0.056\,B - 0.068\,B^2)\,MAX\,306 \quad (3.3)$$
$$+ \varepsilon_t\,(1 + 0.36\,B)/(1 - 0.056\,B)$$

In this fitted model, the error process is an ARMA (1,1) model. Model adequacy is verified using the modified Q_0 statistic. The forecast and the observed peak values of wavelength 291.5 nm are shown in Figure 3. The forecasts are not as close as the auto-regressive model. However, auto-regressive models assume knowledge of explanatory variables for out-of-sample forecasts. Thus, the transfer function model is more useful than the regression method since it can provide forecasts for both the input and output series.

Lack of seasonality in the data set is definitely a drawback and also the very limited nature of the data could account for the poor forecasting results. For preliminary results with four months' data, the forecasts are reasonably good.

62

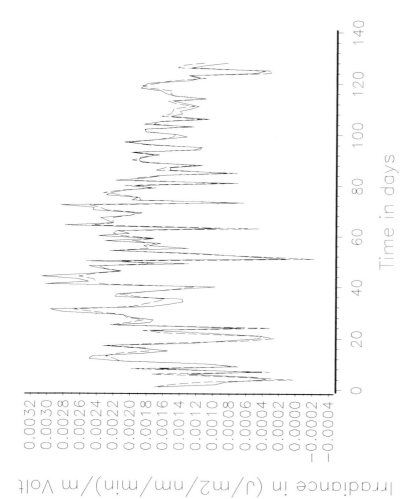

Figure 2: Predicted and observed values for Maximum Daily Irradiance at 291.5nm (using auto-regressive model, with 305.9 nm as predictor).

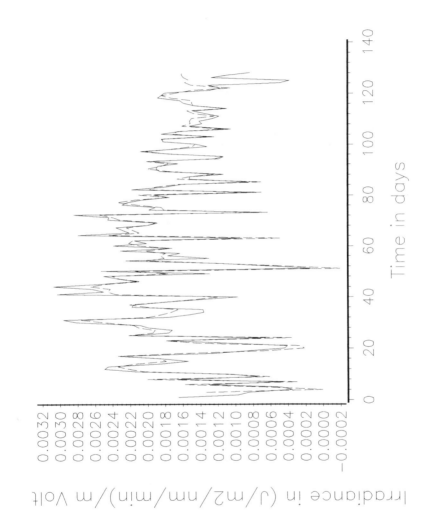

Figure 3: Predicted and observed values for Maximum Daily Irradiance at 291.5nm (using transfer function model, with 305.9 nm as predictor).

Analysis of Hourly Irradiance

From Figure 1, the daily cycle is obvious since at all wavelengths, the intensity reaches a maximum around noon and falls to low values in the morning and evening. The significant cyclical component of the solar hourly irradiance must be adjusted before any statistical analysis is done. The method suggested by (*10*) is adopted in this paper. The hourly means of irradiance at all wavelengths for each month is calculated and treated as the steady periodic component of irradiance for each wavelength. This periodic component of solar irradiance is assumed to be deterministic. Residual time series are formed by subtracting these steady, periodic components from the originally measured hourly series. Hence, the hourly irradiance residuals, instead of original hourly data, are used in the following analysis.

The dependent variable or output variable is RES292 (from 291.5 nm), the explanatory variable or input variable is RES306 (from 305.9 nm), both of which are hourly irradiance residuals. The last day of the available data is left out of the model-building process to check the forecasting ability of the built model.

Auto-Regressive Method. In order to detect the lags that cause serious serial correlation, the auto-correlation for the residuals of the regression of RES292 on RES306 is studied and the finl model is given as:

$$RES\,292_t = -1.8 \times 10^{-7} + 0.000478 \times RES\,306_t + \varepsilon_t \qquad (4.1)$$
$$\varepsilon_t = -1.173\,\varepsilon_{t-1} + 0.722\,\varepsilon_{t-2} - 0.196\,\varepsilon_{t-3} - 0.137\,\varepsilon_{t-23} + \nu_t$$
$$R^2 = 0.96$$

Diagnostic checks, discussed earlier, indicated (4.1) adequately fitted the data. The forecasts for the irradiance are formed by the sum of the forecasts for the residuals and the steady periodic component. The forecasts generated by the above regression model along with the observed irradiance for wavelength 291.5 nm are plotted in Figure 4, and the lower and upper 95% confidence limits along with the observed hourly irradiance for wavelength 292 nm are given in Figure 5. The observed irradiance lies well within the constructed 95% confidence limits for the predicted as well as the out-of-sample forecasts.

Transfer Function Model. The transfer function model is used with RES306 is the input series and RES292 is the output series. The auto-correlation and partial auto-correlation for RES306 are studied. The estimated model is expressed as follows:

$$\varepsilon_t = (1 - 1.08419\,B + 0.23312\,B^2 + 0.04522\,B^3)(RES\,306_t - 0.0002238) \qquad (4.2)$$

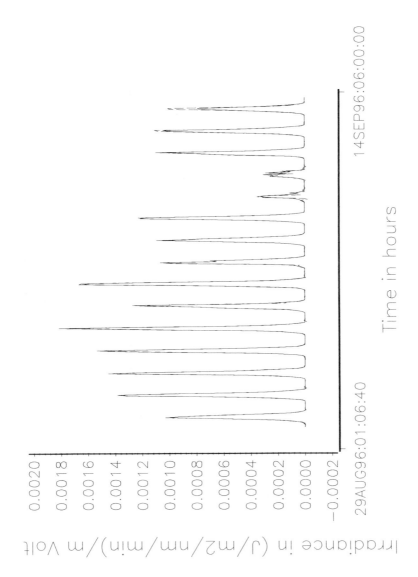

Figure 4: Forecasted and observed values for Hourly mean Irradiance at 291.5 nm (using auto-regressive model, with 305.9 nm as predictor)

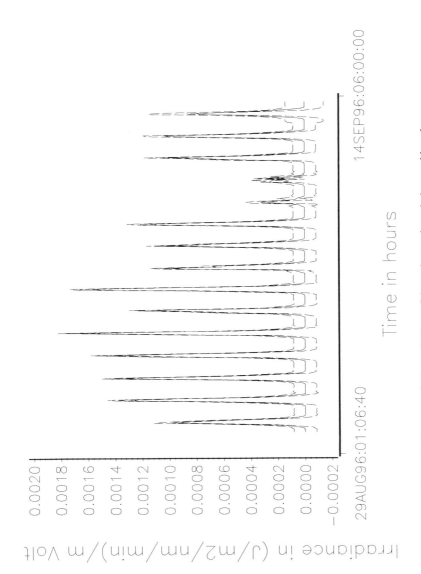

Figure 5: Lower and Upper 95% confidence intervals and observed hourly mean Irradiance at 291.5 nm (using auto-regressive model, with 305.9 nm as predictor)

Diagnostic checks, discussed earlier, were performed to verify that (4.2) fitted our data. The final transfer function model is given as:

$$RES\,292 = -3.85^{-8} + (0.00048/(1-0.0085\,B+0.0316\,B^2))\,RES\,306 \quad (4.3)$$
$$+\,\varepsilon_t/(1-1.175\,B+0.726\,B^2-0.197\,B^3-0.138\,B^{23})$$

The error process for the transfer function model is found to be an AR process, which included the same lags as the above auto-regressive method.

Forecasted lower and upper 95% confidence limits are generated using the fitted model in Equation (4.3) and are plotted in Figures 6 and 7 after taking into account the steady periodic component. The confidence limits captured the observed values. The 95% confidence limits for the out-of-sample forecasts were wider than the auto-regressive model.

Summary

The purpose of this study was to see if knowledge of irradiance of one wavelength would allow effective predictions of irradiance on another wavelength. Solar spectral irradiance data collected at Ocean City, NJ was used.

The daily maximum data was used as the first step. The predicted and forecasted values for both the methods were reasonably close to the observed values. Auto-regressive method had a $R^2 = 89\%$. The "best" transfer function model was picked based on AIC and Q_0 statistic.

For the hourly data, which provided more detailed information, the predictions and forecasts were much better. 95% confidence limits captured the observed irradiance for both models. Auto-regressive method had a $R^2 = 96\%$ indicating good predictability.

The choice of 306 nm as the explanatory variable was arbitrary. Hence, similar auto-regressive analysis was done using 290.2, 305.9, 307.7, 322.1 and 323.6 nm as the explanatory variable using a saturated model of lag 24. The R^2 obtained were 97%, 96%, 95%, 92% and 91% respectively. This reiterates the observation that the closer the explanatory variable is to the response variable the better the prediction. This indicates wavelengths, which are only 2 nm apart closely resemble each other, and using one it is easy to get very good predictions of the other. However, wavelengths that are 20 nm apart do a reasonably good job in predicting irradiance. Since the R^2 obtained for predicting 291.5 nm based on 323.6 nm is 91%.

In summary, with data for separate wavelengths instead of daily or hourly total irradiance data integrated over all wavelengths, this study showed that the irradiance at all wavelengths are highly correlated and the prediction of one particular wavelength using information on the other one is valid in statistical sense. The forecasts produced by these methods are reasonably good especially for the hourly data. It can be assumed that results for the daily maximum data would improve with the incorporation of more data over time.

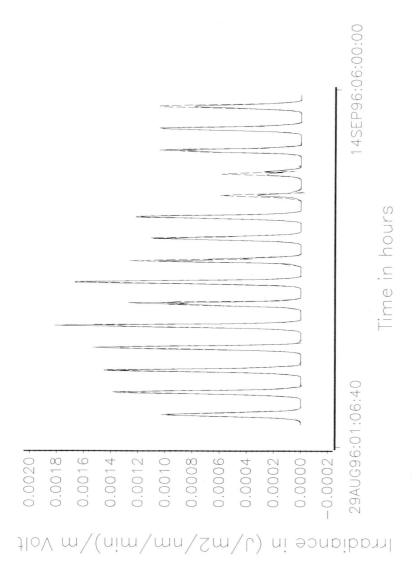

Figure 6: Forecasted and observed hourly mean Irradiance at 291.5 nm
(using transfer function model model, with 305.9 nm as predictor)

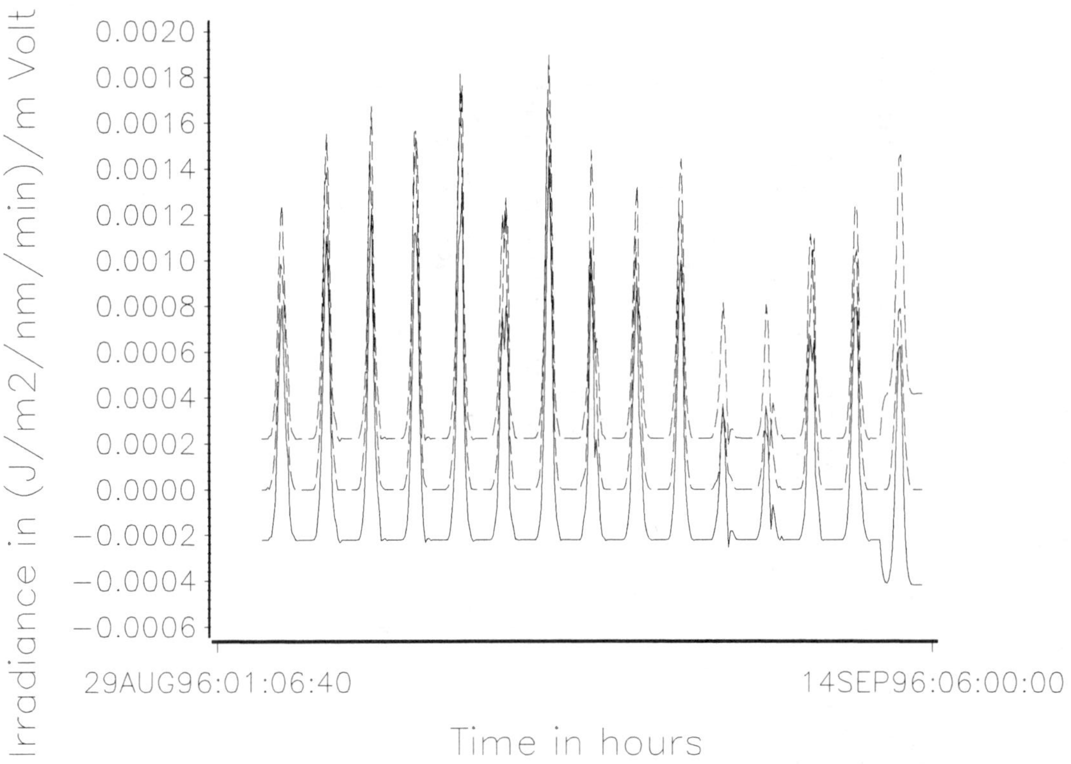

Figure 7: Lower and Upper 95% confidence intervals and observed hourly mean Irradiance at 291.5 nm (using transfer function model, with 305.9 nm as predictor)

Future analysis will show to what extent statistical models vary between sites and between seasons. The results of this analysis provide insight into the design of spectral radiometers in the future. Such design takes into account not only the minimum channels that need to be included to characterize the spectrum, but also the desirability in practice of a certain amount of redundancy in the data set to increase the reliability of detection of small changes in UV over long observation periods.

Literature Cited

(1) Thompson, A.; Early, E. A.; Deluisi, J.; Disterhoft, P. D., W.; Kerr, J.; Rives, J.; Sun, Y.; Lucas, T.; Duhig, M.; Neale, P. J. *National Institutes of Standards and Technology Journal of Research,* **1997**, *102*, 270-322.

(2) Lechner, J. A.; Martin, J. W. *Proceedings of the American Chemical Society Division of Polymeric Materials: Science and Engineering,* **1993,** *69,* 279-280.

(3) Gordon, J. M.; and T. A. Reddy *Solar Energy,* **1988**, *41*,215-226.

(4) Mustacchi, C.; Cena, V.; Rocchi, M.; *Solar Energy,* **1979,** *23*, 47-51.

(5) Vergara-Dominguez, L.;Garcia-Gomez, R.; Figueiras-Vidal, A.R.; Casar-Corredera, J.R.; Casajus-Quiros, F. J. *Solar Energy,* **1985**, *35*, 483-489.

(6) Brillinger, D. R. *Time series data analysis and theory*; Holden Day Inc.: New York, (NY), 1980.

(7) Box E.P.; Jenkins G.M.; Reinsel G.C. *Time series analysis, forecasting and control*; Prentice Hall: Englewood Cliffs, (NJ). 1994;

(8) Wei, W., *Time series Analysis, Univariate and Multivariate Methods*; Addison Wesley Publishers: New York, (NY) 1990.

(9) Judge, G. G.; Hill R.C.; Griffiths W.E.; Lutkephol H.; Lee T.C.; *Introduction to the Theory and Practice of Econometrics*; 2nd edition, John Wiley & Sons, Inc.: New York, (NY), 1988.

(10) Boland, J. *Solar Energy,* **1995**, *55*, 377-388.

Chapter 6

Predicting Temperatures of Exposure Panels: Models and Empirical Data

Daryl R. Myers

National Renewal Energy Laboratory, Center for Renewable
Energy Resources, 1617 Cole Boulevard, Golden, CO 80401

This paper discusses meteorological and temperature measurements and heat transfer models for predicting panel temperatures as a function of solar irradiance, orientation, ambient temperature, windspeed, and wind direction during outdoor exposure testing. Sources of uncertainty associated with physical measurements of meteorological and temperature data, and the propagation of uncertainty into model results are briefly addressed. Equations for making estimates of panel temperatures, based on fundamental principles of time series analysis, and radiative, conductive, and convective heat transfer from empirical data will be presented. We will also discuss models and data used to predict or simulate panel temperature responses in various meteorological climates.

The thermal properties of materials and coatings and the response of these material to the thermal environment are significant components of the degradation mechanisms for materials. Two models for the rates at which reactions occur as a function of temperature are those of *Arrhenius*:

$$R = A e^{-(\frac{E}{kT})}$$

and *Eyring:*

$$t = \frac{A}{T} e^{\frac{B}{kT}}$$

where R is the reaction rate, t is a "nominal" lifetime , k is the Boltzman Constant, A and B are characteristic of the material or failure mechanisms, T is the absolute temperature, and E is the activation energy for the reaction. (*1*) It is this exponential dependence of the rate and lifetime on temperature that make it of critical interest.

Correlating changes in material properties with environmental stresses induced naturally (through outdoor exposure) or artificially (through indoor, possibly accelerated, testing) with a given level of confidence requires knowledge of the uncertainties associated with measurement data, and good experimental design. Predicting material responses to a variety of conditions requires estimates of the accuracy or uncertainty of the modeled environmental conditions and the material properties under investigation.(2,3)

Measurement Issues

Measurements of changes in material properties and temperature data are the basis for studying the effects of thermal energy on material properties. Understanding the measurement processes and their associated uncertainties is needed when analyzing and interpreting the data. The following comments emphasize that the precision of model calculations and results will often far exceed our measurement capabilities.

The physical realization of the theoretical concept of temperature, or thermal energy associated with any material has long been a scientific challenge. Today's technological tools of electronic data loggers and sensors, make its easy to assemble a temperature measurement system that can provide either relatively accurate, or highly misleading temperature data.

The basic issues associated with the accuracy of temperature measurement include the physical principle involved (changes in state or electrical properties), and sources of error due to installation or instrumentation. For any temperature measurement system the temperature of the sensor,, not necessarily the temperature of the device or medium the sensor is attached to is recorded (4).

The International Practical Temperature Scale (5) is realized by national standardizing laboratories with an accuracy of approximately 0.0005 Kelvin (K). The scale is founded on extensive sensor characterization and international determinations of scale reference points (freezing, boiling, or melting points of various materials). Secondary temperature standards used in corporate and research metrology (calibration) laboratories can be accurate to 0.002 K. Commercial calibration standards for temperature are accurate to 0.02 K. Commercial sensor elements are quoted with accuracies of 0.2 to 2.0 K over the range of -40 °C to +100 °C. Table I summarizes the level of accuracy to be expected from this range of sensors.

Table I. Temperature Sensor Accuracies -40 ° C to +150 °C

Sensor Type	Typical Accuracy
Standard Platinum Resistance Thermometer	0.001 K (1 milliKelvin)
Platinum Resistance Thermometer	0.02 K
Thermistor	0.2 K to 3.0 K
Thermocouple	0.2 K to 2.0 K
Radiometric methods	1.0 K to 10 K

The above specifications address only the accuracy of sensors. Errors associated with self-heating by excitation currents, thermal gradients, the quality of the thermal contact of the sensor with the material to be measured, linearity, and the accuracy of electronic data collection equipment are ignored. Detailed uncertainty analysis to estimate the magnitude of such error sources is required to quantitatively evaluate the uncertainty in any measurement system.(2,3)

Uncertainties quoted in Table I may increase by up to a factor of two, even in well characterized measurement systems, in the presence of these additional sources of error. Thus the accuracy of real world temperature measurements can estimated as approximately 0.5 °C to 1.0 °C. The World Meteorological Organization (WMO) specifies the accuracy requirements for synoptic temperature observations of ambient temperature at 0.5 °C (6). The combined uncertainty in very good meteorological and device temperature measurements would then be 1.0 °C to 2.0 °C.

Uncertainties of this magnitude need to be kept in mind when comparing thermal model computations with measured data. Modern computer software and data analysis products often display an inordinate number of significant digits which cannot be taken at face value. An appreciation of measurement accuracy combined with engineering judgment should be used in evaluating model results.

Heat Transfer: Theoretical Considerations

The three fundamental means of exchanging thermal energy, or heat transfer, are conductive, convective, and radiative transfer (7,8). The first two involve physical media, namely solids and fluids (either liquids or gases). The third is based on the transmission, absorption, and reflection of energy in the form of photons. We outline the basic principles of each form of energy exchange to further the understanding of the models discussed later.

Heat Transfer Mechanisms. When there is a temperature gradient within a solid body, the transfer rate, q, or *conduction*, of energy through the body is function of the thermal conductivity, k, the area, A, and the temperature gradient normal to the area. This is *Fourier's Law*, expressed as:

$$q = -kA\frac{\partial T}{\partial X}$$

where the minus sign reflects the second law of thermodynamics requiring heat transfer to be from warmer to cooler regions.

When the temperature within a body is changing with time, the rate at which energy is transferred Q_k, is a function of the mass, m, the heat capacity, c_P, and the rate of change of temperature, ∂T, as a function of position within the material, ∂X, namely

$$Q_k = m \cdot c_p \frac{\partial T}{\partial X}.$$

Convective heat transfer occurs when a fluid (liquid or gas) receives or provides thermal energy to a system. *Newton's Law of Cooling* states that the heat, q, transferred between two a body immersed in a fluid at different temperatures, is proportional to the area of contact and the difference in temperature:

$$q = h \cdot A \cdot (T_s - T_f)$$

where h is the convective heat transfer coefficient, A is the area, and Ts and Tf are the temperatures of the surface and fluid, respectively.

Thermal conductivity of solids varies as a function of temperature, generally as a quadratic in the difference between a reference temperature and the body temperature. For "low" temperatures (within one hundred degrees of ambient), the relationship is generally linear: $k = k_0 (1 + b (T-T_{ref}))$ where k_0 is the thermal conductivity at a reference temperature. In applications around ambient, tabulated values of thermal conductivity are adequate.

The transfer of energy through substrates to coatings, or in the reverse direction, requires application of the principles of conductivity. When several layers of material separate two different temperature regimes (in the simplest case, at steady state conditions), superposition holds, and a sequential application of Fourier's law is appropriate. Integration of Fourier's law over the thickness of a material layer results in an expression of the heat as proportional to the difference in temperature (thermal 'potential difference') and inversely proportional to the thermal conductivity (thermal 'resistance'); namely

$$q = -kA \frac{T_2 - T_1}{x_2 - x_1} = -kA \frac{\Delta T}{\Delta x} .$$

For layers of materials in 'series', where the heat energy flowing out of one layer enters the adjacent layer the continuity of the heat flow allows summation of thermal 'resistances' analogous to Ohm's law for electrical resistances in series. In a two-layer sheet of material situated in an environment with temperature T1 at side one of layer one, and T3 on side two of layer two, assuming thermal conductivities are k1 and k2, and thickness x1 and x2, at the boundary between the layers,

$$q_1 = \frac{T_2 - T_1}{\Delta x_1 / A k_1} = \frac{T_3 - T_2}{\Delta x_2 / A k_2} = q_2 .$$

Combining the two terms, the total heat flow between the regions separated by the layers is:

$$q_{total;} = \frac{T_3 - T_1}{(\Delta x_1 / A k_1) + (\Delta x_2 / A k_2)} .$$

The extension to multiple layers is clear.

For outdoor exposure testing of coatings and thin laminates, especially of small samples, it is likely that thermal equilibrium is rapidly achieved due to the small distances (x) and areas (A) involved. For larger samples, temperature differences as a function of position (i.e., temperature gradients) may complicate the situation.

The above equation shows that a layer of low conductivity material on the back of a panel irradiated on the front reduces the heat flow, and the panel temperature rises as thermal energy is stored.

While liquids are widely used as conductive heat transfer media, they are not addressed here, except to mention that their thermal conductivities are generally linear functions of temperature, and independent of pressure.

Fluid Mechanics, Convective Heat Transfer, and Air. The ability of air to impart or carry away thermal energy in exposed panels is a result of air's thermal and fluid mechanical properties, which are used in modeling heat transfer (*8*). Free convective heat transfer is due to bouyancy forces driven by density differences in the fluid occuring due to contact with a surface at a different temperature. Forced convective heat transfer occurs when an external force (in outdoor cases, wind) moves the fluid past a surface at a temperature different from the fluid. Table II displays the thermal properties of the fluid of interest to us, air, at typical atmospheric pressure.

Table II. Thermal Properties of Air at Atmospheric Pressure

Parameter	Symbol	Value
Thermal Conductivity	k	$5.53 \cdot 10^{-6}\, kcal\,/\,m\,/\,s\,/\,K$
Density	ρ	1.17 kg/m³ @ 40°C
Heat Capacity @ atm press.	C_P	0.24 kcal/kg/K
Thermal Diffusivity	$\alpha = \dfrac{k}{\rho C_p}$	0.0929 m²/hr
Kinematic Viscosity	$v = \dfrac{\mu_m}{\rho}$	1.86 m²/sec
Free Flow Heat Transfer Coefficient	h	0.001 - 0.003 kcal/ m²/sec/°C
Forced (*) Flow Heat Transfer Coefficient	h	0.012 - 0.020 kcal/ m²/sec/°C

Useful conversion factors are 1 kcal/ m²/sec/°C = 738 British Thermal Units (Btu)/ft²/hr = 4184 W/ m²/K. The parameter μ_m is the dynamic viscosity, kg/m-sec (*9*).
(*) Note 'forced' in this context refers to natural wind streams with velocities driven by meteorological pressure gradients, as opposed to fans and turbines moving fluids in pipes.

Parameterization Variables. To characterize the energy transport properties of fluids and gases, many parameterization variables have been developed. The most important of these distinguish the type of flow (laminar or turbulent), fluid transport or motion properties, relations between conductive and convective heat transfer coefficients, and buoyancy and viscosity. Table III displays common parameterization variables, their definitions, and the fluid or flow property they represent.(see *8,9*)

Table III. Transport Property Parameters for Air at Atmospheric Pressure

Parameter	Symbol	Property
Reynolds Number	$Re = \dfrac{\rho V L}{\mu_m}$	Laminar or Turbulent Flow
Prandtl Number	$Pr = \dfrac{C_p \mu_m}{k}$	Fluid Motion/Transport
Nusselt Number	$Nu = \dfrac{hL}{k}$	Convective (h) to conduction (k) ratio. (conduction at interface)
Grashof Number	$Gr = \dfrac{g\beta \Delta T L^3}{v^3}$	Buoyancy/Viscosity ratio
Stanton Number	$St = \dfrac{Nu}{Re \cdot Pr}$	Heat Transfer Related to Skin Friction

V is velocity, L is length, v is the kinematic viscosity defined in Table II, and $\beta = \dfrac{1}{v}\dfrac{\partial v}{\partial T}$ is the volumetric expansion ratio as a function of temperature.

The typical range of values for these parameters under natural meteorological conditions are: Re > 300000, Pr = 0.70, $100 \leq Nu \leq 2000$, $2 \cdot 10^7 \leq Gr \leq 3 \cdot 10^{10}$. All of these parameters and their applications are extensively described in most textbooks on heat transfer. Simplified engineering equations relating the element of primary interest here, the heat transfer coefficient of air flowing around a flat plate have been developed both empirically and theoretically (7). In particular, for air flowing around a flat plate tilted at an angle θ, we have

$$\frac{hL}{k} = Nu = C(Gr \cdot \cos(\theta) \cdot Pr)^a$$

or

$$h = \frac{k \cdot C(Gr \cdot \cos(\theta) \cdot Pr)^a}{L} = \frac{Nu \cdot k}{L}.$$

Table IV displays combinations of the a, C, and L parameters for determining a (wind) forced convective heat transfer coefficient for air under turbulent and laminar flow situations.

Table IV. Fluid Parameters for Heat Transfer Coefficients of Flowing Air

Flow Condition	Gr Pr	L	C	a
Laminar	$10 \cdot 10^5 - 2 \cdot 10^7$	0.5(L1+l2)	0.54	0.25
Turbulent	$2 \cdot 10^7 - 3 \cdot 10^{10}$	0.5(L1+L2)	0.14	0.33

For GrPr = $2 \cdot 10^7$, L =0.5, θ= 30 °, hforced = 0.016 kcal/m²-s. As a rule of thumb, wind forced convection heat transfer coefficients will be in the range of 0.012 to 0.030

kcal/m²-s. For free convection, as occurs in calm air, or from the surface of a panel in a 'wind shadow',

$$h_{free} = \frac{0.21 \cdot (Gr \cdot \cos(\theta) \cdot \text{Pr})^{0.33}}{L}.$$

which is generally in the range of 0.001 to 0.003 kcal/m²-s. (8) Free and (wind) forced convective heat transfer are considered in a thermal model developed by Fuentes (10) for predicting photovoltaic panel temperatures during outdoor exposure testing (11), which we discuss below.

Radiative Transfer . The exchange of thermal energy between two objects of different temperatures, T_1 and T_2, via photons is referred to as radiative heat transfer. This type of transfer requires no intervening medium (i.e., it can occur in a vacuum). The energy transfer is proportional to the fourth power of the temperatures of the interacting bodies. This is expressed in the form of the *Stefan -Boltzman Law*

$$q = \varepsilon\sigma (T_2^4 - T_1^4)$$

where the emissivity, ε, is less than 1, and σ is the Stefan-Boltzman constant $5.667 \cdot 10^{-8}$ Watts/m²/K^4 (K=degrees Kelvin, 273.15 K = 0 °C and 1 K = 1 °C).

The properties of absorptivity, α, transmissivity, τ, reflectivity, ρ, and emissivity, ε, of materials are material dependant. The law of conservation of energy requires that α, τ, and ρ sum to unity (1), since they represent different proportions of the total radiant energy. Note that for solid materials, $\alpha + \rho = 1$, since $\tau = 0$, and for gases (transparent materials) $\alpha + \tau = 1$, as $\rho \sim 0$.

A totally absorbing and perfectly emitting source is the Planck Blackbody. The emissivity of a material is defined as the ratio of the total radiation emitted by the body to that of a perfect blackbody at the same temperature. *Kirchoff's Law* states that the ratio of the emissivity and absorptivity of a body in thermal equilibrium with it's surroundings is 1, so $\alpha = \varepsilon$. This is strictly true on a wavelength by wavelength basis even if the body of interest is surrounded by objects at different temperatures.

A simplifying assumption often made is the gray body assumption, that the emissivity (and absorptivity) of materials is constant independent of wavelength, and the emitted radiation is some fraction of that emitted by a blackbody. This greatly simplifies the computation of energy exchanges between objects. An expression for the steady state energy balance for a horizontal plate exposed to the sun, sky, and ground out of doors is given in Saunders, Jensen and Martin (12) :

$$A \cdot H \cdot \alpha = 2hA(T_p - T_a) - \varepsilon A\sigma(T_s^4 - T_p^4) - \varepsilon A\sigma(T_b^4 - T_p^4)$$

where α is the plate absorptivity, H is the total solar radiation, h is an 'average' convective heat transfer coefficient, ε the panel emissivity, T_a ambient temperature, T_b ground temperature, T_s sky temperature, T_p the panel temperature, and σ is the Stefan-Boltzman constant.

Modeling Panel Temperatures.

An application of the principles described above is the question of how photovoltaic (PV) modules operate outdoors. Module electrical performance is known to be a function of the module temperature. Menicucci and Fernandez (*11*) developed the PVFORM simulation program at the Sandia National Laboratories to model hourly PV array performance throughout a year of 8760 hours. PV module performance is a function of many variables, the most important of which are the module conversion efficiency, solar radiation, and module temperature. We extracted the subroutine used to compute the thermal performance of a PV module and used it to compute one year of modeled panel temperatures. See the Appendix for a brief synopsis of the model.

We compared the model results with measured data at the Outdoor Test Facility at the National Renewable Energy Laboratory (NREL), latitude 39.74 N, Longitude 105.178 W, at an elevation of 1740 m. The module was mounted at a 50 degree south facing tilt, and instrumented with type-T copper constantan thermocouples in three places across the module (two edges and in the center). The panel temperature sensors, ambient temperature, and solar radiation data (as well as electrical performance parameters we will not discuss) were sampled at 10 second intervals, and averaged over a one hour period. The estimated uncertainty in the temperature and irradiance measurements are ± 1°C, and $\pm 5\%$ (50 W/m² out of 1000 W/m² full scale).

The Fuentes thermal performance model incorporates the computation of the radiative and convective exchange between the ground, the sky, the panel, and the surrounding air using the parameterization variables discussed above. Figure 1 is a time series plot of the measured panel temperature (center thermocouple only) and the modeled panel temperature. For clarity, the modeled temperature (the top curve) is the modeled data +50°C, and the bottom curve is the actual measured panel temperature.

The mean daylight ambient temperature for the year was 12.9 °C with a standard deviation of ± 9.8 °C. The mean of the measured panel temperatures was 17.5°C ± 11.0 °C, and the mean of the modeled panel temperature was 15.2 °C $\pm$ 12.0 °C. The difference between measured and modeled means of 2.3°C is about twice the estimated measurement uncertainty. Errors in the estimates of module emissivity (absorptivity) and thermal conductivity, poor thermal contact between the module temperature sensor, or differences in the characteristics of the module and air temperature sensors, could be the source of the biases observed.

Figure 2 is time series plot of the difference between model and measured panel temperature. Larger scatter, and apparent sinusoidal pattern of offsets occur in the early and late parts of the time series, which represent spring and fall months. These characteristics may be due to the way sky temperature was computed, or the more variable meteorological conditions which occur in spring and fall.

The mean difference between the model data and the measured data is -1.8 °C. The panel parameters used in the calculations were an emissivity of 0.63, an absorptivity of 0.63 (about the same as for rough steel or glass), a specific heat capacity of 11000 Joules/m²-K, a height of 1 meter, and tilt angle of 50°.

The Fuentes estimate of the sky temperature might be improved using an estimate such as that of Martin and Berdahl (*13,14*):

$$T_s = T_a[0.711 + 5.6 \cdot 10^{-3}T_{dp} + 7.3 \cdot 10^{-5}T_{dp}^2 + 0.013 \cdot \cos(15t) + 1.2 \cdot 10^{-4}(P - 1000)]^{1/4}$$

Measured versus Modeled Hourly Panel Tempertures

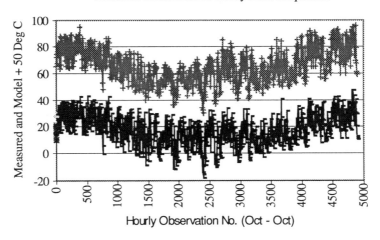

Figure 1. Measured (bottom) and modeled +50°C (top) time series of panel temperatures for one year (October 95 to October 96) at NREL Outdoor Test Facility. High frequency variations are diurnal cycles, overall sinusoidal shape is due to the annual variaton in temperature.

Where T_a and T_{dp} are ambient and dew point temperatures, t is the hour of the day (0-24, 0=midnight), and P is the atmospheric pressure in millibars.

Simulating Meteorological Data.

Thermal properties of material or coating samples can be experimentally determined in the laboratory. Acquiring the necessary meteorological data to play against these parameters can be difficult. Rarely are both the meteorological and solar radiation data needed to run a model calculation available. Many methods have been developed to model both solar radiation and ambient temperature conditions, given only long term averages and ranges (known as means and normals) of such data.

One method for generating time series estimates of meteorological and solar radiation data use Fourier analysis and Autoregressive Moving Average (ARMA) analysis of previously measured data (*15,16*) These approaches compute the present estimate of the parameter from a linear combination of past values, sometimes adding an appropriate random component (white noise).

Saunders, Jensen, and Martin (*12*) use measured hourly data smoothed using moving averages. Each day is treated as a separate data subset, for which the Fourier transform coefficients are derived. The distribution of the coefficients is then examined to determine the statistical or stochastic properties. Long term mean temperatures and standard deviations can then be modified by applying coefficients selected from the distributions. They provide FORTRAN 77 source code for their estimation technique in reference *12*.

Hokoi, Matsumoto, and Kagawa (*15*) , derive the Fourier coefficients by removing annual variations. They normalize previously measured hourly data by the ration sin(ho)/sin(h), where ho is the solar elevation at noon at the midpoint of the year (7/21) and h is the solar elevation at noon on each day. Next, a mean diurnal profile, M(t), for the year for each parameter (solar radiation and temperature) is determined, as

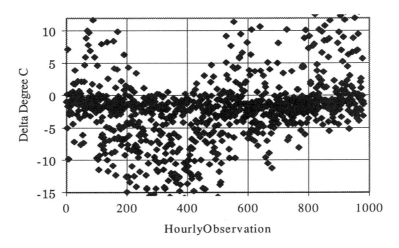

Figure 2. Difference (Modeled minus Measured) panel temperature time series.

well as the associated diurnal profile of standard deviations for each hour. The Fourier coefficients a and b are derived as usual:

$$M(t) = f_o + \sum_j (a \cdot \cos(2\pi \, j\Delta T / 24) + b \cdot \sin(2\pi \, j\Delta T / 24)).$$

Subtracting the mean dirurnal profile from the time series of the data, F(t), results in the random, statistical component: S(t)=F(t)-M(t)

The autocorrelation of the time series data divided by the random component Z(t)=F(t)/S(t) is computed from:

$$G(N) = \frac{1}{n} \sum_i Z_i \cdot Z_{i+N}$$

where N is the lag, or number of time steps by which the normalized time series is shifted, and n is the number of data points in the series. For solar radiation data, nighttime values of 0 are eliminated and not considered. The objective is to choose linear combinations of modeled data that produce the same autocorrelation function as the measured data.

The number of terms used in the linear combination of previous estimates and number of random components used is the order of the ARMA model. Hokoi et al. determined that for solar radiation data, a third order, or ARMA(3,3) model, of the form:

$$S_i - 1.60 \cdot S_{i-1} + 0.64 \cdot S_{i-2} - 0.03 \cdot S_{i-3} = \sigma_i + 0.3 \cdot \sigma_{i-1} - 1.0 \cdot \sigma_{i-2} - 0.16 \cdot \sigma_{i-3}$$

reproduced the statistical properties of solar radiation at Tokyo. The σ_i are selected from a Gaussian distribution with standard deviation equal to that of the data, σ, (0.114 for the data used). The time series of estimated data was constructed by adding in the M(t) term :

$$F_{est} = M(t_i) + S_i = 1.6 \cdot S_{i-1} - 0.64 \cdot S_{i-2} + 0.003 \cdot S_{i-3} + \sigma_i + 0.3 \cdot \sigma_{i-1} - 1.0 \cdot \sigma_{i-2} - 0.16 \cdot \sigma_{i-3}$$

See reference (*16*) for more detailed discussion of this approach and technique.

Similarly, annual and diurnal variation in temperature data can be evaluated. In this case, the cross-correlation of temperature with solar radiation data come into play, contributing additional two additional terms to form an ARMA (3,2,3) model of the form:

$$T_i - 1.673 \cdot T_{i-1} + 0.692 \cdot T_{i-2} - 0.0062 \cdot S_i - 0.00867 \cdot S_{i-1} - 0.00266 \cdot S_{i-2}$$
$$= \varpi_i - 0.586 \cdot \varpi_{i-1} - 0.184 \cdot \varpi_{i-2}$$

where the Ti are the three statistical components of the temperature data, and the Si are the cross correlations for the solar radiation data derived earlier. Random components of the temperature ϖ_i come from a Gaussian distribution with standard deviation ϖ, (0.276 for the Tokyo data). These models, and the Fast Fourier Transform approach of Saunders, Martin, and Jensen reproduce faithfully the stochastic characteristics of the data used in their generation.

Another approach is to randomly select daily maximum and minimum temperatures from a normal distribution for the climate in question. The daily statistical values are arranged in order using site independent distributions for establishing the sequence (17,18). Hourly values are modeled using a cosine shaped curve fit to the daily maximum and minimum values. This technique results in a deterministic value for hourly temperature, with no additional random component. The loss of information is not so important, since random variations tend to cancel out through the year. See (17) for a detailed discussion.

Obtaining Meteorological Data. Techniques described above rely on measured data, or statistics such as means, normals, and extremes, and perhaps standard deviations , which may be estimated from minimum and maximum values (19). The National Climatic Data Center, (NCDC) Climate Services Branch, Federal Building, Ashville, NC 28801-2733, phone 704-271-4800. NCDC provides a wide variety of both solar and meteorological data on Compact Digital-Read Only Memory (CD-ROM) and magnetic media. NCDC also is the office for the World Data Center for Meteorology and exchanges data with foriegn countries and other World Data Centers in Japan, China, and the Commonwealth of Independent States (Russia).

Examples of published solar and climate data summaries are the *Solar Radiation Data Manual for Flat-Plate and Concentrating Collectors* (20) and the *Solar Radiation Data Manual for Buildings* (21) available from the National Renewable Energy Laboratory Document Distribution Center, 1617 Cole Blvd, Golden CO 80401, 303-275-4363. These are summaries of 30 year monthly mean, minimum, and maximum solar radiation, ambient temperatures, windspeed, and relative humidity derived from hourly data for 239 sites across the U.S. (22,23,24)

Conclusion

We discussed typical uncertainties associated with temperature data using various sensors. Accuracies of 1°C to 2 °C represent high quality measurements in the outdoor environment. Heat transfer from the environment to exposure panels, and fluid mechanical properties of air and their role in estimating heat transfer were briefly described. A simple thermal model was evaluated in the light of measured data with estimates of uncertainty. Sources of highly summarized and hourly resolution solar and meteorological data were provided. That data could be used for the basis of ARIMA and Fourier Transform based models for generating time series data for estimating exposure panel temperatures.

Appendix. Description of Fuentes' PV Module Thermal Model.

Fuentes assumes heat balance resulting in this expression for module temperature (T_m):

$$T_m = \frac{(h_c \cdot T_a + h_s \cdot T_s + h_g \cdot T_g + \alpha \cdot (H_0 + \frac{\Delta H}{L})) \cdot (1 - e^L) + \alpha \cdot \Delta H}{h_c + h_s + h_g} + T_{m0} \cdot e^L$$

where

$$L = -(h_c + h_s + h_g) \cdot \Delta t / (m \cdot c) \qquad \text{and}$$

Tmo	module Temperature at time to, K;	m	mass per unit area (kg/m²);
Tm	module temperature at time t, K;	c	panel specific heat (J/kg K);
Ta	ambient temperature K;	Ts	sky temperature K;
Tg	ground temperature, K;		

hc   overall convective heat transfer coefficient (W/m²-K)

hs   radiative heat transfer coefficient to the sky (W/m²-K)

hg   radiative heat transfer coefficient to the ground (W/m²-K)

α module absorptivity

Ho   plane of panel irradiance for previous time step, W/m²

ΔH   change in plane of panel irradiance from previous time step, W/m².

Free and (wind) forced convective heat transfer are considered. The free convective heat transfer coefficient is $h_{free} = 0.21 \cdot (Gr \cdot 0.71)^{0.32} \cdot k / D$ where k is the thermal conductivity of air (2.1695E-4*Tfilm). Tfilm is estimated as the average of Tm and Ta. D is the estimated 'hydraulic diameter' of the plate, assumed to be 0.5 meters. The forced convective heat transfer coefficient is computed as $h_{forced} = b \cdot (Re^c / 0.71^a) \cdot \rho \cdot 1007 \cdot w$ where w is the wind speed at the module height, and a, b and c are selected depending on the Reynolds number (0.67,0.86, and -0.5 for laminar flow, or 0.4, 0.028 and -0.2 for turbulent flow). ρ is the density of air computed at the average of Ta and Tm: 0.003483*101325/Tfilm. The overall heat transfer coefficient, hc, is the cube root of the sum of the cubes of the free and forced coefficients. Sky temperature estimate is computed from the ambient temperature using $T_s = 0.68*(0.0552*T_a^{1.5})+0.32*T_a$.

Ground temperature, Tg, is assumed to be between the panel and ambient temperatures. Initial values of Ta, Ho, and Tmo are used to begin the simulation. Subsequently, the irradiance, ambient temperature, and wind speed data are used to compute the convective and radiative heat transfer coefficients and the new module temperature.

References

1.   Nelson, Wayne. *Accelerated Testing Statistical Models, Test Plans, and Data Analysis.* Wiley Series in Probability and Mathematical Statistics, John Wiley and Sons: New York, NY, 1990; Chapter 2, pp 76-100.

2.   Abernethy, R.B.; Benedict, R.P. *ISA Transactions* **1985**; 2 ,74.

3.   International Organization for Standardization *BIPM, IEC, IFCC, ISO, IUPAC, IUPAP, OIML,Guide to the Expression of Uncertainty in Measurement,* , 1993; ISO, Geneva, Switzerland.

4.   Houghten, F.C.; Olson, H.T. In *Temperature Its Measurement and Control in Science and Industry,* American Institute of Physics, Reinhold Publishing Corporation, New York, NY, 1941; pp 855-861

5.   Preston-Thomas, H.; Quinn, T.J. In *Temperature Its Measurement and Control in Science and Industry,* American Institute of Physics, Reinhold Publishing Corporation, New York, NY, 1992; pp 63-67.

6.   World Meteorological Organization, *Guide to Meteorological Instruments and Methods of Observation,* Fifth Ed., WMO No. 8, Secretariat of the World Meteorological Organization, Geneva, Switzerland, 1983; Chapter 4, pp 4.1-4.19

84

7.   American Society of Heating, Refrigerating, and Air-Conditioning Engineers, *ASHRAE* Handbook *1977 Fundamentals*, ASHRAE, New York, NY, 1977; Chapter 2, pp 2.1-2.34
8.   Karlekar, B.V.; Desmond, R.M. Engineering *Heat Transfer*, West Publishing Co. St. Paul, MN, 1977; Chapter 7, pp 294-317.
9.   Tuma, J.J., *Handbook of Physical Calculations*, McGraw Hill Book Company, New York, NY, 1976; Chapter 6, pp 142-158.
10.   Fuentes. M.K., *A Simplified Thermal Model of Photovoltaic Modules,* Sand79-1785, Sandia National Laboratories, Allbuquerque, NM, 1980.
11.   Menicucci, D.F.; Fernandez, J.P., *Users Manual for PVFORM: A Photovoltaic System Simulation Program for Stand-Alone and Grid-Interactive Applications.* Sand85-0376, Sandia National Laboratories, Allbuquerque, NM, 1991.
12.   Saunders, S.C; Jensen, M.A., Martin, J.A.;*A Study of Meteorological Processes Important in the Degradation of Materials Through Surface Temperature.* NIST Technical Note 1275, National Institute of Standards and Technology, Gaithersburg MD, 1990.
13.   Martin, M.; Berdahl, P., *Solar Energy*, **1984**;Vol 33, No. 3/4 pp 241-252
14.   Martin, M.; Berdahl, P., *Solar Energy*, **1984**;Vol 33, No. 3/4 pp 321-336
15.   Hokoi, S; Matsumoto, M.; Kagawa, M. *ASHRAE Transactions.* **1990**, Vol . 96, Pt. 2
16.   Bennet, R.J., *Spatial Time Series Analysis,*Pion Limited, London, 1979, Chapter 2, pp.77-81.
17.   Knight,K.M; Klein, S.A.; Duffie, J.A., in *Solar '90, Proceedings of the 1990 Annual Conference American Solar Energy Society;* Burley, S.; Coleman, M.J., Ed. American Solar Energy Society, Boulder, CO. 1990
18.   Degelman, L.O., *ASHRAE Transactions, Symposium on Weather Data Seattle, WA.* June 1976 pp 435
19.   Natrella, M.G., *Experimental Statistics,* National Bureau of Standards Handbook 91, National Bureau of Standards (now National Institute of Standards and Technology) Gaithersburg, MD, 1963. Chapter 2-2, pp 2.6
20.   NREL, *Solar Radiaiton Data Manual for Flat Plate and Concentrating Collectors,* NREL/TP-463-5607. National Renewable Energy Laboratory, Golden, CO. 1994
21.   NREL, *Solar Radiaiton Data Manual for Buildings,* NREL/TP-463-7904. National Renewable Energy Laboratory, Golden, CO. 1995.
22.   Maxwell, E.L., in *PV Radiometrics Workshop Proceedings,* Myers, D., Ed. NREL/CP-411-20008, National Renewable Energy Laboratory, Golden, CO. 1995
23.   Stoffel, T.S., Ibid
24.   Marion, W., Ibid.

Chapter 7

Predicting the Temperature and Relative Humidity of Polymer Coatings in the Field

D. M. Burch[1] and J. W. Martin[2]

[1]Building Environment Division, Room B320, Bldg. 226
[2]Building Materials Division, Room B348, Bldg. 226, Building and Fire
Research Laboratory National Institute of Standards and Technology,
Gaithersburg, MD 20899

This paper explores the applicability of the NIST Moisture and Heat
Transfer Model (called MOIST) to predict the temperature and relative
humidity (or moisture content) of a polymer coating exposed to outdoor
weather conditions. The rate of degradation of a polymer coating is
caused primarily by climate stress of sunlight, temperature, and water
(dew, humidity, and rain). For a coating that fails through a loss of
appearance, the primary degradation agent has been shown to be
ultraviolet radiation with temperature and relative humidity serving as
accelerants. The ability to accurately predict the temperature and
relative humidity of a polymer coating will permit a more precise
evaluation of photo-degradation accumulation process and its effect on
service life.

The MOIST model was shown to provide reasonable predictions
of the temperature and relative humidity of polymer coatings exposed to
outdoor conditions in the field. The model predicted wetting of polymer
coatings during periods having dew condensation.

Martin (*1*) has suggested that a quantitative service-life model (currently used in the
medical, biological, and agricultural communities) may be successfully applied to
predict photolytic material damage in polymer coatings exposed to ultraviolet (UV)
solar radiation in the field. The material damage, Γ, of a polymer coating due to
photodegradation is related to total effective UV dose, D_{tot}, though a damage function.
Various forms of the damage function (linear, power law, and exponential functions) are
described by Martin (*1*).

The total effective UV dose may be evaluated from the following double integral
expression:

U.S. government work. Published 1999 American Chemical Society

$$D_{tot} = \int_0^t \int_{\lambda_{min}}^{\lambda_{max}} I_o(\lambda,t)(1 - e^{-A(\lambda)})\phi(\lambda,T,rh)d\lambda dt \qquad (1)$$

where:

$I_O(\lambda,t)$ = spectral UV irradiance to which a material is exposed at time t (W/cm^{-2})

$(1-e^{-A(\lambda)})$ = spectral absorptivity of the material (dimensionless)

$\phi(\lambda,T,rh)$ = spectral quantum yield (dimensionless)

In the inner integral, the functions are being integrated over wavelengths ranging from the minimum, λ_{min}, to maximum, λ_{max}, photolytically effective wavelengths. In the outer integral, the cumulative dose is found by integrating the functions over a specified time period.

In applying equation 1, the spectral UV irradiance is measured in the field. The spectral absorptivity and spectral quantum yield are measured by conducting a series of laboratory experiments. The spectral quantum yield is defined by:

$$\phi(\lambda) = \frac{No.\ of\ molecules\ undergoing\ \deg radation\ at\ wavelength\ \lambda}{No.\ of\ quanta\ at\ wavelength\ \lambda\ absorbed\ by\ the\ polymer\ coating} \qquad (2)$$

The spectral quantum yield is usually estimated in laboratory measurements from changes in the macroscopic properties (loss of gloss, yellowing, loss of appearance, changes in elongation to break or changes in tensile strength).

In the above quantitative service-life model, it is important to note that the spectral quantum yield is a function of temperature, T, and relative humidity, rh. Both of these parameters act as accelerants in the degradation process. Therefore, if we can predict the temperature and relative humidity of a polymer coating, then we may apply the above theory to predict the photolytic damage of a polymer coating when it is exposed to UV radiation in the field.

The National Institute and Standards and Technology (NIST) has recently developed a heat and moisture transfer model, called MOIST, that predicts the temperature and moisture content (or relative humidity) at the material layers of a building construction (2). This model has been verified by way of comparison to a series of NIST laboratory experiments (3). In these experiments, several wall specimens were exposed to temperature variations. Measurements of inside surface heat fluxes and construction layer moisture contents agreed very well with corresponding values predicted by MOIST. These experiments, however, did not include wetting of the exterior wall surface by dew or rain, which is an important aspect of the present study. The applicability of the MOIST model to predicting the temperature and relative humidity at a polymer coating is explored in the paper presented herein.

Description of Model

Some of more important assumptions are:

      ·   heat and moisture transfer are one dimensional;

      ·   snow accumulation on horizontal surfaces, and its effect on the solar absorptance and thermal resistance is neglected; and

· transport of heat by liquid movement is neglected.

Other assumptions are given in the model description.

Basic Transport Equations. The basic transport equations are taken from Pedersen (4) and are briefly presented below. (Carsten Pedersen has changed his name to Carsten Rode). Within a polymer coating (see Figure 1), the moisture distribution is governed by the following conservation of mass equation:

$$\frac{\partial}{\partial y}(\mu \frac{\partial p_v}{\partial y}) - \frac{\partial}{\partial y}(K \frac{\partial p_l}{\partial y}) = \rho_d \frac{\partial \gamma}{\partial t} \tag{3}$$

The first term on the left side of equation 3 represents water-vapor diffusion, whereas the second term represents capillary transfer. The right side of equation 3 represents moisture storage within the material. The potential for transferring water vapor is the vapor pressure (p_v) with the permeability (μ) serving as a transport coefficient. The potential for transferring liquid water is the capillary pressure (p_l) with the hydraulic conductivity (K) serving as the transport coefficient. The signs on the first two terms are different because water vapor flows in the opposite direction of the gradient in water vapor pressure, and capillary water flows in the same direction of the gradient in capillary pressure. Other symbols contained in the above equation include the dry density of the material (ρ_d), moisture content (γ), distance (y), and time (t). The sorption isotherm (i.e., the relationship between equilibrium moisture content and moisture content) and the capillary pressure curve (i.e., the relationship between capillary pressure and moisture content) were used as constitutive relations in solving equation 3.

The hydraulic conductivity (K) in equation 3 is related to the liquid diffusivity (D_γ) by the relation:

$$K = -\frac{\rho_d D_\gamma}{\frac{\partial p_l}{\partial \gamma}} \tag{4}$$

The term in the denominator of the right side of the equation is the derivative of the capillary pressure with respect to moisture content.

Within the polymer coating (see Figure 1), the temperature distribution is calculated from the following conservation of energy equation:

$$\frac{\partial}{\partial y}(k \frac{\partial T}{\partial y}) + h_{lv} \frac{\partial}{\partial y}(\mu \frac{\partial p_v}{\partial y}) = \rho_d(c_d + \gamma c_w)\frac{\partial T}{\partial t} \tag{5}$$

The first term on the left side of equation 5 represents heat conduction, whereas the second term is the latent heat transfer derived from any phase change associated with

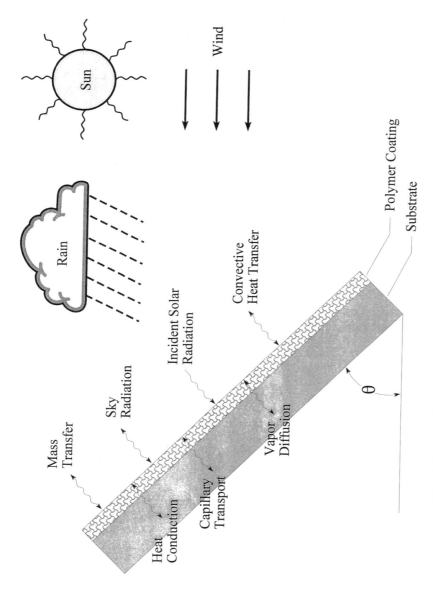

Fig. 1 Schematic polymer coating and substrate showing heat and moisture transport processes.

the movement of moisture. The right side of equation 5 represents the storage of heat within the dry material and accumulated moisture within the materials pore space. Symbols in the above equation include temperature (T), the dry specific heat of the material (c_d), the specific heat of water (c_w), and latent heat of vaporization (h_{lv}).

In equation 3, the water vapor permeability (μ) and the hydraulic conductivity (K) are strong functions of moisture content. In equation 5, the thermal conductivity (k) is also a function of temperature and moisture content.

Boundary Conditions. At the exposed boundary of the polymer coating (see Figure 1), the rate of water available from rain plus the water-vapor transfer rate from the surrounding atmosphere is set equal to the water-vapor transfer rate into the surface plus the liquid transfer rate into the surface, or:

$$\dot{W} + m(p_o - p_s) = -\mu\frac{\partial p_v}{\partial y} + K\frac{\partial p_l}{\partial y} \tag{6}$$

Here $\dot{W}$ is the rate of water availability from rain, m is the mass transfer coefficient, p_o is the outdoor water vapor pressure, and p_s is the surface water-vapor pressure. Rain wetting is not included in the current MOIST, but it is shown above for the sake of completeness. Both the derivatives are evaluated at the exposed surface of the polymer coating. The last term has a positive sign because capillary transport occurs in the same direction as the gradient in capillary pressure.

Also at the exposed surface of the polymer coating, the rate of absorbed solar radiation is equal to sum of the rate of heat conduction into the surface, the heat loss rate from the surface to the outdoor air by convection, and the rate of radiation exchange between the surface and the sky, or:

$$\alpha H_{sol} = -k\frac{\partial T}{\partial y} + h_{c,o}(T - T_o) + h_{r,o}(T - T_{sky}) \tag{7}$$

All quantities are evaluated at the exposed surface. Here $h_{r,o}$ is the radiative heat transfer coefficient defined by the relation:

$$h_{r,o} = E\sigma(T_s + T_{sky})(T_s^2 + T_{sky}^2) \tag{8}$$

where E is the emittance factor which includes the surface emissivity and the view factor from the outdoor surface to the sky, T_s is the surface temperature, and T_{sky} is the sky temperature. When a surface views the sky, the surface exchanges thermal radiation with clouds and green house gases contained in the atmosphere. In these calculations, the sky is treated as an equivalent black body at temperature, T_{sky}.

At the interface between the polymer coating and its substrate, the temperature, water-vapor pressure, and capillary pressure are assumed to be continuous.

Solution Procedure and Model Verification. A FORTRAN 77 computer program, called the MOIST 3.0, has been prepared to solve the above system of equations. It was necessary to modify MOIST to set the indoor (house) temperature equal to the outdoor temperature to reflect the fact that the ambient air temperature beneath the panel would coincide with the outdoor air temperature. In this computer model, finite-difference equations are used to represent the basic moisture and heat transport equations (equations 3 and 5).

Necessary Input Data

The purposes of this section were two fold. First, we wanted to identify the parameters that must be specified as input for MOIST. Second, we wanted to give the particular input parameter values used in the illustrative analysis (presented later).

The physical parameters are given in Table I.

The moisture transfer parameters are given in Table II. The transport properties (μ and D_y) are normally a function of moisture content, but they were assumed to be constant in the illustrative analysis. The MOIST program also requires the following two moisture storage functions: the sorption isotherm (i.e, a relationship between moisture content and relative humidity) and the capillary pressure curve (i.e., a relationship between capillary pressure and moisture content). The two moisture storage functions were found to have very little effect on the predicted temperature and relative humidity.

The heat transfer parameters are given in Table III. Their particular values were observed to have an unimportance effect on the predicted results in the illustrative analysis.

Table IV summarizes the outdoor climate parameters that are needed for an analysis. In the illustrative analysis given in the next section, the outdoor temperature, relative humidity, and wind speed were obtained from Weather Year for Energy Calculations (WYEC) hourly weather data (7). The total incident solar radiation onto the exposed surface was calculated using the algorithms given by Duffie and Beckman (5). Here the total incident solar radiation onto a horizontal surface given in the WYEC hourly weather data was used as input. The long-wave sky temperature was calculated from the outdoor dew point temperature using an equation developed by Bliss (6).

It is worth mentioning that the solar absorptance, thermal emittance, water vapor diffusivity, and liquid diffusivity are currently not measured by the coatings community. The application of MOIST will require that laboratory experiments be conducted to determine these additional parameters. The water-vapor permeability of the polymer coating may be determined by a series of permeability cup measurement as outlined in Burch, Thomas, and Fanney (11). The liquid diffusivity may be determined by measuring the water uptake of a polymer coating wafer after it is submerged in water.

Illustrative Analysis

In the computer simulations, it was only necessary to use two finite-difference nodes in

Table I. Physical Parameters		
Property	**Value**	**Units**
Polymer Coating Thickness	0.13	mm
Solar Absorptance	0.7	Dimensionless
Thermal Emittance	0.9	Dimensionless
Tilt (from Horizontal)	0	Degrees
Azimuth Orientation	South	Degrees

Table II. Moisture Transfer Parameters		
Property	**Value**	**Units**
Water-Vapor Permeability (μ)	7.3×10^{-14}	kg/s·m·Pa
Liquid Diffusity (D_γ)	1.3×10^{-8}	m²/s
Capillary Saturated Moisture Content (γ_s)	1.0	Dimensionless

Table III. Heat Transfer Parameters		
Property	**Value**	**Units**
Thermal Conductivity (k)	0.64	W/m·K
Density (ρ_d)	1,650	kg/m³
Specific Heat (c_p)	1,256	J/kg·K

Table IV. Outdoor Weather Parameters	
Parameter	**Units**
Ambient Temperature	°C
Ambient Relative Humidity	%
Wind Speed	m/s
Total Incident Solar Radiation	W/m²
Sky Temperature	°C

the polymer coating because it was very thin. However, we had to use a time step of 0.01 hour to achieve a stable solution without unreasonable moisture content fluctuations in the polymer coating. The polymer coating had an extremely fast response time that gave rise to a tendency for the finite-difference solution to overshoot and undershoot the "true" solution.

Base Case (Miami 7-Day Summer Period). The MOIST model was first used to investigate the relative humidity and temperature at the exposed and inner surface of the polymer coating applied to a panel during a 7-day summer period in Miami, FL. This analysis period was the first week of August. The surface relative humidities are given in Figure 2a, the surface temperatures in Figure 2b, and the solar flux incident onto a horizontal surface in Figure 2c. From the solar flux data, this analysis period appears to be predominantly clear without cloudy days. The exposed surface temperature ranges from a low of 24 °C to a high of 50 °C. The outdoor relative humidity ranges from a low of 20% to a high of 95%.

From Figure 2b, it is seen that the temperature at the exposed surface is always above the outdoor dew point temperature. This means that dew condensation does not occur during cool night periods. As a result, the surface relative humidity of the polymer coating is below liquid saturation (i.e., rh < 97%). It should be pointed out that the outdoor convective heat transfer for this base case simulation corresponds to a 2.2 m/s wind speed, which substantially reduces the amount of surface temperature depression due to radiation exchange with a cold night sky. It will be later shown (see Figure 10)
that, when the outdoor convection is reduced to a still air condition, the surface temperature decreases below the outdoor dew point temperature. Under this condition, dew condensation wets the exposed polymer surface.

Effect of Climate (Miami Winter versus Miami Summer). The MOIST model was next used to generate a similar set of results for a 7-day winter period in Miami, FL (see Figure 3). The first week of February was used as the winter analysis period. The surface relative humidities span a similar range as the previous results, but the surface temperatures are approximately 10 °C lower. The lower surface temperature during the winter is believed to give rise to less material degradation. Note that the exposed surface temperature is again seen to be above the outdoor dew point temperature during cool night periods, thereby indicating that dew condensation does not occur.

Effect of Geographic Location (Miami, FL versus Phoenix, AZ). The MOIST model was next used to generate a similar set of results for a 7-day summer period in Phoenix, AZ (see Figure 4). When the Phoenix summer period is compared to the previous Miami summer period, the panel temperature is seen to be about 10 °C warmer than the panel temperature in Miami. These differences are about the same as those reported by Fischer and Ketola (8). However, the panel surface relative humidities in Phoenix are considerably lower than those in Miami (see Figure 4a), due to considerably lower absolute humidities in Phoenix.

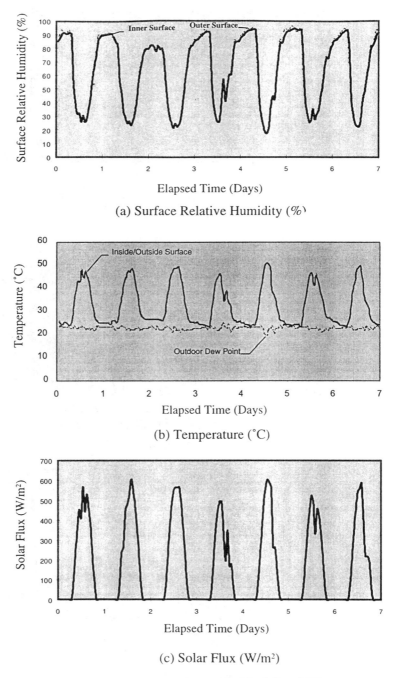

(a) Surface Relative Humidity (%)

(b) Temperature (˚C)

(c) Solar Flux (W/m²)

Fig. 2 Summer 7-day period in Miami FL.

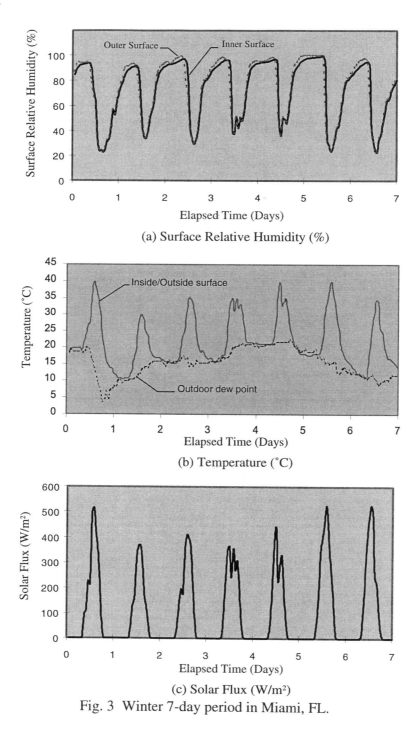

(a) Surface Relative Humidity (%)

(b) Temperature (°C)

(c) Solar Flux (W/m²)

Fig. 3 Winter 7-day period in Miami, FL.

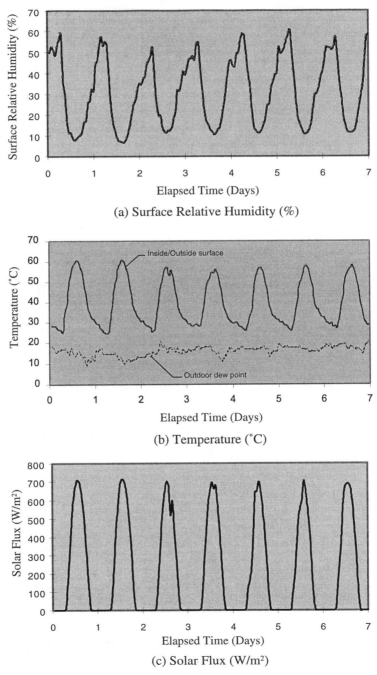

(a) Surface Relative Humidity (%)

(b) Temperature (°C)

(c) Solar Flux (W/m²)

Fig. 4 Summer 7-day period in Phoenix, AZ.

Selection of Base Case. Of the three sets of previous results (Figures 2-4), the seven day summer period of Miami (FL) had the highest combination of temperature and relative humidity. It was believed to represent the worst case of the three previous conditions. It was used as a base case for the sensitivity analysis presented below. In the sensitivity analysis, one parameter is varied at a time, and its effect on the temperature and relative humidity of the inner polymer surface is investigated. The first 24-hour period (August 1) of the 7-day Miami summer period is plotted in order to better see the effect of each of the parameters. The base case results are plotted for the first 24-hour period in Figure 5.

Effect of Substrate. We considered the following three substrates: 1.3 cm plastic (base case), 0.32 cm steel, and 1.3 cm wood. The surface temperature results are given in Figure 6b. The peak surface temperatures are highest for the wood substrate. This is because the wood substrate has the highest thermal resistance and therefore the smallest backside heat loss. Under this condition, a smaller amount of the solar gain is conducted out of the backside of the panel. The opposite argument can be made for the exposed panel with a metal substrate which had the lowest mid-day temperature.

From the surface relative humidity results (see Figure 6a), the panel with the wood substrate has less extreme relative humidity fluctuations. The values are lower at night and higher during the day. The wood substrate provides considerably larger moisture storage capacity than the other two substrates. The other two substrates are for all practical purposes impermeable to moisture transfer and are therefore unable to store moisture. The moisture storage capacity offered by the wood substrate acts like a flywheel and dampens the relative humidity fluctuations.

Effect of Paint Permeance. We next examined the effect of paint permeance. The following two paint permeance were considered: 57.0×10^{-11} kg/s·m²·Pa and 5.7×10^{-11} kg/s·m²·Pa. The first paint permeance corresponds to a latex paint system, while the second one (which is one tenth the first value) corresponds to an oil-base paint system. The results are given in Figure 7. The paint permeance is seen to have very little effect on panel temperature (see Figure 7b). With regards to the surface relative humidity (see Figure 7a), the simulation with the smaller paint permeance is seen to lag behind the other simulation. Here the smaller paint permeance is slowing down the transfer of moisture into the paint layer. As a result, the relative humidity at the inner surface exhibits a slower response.

Effect of Solar Absorptance. The effect of solar absorptance is shown in Figure 8. Three solar absorptances were considered: a light-colored surface ($\alpha = 0.4$), a medium-colored surface ($\alpha = 0.7$), and a dark-colored surface ($\alpha = 1.0$). It is interesting that the peak panel temperature for the dark colored surface is very close to the maximum black panel temperature reported in August by Fischer and Ketola (8). The observed variations in surface relative humidity are a direct consequence of surface temperature variations. Higher temperature produces faster drying of the polymer coatings, thereby giving rise to lower relative humidities.

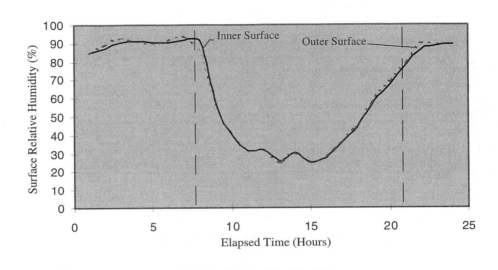

(a) Surface Relative Humidity (%)

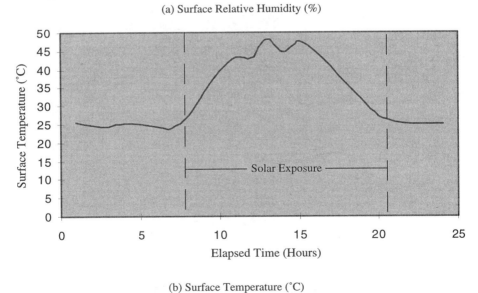

(b) Surface Temperature (°C)

Fig. 5 Summer 24-hour period in Miami, FL.

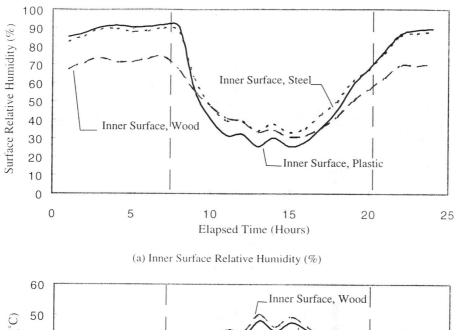

(a) Inner Surface Relative Humidity (%)

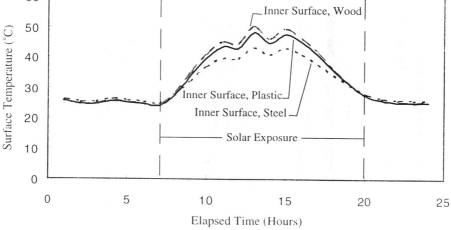

(b) Inner Surface Temperature (°C)

Fig. 6 Effect of Substrate

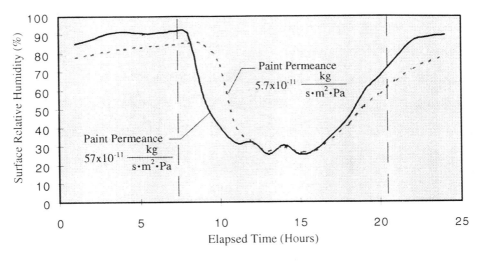

(a) Inner Surface Relative Humidity (%)

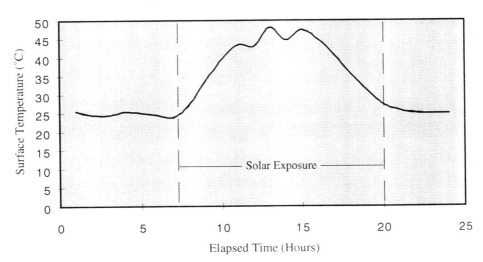

(b) Inner Surface Temperature (°C)

Fig. 7 Effect of Paint Permeance

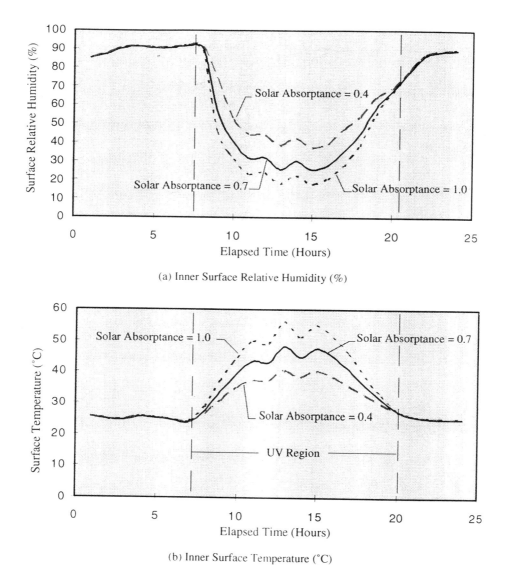

(a) Inner Surface Relative Humidity (%)

(b) Inner Surface Temperature (°C)

Fig. 8 Effect of Solar Absorptance

Effect of Tilt. MOIST simulations were carried out for panels having the following tilts: horizontal (0°), latitude (25.8°), 45°, and vertical (90°). The surface temperature results are given in Figure 9b. The vertical exposure panel is seen to have the lowest peak temperature during the day due to smaller incident solar radiation. However, during cool night periods, the vertical panel has a higher temperature because this panel views a considerably smaller portion of the sky. Therefore, it is cooled less by radiation exchange with the cold night sky. The panels sloping at 25.8°, 45°, and 90° have the highest peak temperature during the day. It is interesting that the panels sloped at 25.8° and 45° have very nearly the same temperatures during the day, thereby indicating that the incident solar radiation onto these two panels is approximately the same. This later result is consistent with the previous findings of Walker (9) who reported that panels tiled at 45° receive very nearly the same amount of incident solar radiation as panels tilted to latitude.

The surface relative humidities are given in Figure 9a. The surface relative humidity curves have the opposite trend as the surface temperature curves. That is, curves having the highest peak temperatures have the lowest surface relative humidities. Higher temperatures produce faster drying of the polymer coatings, thereby resulting in lower surface relative humidities. The opposite is true at night. Here the fundamental driving force for moisture transfer is the temperature difference between the outdoor air and cold panel surface. Colder panels have larger temperature differences and therefore support larger moisture transport from the outdoor environment to the panel.

Effect of Outdoor Convection. MOIST simulations were also carried out for the following outdoor convection conditions: still air ($h_{c,o}$ = 2.8 W/m²·°C) and windy ($h_{c,o}$ = 34.1 W/m²·°C). The temperature curves are given in Figure 10b. As expected, the panel exposed to windy condition had considerably lower temperatures during solar exposure. The resulting high convective heat transfer rate removed much of the solar radiation absorbed by the panel. It is interesting to observe that the panel exposed to still air conditions was cooled below the outdoor dew point temperature by radiation exchange with the cold night sky. As a result, dew condensation occurred on this panel.

The surface relative humidities results are given in Figure 10a. The relative humidity at the lower surface of the polymer coating rose above 97% during periods when dew condensation occurred. A relative humidity above 97% indicates the presence of liquid water in the large pores of a material. Q-Panel LabNotes (April, 1997) have reported that on the average panels are wet about 30% of the time. This means that panels are exposed to eight hours of wetness per day, or about 2,900 hours of wetness per year. Panel wetness is caused by wetting by rain and dew condensation. Obviously, wetting by rain alone can not explain this very high amount of wetting. Wetting is believed to be a major mechanism causing degradation of polymer coatings. Also of interest is the very large diurnal swing in relative humidity (i.e., 10%< rh< 99%) under the still air condition.

It is worth mentioning that the panel exposed to the still air condition is wet during a 2-hour morning period when it receives ultra-violet radiation. The authors believe that wetness may act as an accelerant for ultra-violet degradation. The effect of

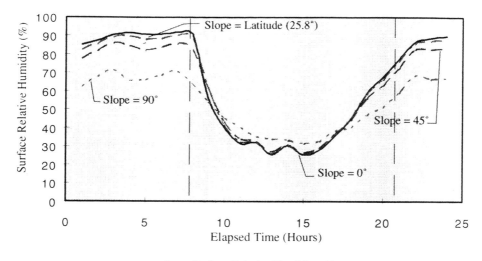

(a) Inner Surface Relative Humidity (%)

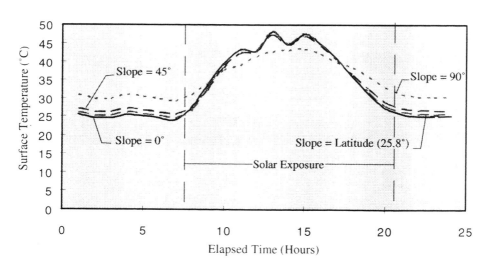

(b) Inner Surface Temperature (°C)

Fig. 9 Effect of Tilt

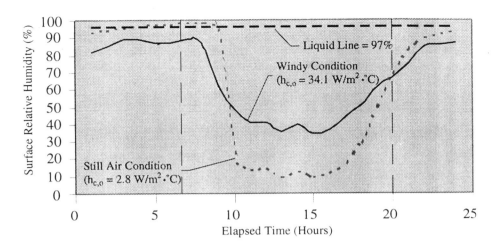

(a) Inner Surface Relative Humidity (%)

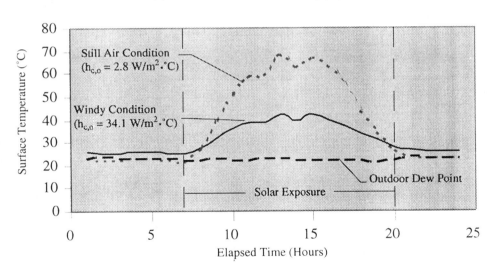

(b) Inner Surface Temperature (°C)

Fig. 10 Effect of Wind Speed

cumulative UV dose received under wet conditions on the coating degradation warrants further investigation.

In the current MOIST model, a constant convection coefficient is specified for an entire simulation. Since dew condensation wetting is such an important wetting mechanism, it would be desirable to add algorithms to MOIST which would permit the convective coefficient to be calculated as a function of the wind speed. In addition, the current MOIST model does not include rain wetting which also would be desirable to add.

In carrying out the present study, some enhancements to the dew condensation algorithms have come to the attention of the authors. In the current MOIST model, the outdoor air is coupled to the first material node by a water-vapor conductance which includes both the air boundary layer conductance and a material conductance between the surface and the first finite-difference node. A more accurate representation of the physics would be make the first node a surface node coupled directly to the outdoor air by an air mass transfer coefficient. This surface node is turn would be coupled to the first material node by a liquid conductance. In this formulation, liquid water would condense at the outside expose surface during periods having dew condensation. This liquid water, in turn, would be readily transported to the first material node by a capillary (liquid) conductance. This enhancement will provide considerably more accurate moisture absorption predictions during wetting periods. In addition, it will better accommodate the addition of rain wetting to the MOIST model.

Simplified Model for Predicting Polymer Coating Surface RH. Let us assume that we have a good thermal model that can accurately predict the panel temperature. There are some very good thermal models in the public domain (e.g., BLAST, TARP, and DOE2, etc.). These models have the capability to accurately predict the polymer coating temperature as a function of time.

Neglecting periods of wetting by dew condensation and rain, let us assume that the polymer coating is in a state of moisture equilibrium with the outdoor environment at each hour of the day. This assumption may be valid because the polymer coating is very thin and therefore has a fast moisture response time. Under this condition, the dew point temperature at all locations within the polymer coating must be equal to the outdoor dew point temperature. We can predict the polymer coating temperature from the thermal model. From simple psychrometric relationships, we may calculate the relative humidity in the coating.

A comparison between the above simplified model and the considerably more detailed MOIST model is given in Figure 11. Here each point is an hourly calculation. The simplified model results are on the vertical axis, while the corresponding MOIST predictions are given on the horizontal axis. The 45° sloping line depicts perfect agreement between the simplified model and the MOIST model. The agreement between the two models is within ± 10% rh. The results indicate that during periods, when the polymer coating is not wetted by dew condensation or rain, the polymer coating is in an approximate state of moisture equilibrium with the outdoor environment.

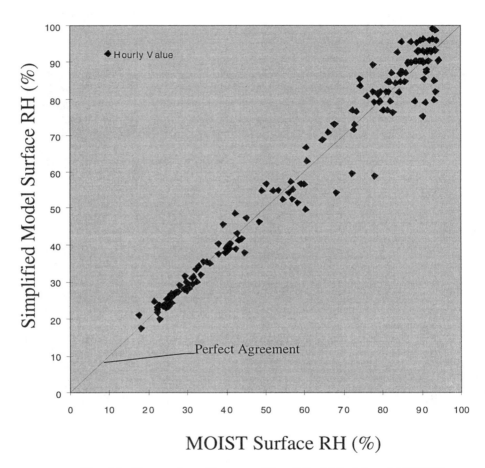

Fig. 11 Comparison Between Simplified Model and MOIST

The above results indicate that the simplified model only provided fair relative humidity predictions (± 10% rh). Therefore, the MOIST model should be used to predict the relative humidity of a polymer coating. However, since the simplified model did provide a strong correlation to the more exact MOIST model, parameters used in its formulation are the dominant parameters governing its moisture performance. These parameters include the panel temperature and the outdoor dew-point temperature.

Summary and Conclusions

The MOIST model was shown to provide reasonable predictions of the temperature and relative humidity of polymer coatings exposed to outdoor conditions in the field. The model predicted wetting of polymer coatings during periods having dew condensation. It was recommended that the following enhancements be made to the MOIST model: (1) incorporate algorithms that predict the convective heat transfer rate at the panel surface as a function of the outdoor wind speed; (2) improved the wetting dew condensation algorithms; and (3) include rain as a wetting mechanism. After the enhancements are implemented, it is recommended that the revised model be verified by way of comparison to a series of laboratory or field experiments.

Literature Cited

1. Martin, J.W. "Quantitative Characterization of Spectral Ultraviolet Radiation-Induced Photodegradation in Coating Systems Exposed in the Laboratory and the Field." *Progress in Organic Coatings.* **1993**, Vol 23, pp. 49-70.
2. Burch, D.M. and Chi, J. "MOIST: A PC Program for Predicting Heat and Moisture Transfer in Building Envelopes (Release 3.0)." *NIST Special Publication 917.* **1997**, September.
3. Zarr, R.R.; Burch, D.M.; and Fanney, A.H. "Heat and Moisture Transfer in Wood-Based Wall Constructions: Measured Versus Predicted." *NIST Building Science Series 173.* **1995**, February.
4. Pedersen, C.R. (now Rode, C.R.). "Combined Heat and Moisture Transfer in Building Constructions." *Report No. 214.* Technical University of Denmark, **1990**, September.
5. Duffie, J.A. and Beckman, W.A. *Solar Engineering of Thermal Processes.* Second Edition, John Wiley & Sons, Inc. **1991**.
6. Bliss, R.W. "Atmospheric Radiation Near the Surface of the Ground." *Solar Engineering.* **1961**, Vol 5, No 103.
7. Crow, L.W. "Development of Hourly Data for Weather Year for Energy Calculations (WYEC)." *ASHRAE Journal.* **1981**, Vol 23, No 10, pp. 34-41.
8. Fischer, R.M. and Ketola, W.D. "Surface Temperatures of Materials in Exterior Exposures and Artificial Accelerated Tests." *Accelerated and Outdoor Durability Testing of Organic Materials.* **1994**, pp. 88-111.
9. Walker, P.H. "Importance of Position in Weather Tests." *Journal of Industrial and Engineering Chemistry.* **1924**, Vol 16, No 5.

10. Q-Panel LabNotes. "Materials Exposed Outdoors are Wet for Longer than You Think." **1997,** April.

11. Burch, D.M.; Thomas, W.C.; and Fanney, A.H. "Water Vapor Permeability Measurements of Common Building Materials." *ASHRAE Transactions.* **1992,** Vol 98, Part 2.

Nomenclature

Symbol	Units	Definition
c_d	J/kg·°C	Dry specific heat
c_w	J/kg·°C	Specific heat of water
D_γ	m²/s	Liquid diffusivity
D_{tot}	J/cm²	Total effective ultra-violet dose
E	dimensionless	Emittance factor
h_{lv}	J/kg	Latent heat of vaporization
$h_{c,o}$	W/m²·°C	Convective heat transfer coef. at outside surf.
$h_{r,o}$	W/m²·°C	Radiative heat transfer coef. at outside boundary
H_{sol}	W/m²	Incident solar radiation onto a surface
I_o	W/cm²	Spectral ultraviolet irradiance
k	W/m·°C	Moist thermal conductivity
K	kg/s·m²·Pa	Hydraulic conductivity
m	kg/s·m²·Pa	Water-vapor mass transfer coefficient
p_l	Pa	Capillary pressure
p_v	Pa	Water vapor pressure
$p_{v,i}$	Pa	Water vapor pressure of indoor air
rh	%	Relative humidity
t	s	Time
Symbol	Units	Definition
T	°C	Temperature
T_s	°C	Surface temperature
T_{sky}	°C	Sky temperature
T_o	°C	Outdoor air temperature
y	m	Distance
$\dot{W}$	kg/h	Moisture generation rate
γ	kg/kg	Moisture content on dry mass basis
γ_s	kg/kg	Capillary saturated moisture content
λ	cm	wavelength
λ_{max}	cm	maximum photolitically effective wavelength
λ_{min}	cm	minimum photolitically effective wavelength
ρ_a	kg/m³	Air density
ρ_d	kg/m³	Dry material density
σ	W/m²·°C⁴	Stefan-Boltzmann constant
ϕ	dimensionless	spectral quantum yield
μ	kg/s·m·Pa	Water-vapor permeability

Chapter 8

Monitoring and Characterizing Air Pollutants and Aerosols

S. E. Haagenrud

Norwegian Institute for Air Research,
Postboks 100, N-2007 Kjeller, Norway

For industry to respond to the standards and requirements for more durable and sustainable organic coatings, extensive data and knowledge in the field of service life prediction of such materials needs to be compiled or generated. The service life prediction models for coating materials are based upon knowledge of degradation mechanisms, dose-response functions and the exposure environment. Much of the required data could be provided by extensive co-operation with the meteorological and environmental research community.

The systematic means for characterization, classification and mapping of environmental degradation factors are described, and an overview of international environmental programs for measuring, modelling and mapping of air quality on macro, meso and local levels are given. A critical deficiency is the lack of methods, data and models for micro-environmental mapping. Some methods and tools are described.

Geographical information systems (GIS) for integrating, processing and presenting data in a user-friendly way exist, and will greatly enhance data accessibility.

Action Needed to Safeguard our Built Environment
In the developed countries, the building stock and infrastructure constitute more than 50 per cent of each country's real capital. This built environment is in a bad state. After the "build and let decay" age during the last 30 years, the concern is not only the environmental impact on the cultural heritage. Generally, the damages to building materials and constructions have become enormous economic, cultural and environmental problems.

The wasteful consumption of energy and materials linked to the degrading built environment makes this a major environmental problem in the context of sustainable development.

© 1999 American Chemical Society

To safeguard our built environment, action is urgently needed. In principle, there are two possibilities – and both should be pursued in parallel. Firstly, society should try to improve the exposure environment surrounding the materials, and secondly, better products, processes, methods and standards should be developed. The first action is being pursued by the environmental research area via cost-benefit analysis for degradation of materials and buildings, while the second issue is the concern of many research&development (R&D) and standardisation programmes around the world (1), (2) and (3).

Standards for Service Life Planning requires Environmental Characterisation

An international standard for prediction of service life of building materials, components and buildings is currently in the process of being generated within *ISO/TC59/SC14* (4). This group was set up from the joint initiative for standardisation of service life methodologies by the EUREKA umbrella project EUROCARE and CIB/RILEM, towards CEC and CEN in 1991. It was based upon the generic RILEM recommendation for prediction of service life (PSL). (5). In Europe the entry into force of the Construction Products Directive (CPD) also creates an urgent and increased need for standards addressing the issue of durability (6).

It is interesting to see that the European Organisation for Technical Approval (EOTA), in their Guidance Paper on "Assessment of Working Life of Products" to the Convenors of the Technical Committees has adopted the service life methodology of RILEM (PSL), the ISO 6241, the damage function approach and emphasizing the specific need for characterising the exposure environment on the geographic scales of Europe (EOTA, 1996).

The ISO work uses the Factorial Approach (Draft 1) to estimate the design life of a building component. It starts from the Reference Service Life (RSL), which is modified for Quality of materials etc (Table I).

Table I. Method for estimating service life of components using factors to represent agents

$ESLC = RSLC * A * B * C * D * E * F$		
where		
ESLC	:	Estimated Service Life
RSLC	:	Reference Service Life of the components (PSL methodology
A	:	Quality of components
B	:	Design level
C	:	Work execution level
D1	:	Indoor environment
D2	:	Outdoor environment
D3	:	Subterrean
E	:	In use conditions
F	:	Maintenance level

The establishment of RSL should be based on the RILEM PSL model. The PSL and this new standard are based upon knowledge of the exposure environment, the degradation mechanisms and *dose-response functions*. For industry to respond to the standards and requirements great emphasis is put on industry's ability to adapt, further develop and implement this methodology. A lot of data and knowledge needs to be compiled or generated. As shown in the following, much of these data could be provided through co-operation with the meteorological and environmental research communities.

Model for degradation and barriers to service life prediction.

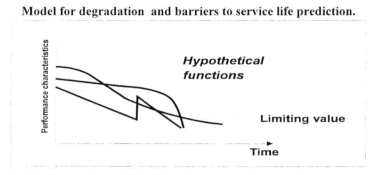

Figure 1: Performance over time functions.

Materials degradation and loss of characteristic properties, as described by performance over time functions (Figure 1), are in most cases due to chemical and physical deterioration or corrosion. The corrosion can be expressed by the mathematical model consisting of a power function of degradation factors and elapsed time:

$$M = a \cdot t^b \tag{1}$$

where   M = corrosion at time t;
        a = rate constant, which can be expressed by the deposition of pollutants or other degradation agents to the surface;
        b = power exponent governed by diffusion processes, where b ~½ for the case of corrosion products forming a protective layer through which fresh reactants must diffuse.

In the environmental area numerous studies have been performed aiming to find the corrosive effect of air pollutants and to establish the relationship between materials decay and the environmental degradation factors. Although generating a lot of useful data and knowledge about the effects, the studies lack a homogeneous approach in terms of measurements, time frames and data-analytical procedures. However, it should be pointed out that the dose response functions which are currently available are very limited in terms of *choice of degradation indicators and establishment of performance requirements*. A survey of many of these studies and their reconciliation is given in (7).

The degradation rate functions are not directly suitable for service life assessments. To transform the degradation into service life terms, performance requirements or **limit states** for allowable degradation before maintenance or complete renewal of material or component, have to be decided. The dose-response function then transforms into **damage function**, which is also a performance over time function, from which a service life assessment can be made.

The establishment of the limit state is complicated, and can be discussed both from a technical, economic and environmental point of view. Within the building society the first two aspects have so far been dominating (6) and (8). However, within the environmental area the discussion has started on fixation of the limit state from the "sustainable requirement" point of view (9). A convergence of these requirements would have great interest and impact in the building sector .

A major barrier for further progress concerning the durability and service life aspects within the building community, is the lack of knowledge of and implementation of the damage function approach.

Tomiita (10) has used the mathematical damage function approach (eq. 1), and developed it into a cumulative damage model as a computer software application. He has thus been able to use the right type and form of the environmental degradation factors, and therefore also to predict and compare degradation rates from long term field exposure and laboratory exposures.

The same approach is advocated by Martin et al. (11) in discussing and proposing methodologies for predicting the service lives of coating systems.

Another major barrier to reliable predictions of service life and/or maintenance intervals is insufficient knowledge of the relevant exposure environment on the various geographical scales.

There exist no common and exact definitions of the different scales, but one frequently used basis of classification for climate and environment is the division into macro, meso, local and micro scales (Figure 2):

By *macro* is normally meant the gross meteorological conditions described in terms like polar climate, subtropical climate and tropical climate. The descriptions are based on measurement of meteorological factors such as air temperature, precipitation etc.(IEC 721-2-:1982).

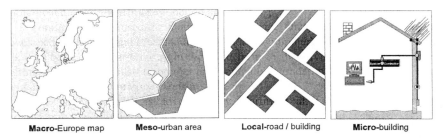

| **Macro**-Europe map | **Meso**-urban area | **Local**-road / building | **Micro**-building |

Figure 2: Exposure environment on different geographical scales.

When describing *meso* climate, the effects of the terrain and of the built environment are taken into account. The climatological description is still based on the standard meteorological measurements.

By *local* scale is meant the local conditions in the building proximity, such as for example in the streets to the building. The ***micro*** climate describes the meteorological variables in the absolute proximity of a material surface. The micro climate or micro environment is crucial to a materials' degradation.

Substantial knowledge and data exist on the environmental exposure conditions on the *macro* and *meso* level, a *third* barrier is just the adaptation of data and knowledge to the *local* and *micro environmental* conditions. The complexities of a structure can result in very different climatic and environmental conditions on a single structure and greatly affecting damage (7), The dose-response functions are primarily established under more or less controlled experimental conditions, and a *major task* would be the transition to real constructions. Measuring and modelling methods for *micro environmental* loading and materials degradation have to be developed and extensive measurements carried out.

UN ECE International Co-operative Programme (ICP)

Many studies have been performed world-wide to establish damage functions (7).
Due to lack of space only some of the functions from the UN ECE are presented here

The most extensive and best designed test programme in the environmental research area is the International Co-operative Programme (ICP) within the United Nations Economic Commission for Europe (UN ECE). The programme which started in September 1987, aims to evaluate the effect of airborne acidifying pollutants on corrosion of structural metals, stone materials, paint coatings on steel and wood, and electric contact materials, and involves exposure at 39 sites in 12 European countries and in the United States and Canada (Figure 3).

Samples have been withdrawn after 1, 2 and 4 years exposure and dose-response functions have been developed (12) for carbon steel, zinc, aluminium, copper, bronze and calcareous stone. The equations should at present be seen as provisional and may be subject to further elaboration when the results from the 8 year exposure will be available in 1996.

For unsheltered exposure most of the dose-response functions have the same form

$$\text{ML or MI} = a + b \text{ TOW } [SO_2][O_3] + c \text{ Rain amount } [H+] \qquad (2)$$

where

ML, MI = mass loss resp. mass increase
SO_2 = air concentration of sulphur dioxide = $\mu g/m^3$
NO_2 = air concentration of ozone = $\mu g/m^3$
TOW = time of wetness in fraction of a year
$H+$ = acidity in preciptation (meqv/l)

This is the first time a synergistic effect of O$_3$ and SO$_2$ has been indicated in a field exposure. There is, however, also a very complex interaction between O$_3$, SO$_2$ and also NO$_x$. The further study of this is the topic of a revised field exposure.

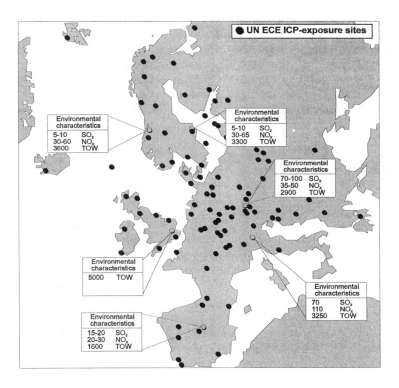

Figure 3: UN ECE ICP sites.
(Units: SO$_2$ and NO$_2$ = µg/m³ , TOW = hours/year.)

Key Environmental Degradation Factors and Appropriate Damage Functions

Various systems have been used to classify degradation agents. The standard ISO 6241-1984 (E) presents a detailed list of agents relevant to building performance and requirements. This systematic classification implies that the agents are *listed according to their own nature* as *Mechanical, Electro-magnetic, Thermal, Chemical and Biological agents*, and to *their origin* (external-internal to the building, atmosphere, ground etc.), and **not** *to the nature of their action* on the buildings or components. The agents that apply in any particular situation, and their magnitudes, will depend on the building's situation, form, intended use and the way it is designed to perform. The ISO6241-1984 (E) systematics will also be used in the new ISO Design Life Standard.

In order to characterise and report the right *type* and *form* of the environmental degradation factors, they have to be related to the degradation mechanism and dose-response functions for the specific materials in question (13) and (11). This will facilitate comparisons to be made between field and laboratory measurements.

Mechanical agents.

Thermal expansion-Daily temperature difference. All external construction materials experience both diurnal, seasonal and annual temperature fluctuations (14), (11) and (15). The fluctuations of temperature induce movements in the joints between building elements or cracks in concrete. Sealants and exterior finishing materials suffer from repeating tensile and compressive and shear stresses and are fatigued.

One special temperature variation that may greatly affect, for example, the micro cracking of coatings is rapid drop in surface temperature. A dark coating on an insulated sheet metal facade with a surface temperature of approximately 70 °C, when exposed to cold heavy rainfall may lose 50–60 °C in a few minutes.

The majority of available temperature data originate from measurements of the ambient atmospheric temperature at meteorological stations. However, the relations between data on ambient temperature, surface temperature and temperature in the bulk of materials are very complex.

The *daily temperature difference* of building elements can be treated as a *mechanical deterioration index*. Tomiita (13) collected daily maximum/minimum BPTS (Black Panel Temperature Surface), TP_{max}/TP_{min} (°C), and expressed them as functions of climatic data.

By inserting the meteorological data observed at 66 points during 1976-1985 the daily maximum/minimum BPT was estimated. The yearly averages of the daily differences of BPT was classified and mapped for Japan. By multiplying the length of a building element by its thermal expansion rate, the daily movement in the joint or crack can be calculated.

Moisture. The presence of moisture enables physical, chemical or biological degradation to take place. Moisture therefore acts as a mechanical, thermal (frost) and chemical agent.

Driving rain. The quantity of water falling on the vertical faces of buildings is related to the combined effects of wind and rainfall (see Figure 4) (16).

Most organic materials and many inorganic materials absorb moisture to varying degrees. The direct effect of water alone on a material item can be as follows:

(a)   volumetric expansion;
(b)   change in mechanical properties (e.g. strength);

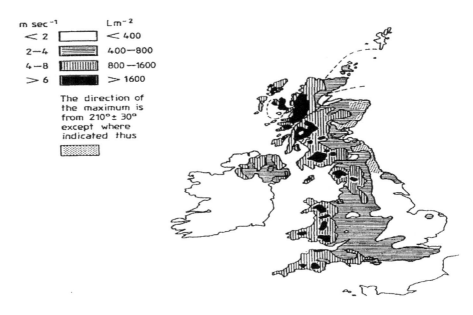

Figure 4: Maximum directional annual driving rain index map for UK.

(c)   development of bending and twisting forces;

(d)   change in electrical properties;

(e)   change in thermal properties;

(f)   change in appearance.

Quantitative dose-response functions exist between driving rain and some of these characteristics. In Germany 140 litre/m^2 is given as the maximum amount of driving rain that a timber-framed houses can withstand without special protection for sealants (17).

On the macro and meso scale general meteorological data on the various forms of water are available from national meteorological institutes.

Electromagnetic agents

Solar radiation. The energy that reach the earth's surface from solar radiation is concentrated in certain wavebands. Figure 5 shows the variation in the energy received at different wavebands from direct sunlight. A portion of solar radiation is absorbed or reflected by the earth's atmosphere. This portion varies from one waveband to another and is affected by cloud cover. The degree of shading from direct sunlight also affects the quantity of the radiation received in each waveband.

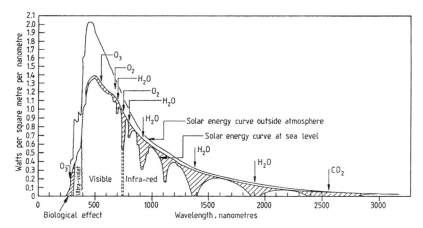

Figure 5: Curves showing the distribution of energy in the solar spectrum outside the atmosphere and at sea level (from BS 7543:1992).

Ultraviolet radiation (UVR)-290 nm to 400 nm. A large proportion of the UV waveband is scattered by the atmosphere, the remainder reaching the earth's surface has no adverse effect on inorganic materials. It is however, very important in its potential for deterioration of some organic materials. Although UVR constitutes no more than 1–7% of the total solar radiation intensity, in practice it determines the service life of polymeric materials in outdoor use. The penetrating power of the UV - radiation is not great, and the action is consequently confined to surface layers exposed to radiation (18).

The UVR-intensity varies with the atmosphere, local weather, air pollution, time of day etc. Due to the environmental ozone-UV problem there is now extensive intensity measurements being carried out on a global scale (19).

Measurements of total UV radiation over the spectral range are of limited value because of the sensitivity of materials to specific wavebands. For example polystyrene shows a maximum sensitivity at about 318 nm whereas polypropylene peaks about 370 nm. As a general rule it has been found that radiation shorter than about 360 nm tends to cause yellowing and embrittlement. Radiation of a longer wavelength tends to cause fading (20).

Many organic dyestuffs are degraded by UV radiation, which may affect appearance.

UV radiation can also play an important role in initiating a degradation reaction which is then propagated under suitable conditions of moisture and temperature, e.g. the yellowing and surface denudation of glass fibre reinforced polyester roofing sheets caused by the combined actions of UV radiation and moisture.

Estimates of the UVR environment for a certain geographic location can be made from meteorological data on global solar radiation or number of sunshine hours.

Tomiita (13) has used a photodiode sensor (range 305–390nm), to measure the hourly solar UV energy on a horizontal plane in outdoor exposure in Japan for a year, and expressed it as a function of the entire range of solar radiation and solar altitude. The results were classified and mapped.

Martin et al. (11) propose in a similar way a metric for UV dosage.

Chemical agents

Water and temperature complex-the time of wetness concept. The term Time-of-wetness (TOW) is much used, and often misleadingly. To avoid misunderstanding the original definition which originates from the field of atmospheric metal corrosion should be remembered. Here TOW is defined as the *period of time* during which the material surface is subjected to *enough moisture for the corrosion rate to be significant*. This lead to the concept of a *time above a critical wetness, i.e. that both the level and duration of wetness is important. Therefore, to avoid misconception, the term should generally include the word "critical", to be called something like "the time of critical wetness"* (TOWcrit), *and should be defined as a material specific property.*

From empirical observations, the term TOW for metals is defined as the time when the relative humidity is greater than 80% at temperature above 0°C. This definition is used in ISO standard 9223 "Corrosion of metals and alloys – corrosivity of atmosphere – classification", where the TOW is calculated together with selected climatological characteristics of the macroclimatic zones of the world. This classification is extracted from the standard IEC 721-2-1:1982..

Haagenrud (21); Tomiita (13); Morcillo and Feliu (22) have calculated the TOWcrit according to the ISO criterion from available meteorological data. Tomiita and Kashino (23) calculated also the number of wet-dry cycles causing condensation. These values have been mapped for Japan.

The level and duration of moisture in the immediate vicinity of the material surface is also crucial to the degradation of *non-metallic materials*. Condensed moisture on coating surfaces over long periods of time is, for example, considered to have a more deteriorating effect than shorter periods of rain (24) and (18).

The humidity impact is also observed for *porous materials* like wood, stone, rendering and concrete, underlining the extreme importance of the fact that the concept also accounts for the *wetness inside* porous materials. For example, for wood the degradation process of rottening takes place after prolonged period of time above a moisture content of 15–20 weight % (25).

Monitoring methods and data for TOWcrit is very important and should be the subject of extensive R&D. The WETCORR instrument can be used in such applications (26) and (27).

Ozone. Ozone is one of the most important oxidising constituents and reacts with many polymers such as polythene, polybutadiene, polystyrene etc. The effects are normally discoloration and embrittlement. Ozone attacks double bonds in the polymer, causing chain scission and crosslinking reactions. The degradation reactions occur via the formation of peroxy radicals, which photochemically may be excited to other free radicals. The reactions take place on the material surface.

There are great variations in the occurrence of ozone in the atmosphere. At high altitudes, more than 20 km above earth level, the amount is more than one thousand times higher than at terrestrial levels.

Ozone may exist in nature as an effect of solar radiation on oxygen in the presence of air pollutants. This has caused major concern in the last few years, and evidence for damages to health and vegetation has triggered off a European Council Directive (92/72/EEC) for monitoring and warning of ozone. Data are available from 461 monitoring sites in 14 Member States, for 1994.

The various ozone threshold values as defined in the Directive are shown in the Table II.

Table II. Thresholds for ozone concentrations in air.

Thresholds	Values
Health protection threshold	110 $\mu g/m^3$ for 8-hours mean
Vegetation protection threshold	200 $\mu g/m^3$ for 1-hours mean
	65 $\mu g/m^3$ for 24-hours mean
Population information threshold	180 $\mu g/m^3$ for 1-hours mean
Population warning threshold	360 $\mu g/m^3$ for 1-hours mean

If ozone threshold values for damage to building materials could be established, which seem likely from the UN ECE dose-response functions, the necessary data is readily accessible from these public sources.

Biological agents-Fungi and bacteria. Living organisms such as fungi and bacteria are important environmental degradation agents of organic building materials, but they may also affect inorganic materials such as calcareous stones, sandstone and even metals (28). While biological factors are not weathering factors, biological attack of exterior building materials depends highly on weather conditions. Fungi causing wood decay, and wood boring insects are ubiquitous, and they can cause damage in most parts of a building.

Mattson (25) has given a systematic survey of microbiological attack and its causes on timber . *Temperature, moisture content of timber* and *nutrients* are the predominant factors when considering its susceptibility to both fungal and insect attack. The *duration of time* for their impact is also of great importance.

For fungi to grow on wood a moisture content of 23% or more is necessary. If growth has started, it can continue at lower moisture levels. Table III show moisture-temperature requirements for some typical damaging fungi in Norway (25).

Table III: Moisture temperature regimes for typical damaging fungi in Norway (24).

Type of Fungi	Moisture in Wood	Temperature Optimal	Lethal
Corticiacea spp.	50-70%	ca. 30 °C	ca. 50 °C
Blue stain fungi	> 30%	ca. 25 °C	ca. 45 °C
Dry rot fungus(Serpula lacrymans	20-55%	ca. 23 °C	ca. 35 °C
"Poria" fungi, white pore fungus (Antrodia serialis, A, sinuosa, A, xantha, Fibroporia waillantii)	35-55%	ca. 28 °C	ca. 45 °C
Cellar fungi	30-50%	ca. 23 °C	ca. 40 °C
Mould fungi	20-150%	ca. 20-45 °C	ca. 55 °C
Dacrymytces stillatus	20-150%	ca. 23 °C	ca. 35 °C
Gloeophyllum sepiarium	30-50%	ca. 35 °C	ca. 75 °C

The information above on fungus activity fits very well with the climatic index developed by Scheffer for the U S (29), as follows:

$$\text{Climate index} = \frac{\sum_{Jan.}^{Dec} [(T-2) \ (D-3)]}{17} \tag{3}$$

where T is the mean monthly temperature in °C and D is the mean number of days in the month with 0.25 mm or more precipitation. This climatic index, which is also a damage function, for decay of wood is based on US public bodies requirements for estimation of needs for wood protective measures.

Measurements, Modelling and Mapping of Air Quality

The measuring, testing and evaluation of air quality are assuming growing importance in developed countries as elements of a comprehensive clean air policy geared to sustainable development. A huge bulk of data are therefore generated on the various geographical levels. This concerns point measurements of both emissions and ambient air concentration, and emission surveys of almost every type of sources.

Respectively, the UN Global Environment Monitoring System (GEMS/AIR), the transboundary UN ECE EMEP-programme and the tasks organised under the European Environment Agency in Copenhagen are described, and some data are given.

GEMS/AIR is an urban air pollution monitoring and assessment programme, which evolved from a World Health Organisation (WHO) urban air quality monitoring pilot project, that started in 1973. Since 1975, WHO and the United Nations Environment Programme (UNEP) have jointly operated the programme as a component of the United Nations systemwide Global Environment Monitoring System. GEMS is a component of the UN Earthwatch system.

Since its beginning in 1973, the GEMS/AIR network has included some 270 monitoring sites in 86 cities in 45 countries. GEMS/AIR is the only global programme which provides long-term air pollution monitoring data for cities in industrialised as well as in developing countries. Thus the programme enables the production of assessments on the levels and trends of urban air pollution world-wide.

UV-intensity maps. Ozone depletion leads to increased ultraviolet radiation reaching the earth surface. Excessive exposure to UV rays from the sun leads to sunburn, and, in some cases, to skin cancer.

UV-intensity maps are being produced by GRID-Arendal in co-operation with NILU. Sun-angle and satellite measurements of stratospheric ozone are entered into a model to calculate UV-intensity at noon for the whole earth. Maps are then produced indicating this intensity for selected areas.

Institutions, media and private persons can now order a UV-intensity information package from GRID-Arendal, with colour maps and graphics showing the UV and ozone situation over a selected period of time.

The unit used on the UV-intensity maps produced is the UV-index developed by Environment Canada. The UV-index runs on a scale from 0 to 10, with 10 being a typical mid-summer day in the tropics. A relative scale from low to extreme is also applied. In extreme conditions, with a UV-index higher than 9, light and untanned skin will burn in less than 15 minutes. There is now also information on UV monitoring around the world on home pages on the World Wide Web.

This information show that quite a few countries now also perform spectral measurements which is of course even more interesting from the materials durability point of view.

UN ECE European Monitoring and Evaluation Programme (EMEP). In Europe, most of the research and monitoring activities related to long range transport of air pollutants have been connected to the "acid rain" issue. To study this, the Organisation for Economic Co-operation and Development (OECD) in 1972 launched a co-operative programme with the objective "to determine the relative importance of local and distant sources of sulphur compounds in terms of their contribution to the air pollution over a region, special attention being paid to the question of acidity in atmospheric precipitation." The programme, which ended in 1977, provided the first comprehensive insight into the transport of air pollutants on a continental scale. It made it possible for the first time to quantify the depositions within one country due to emissions in any other country.

The OECD programme was followed by the Co-operative programme for the monitoring and evaluation of long-range transmission of air pollutants in Europe (EMEP). It was organised under the auspices of the United Nations Economic Commission for Europe (UNECE), in co-operation with the United Nations Environment Programme (UNEP) and the World Meteorological Organisation (WMO). Today EMEP is an integral part of the co-operation under the 1979 Geneva Convention on Long-range Transboundary Air Pollution (30).

The main objective of EMEP is to provide governments with information on deposition and concentration of air pollutants, as well as on the quantity and significance of long-range transmission of air pollutants and transboundary fluxes.

The EMEP-activities are divided into two main parts: chemical and meteorological. A Chemical Co-ordinating Centre (CCC), located at the Norwegian Institute for Air Research (NILU), is responsible for a very extensive monitoring and chemical part of the programme. The meteorological part of EMEP is being undertaken at two Meteorological Synthesizing Centres, an eastern centre (MSC-E) in Moscow, and a western centre (MSC-W) at the Norwegian Meteorological Institute in Oslo. Their main task is to design, operate and verify atmospheric dispersion models.

Working Groups on effects are also established within the context of EMEP, under which the WG on materials exist, and which has established the UN ECE ICP on materials Exposure program.

Air quality information dissemination at European Environment Agency (EEA)

After years of preparation the European Environment Agency (EEA) was established in Copenhagen in December 1994 with the main task being to provide the European Community and its Member States with *objective, reliable, and comparable information at a European level enabling the Member States to take the requisite measures to protect the environment, to assess the results of such measures and to ensure that the public is properly informed about the State of the environment.*

The current set of air quality directives in the European Union comprise the compound-specific directives for SO_2, particulate matter/black smoke, NO_2, lead and ozone, of which lead was the first (1982) and ozone the last (1992). These directives require in principle that all exceedances are detected, and thus that monitoring is carried out in all areas where exceedance of the limit values are expected.

122

The Exchange of Information (EoI) decisions (1982 and 1994) provide the framework for making monitoring data from selected sites regularly (annually) available to the Commission. The total number of monitoring sites exceeds 6000 (Figure 6). The first State of the Environment report for Europe (Dobris Report published by the EEA in 1995, took a pan-European approach.

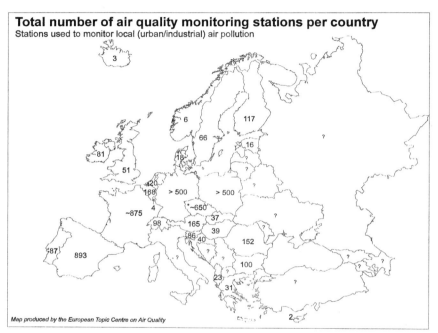

Figure 6: **Number of sites per country for the monitoring of urban/local industrial air pollution.**

On the *local/urban scale*, the Dobris report provided information on air quality in more than 100 European cities, mainly for the years 1985 and 1990. The assessment covered SO_2 and particles ("winter smog compounds"), NO_2 and O_3 ("summer smog compounds"), and also to some extent CO and lead. The data coverage was good for SO_2 and particles (mainly black smoke), less extensive for the other compounds. Data were collected and presented both for high short-term concentrations, and long-term averages.

Air quality information and management systems

Surveillance and management of air quality can now be facilitated and performed via total information systems. The Air Quality Information System, AirQUIS, represents the air pollution part of a modern Environmental Surveillance and

Information System, ENSIS, developed and demonstrated during the Winter Olympic Games, 1994 in Lillehammer (31).

The AirQUIS system was developed by institutions dealing with air pollution, information technology and geographical information systems (GIS). The combination of on-line data collection, statistical evaluations and numerical modelling enable the user to obtain information, carry out forecasting and future planning of air quality. The system can be used for monitoring and to estimate environmental impacts from planned measures to reduce air pollution (Figure 7).

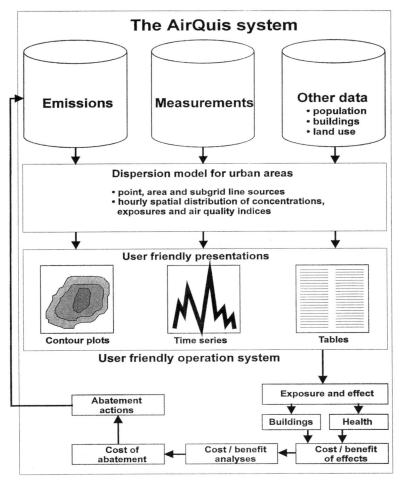

Figure 7: The AirQUIS system.

One main application of the AirQUIS system will be as an effective tool for air quality abatement strategy. The contribution of air pollution from different source categories such as traffic, household and industry to the population and building exposure in an urban area can be calculated based upon data on emissions, dispersion and distribution of buildings and population. Different recommended measures to reduce air pollution can be evaluated due to population and building exposure and cost-benefit or cost-efficiency analyses. A priority list of the selected measures can be developed, taking into account air pollution exposure, health aspects and related costs.

The module for modelling and calculating buildings degradation, service lives and maintenance costs is called CorrCost and is further described in an application used in Oslo (32).

Micro-environmental Characterisation

Temperature and wetness – the WETCORR instrument. The WETCORR instrument is designed for recording of the wetness and temperature condition in the micro environment of constructions, see Figure 8.

The measuring principle makes use of the electrochemical nature of the corrosion processes by measuring the current flow in an electrochemical cell as a function of the thickness of the humidity film bridging the electrodes surface. (33). By selecting various current levels the time above certain humidity levels can be monitored.

The sensor developed so far consists of a small gold cell for measuring time of wetness, TOW^{100}, defined as the time with 100% RH and/or rain, condensation etc., and a temperature sensor for recording the surface temperature. To ensure that the temperature sensor follows the surface temperature, the cell backing is made of aluminium oxide with good thermal conductivity.

Work is going on to develop other types of sensors, such as for example sensors for measuring resulting moisture uptake within wood (EU-project ENV-CT95-0110 (DG 12-ESCY)). Figure 8 show the set-up of such measurements, illustrating also the measurement principle.

UV meter. The solar ultraviolet radiation reaching the ground is controlled by several factors, such as solar elevation, cloud cover, total ozone amount and ground reflection. A decrease in total ozone abundance is expected to lead to an increase in harmful UV radiation if all other factors are kept unchanged.

NILU has developed a multi-channel radiometer for measurement of solar ultraviolet radiation. The instrument measures the irradiance (direct plus diffuse radiation) in the UV-B region (280 nm–315 nm) and in the UV-A region (315 nm–400 nm) in 5 channels. The instrument contains a built-in data logger which can store up to 3 weeks of 1 minute average readings from all channels, as well as the temperature, which is measured close to the detectors. The instrument can also be set to store a 1 minute average data every 15 minutes only. In this mode the storage capacity of the data logger is one year.

The UV-B channels are sensitive to variations in total ozone and variations in cloudiness. UV-A channels, however, are sensitive to variations in cloudiness but not sensitive to ozone variations.

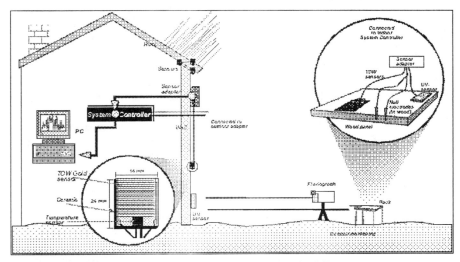

Figure 8: WETCORR measurements on and within building
materials.

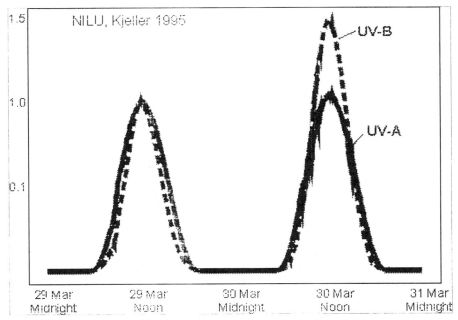

Figure 9: Measurements at NILU show the sensitivity to ozone
variations. The 308 nm level (UV-B) increased by 50%
from March 29 to March 30 due to a 20% decrease in the
total ozone abundance. The sky was clear on both days and
no change was observed in the UV-A channels.

Measurements from March 29 and March 30, 1995 at NILU illustrate the sensitivity to ozone variations, Figure 9. The integrated ozone amount was 417 Dobson Units (DU) on March 29 and 329 DU on March 30, i.e. a 20% decrease. The measured increase in UV radiation from March 29 to March 30 was 100%, 50% and 27% in the 305 nm, 308 nm, and 313 nm channels, respectively. No changes were observed in the UV-A channels since the sky was clear on both days and since UV-A wavelengths are known to be insensitive to ozone.

By using a radiative transfer model combined with such irradiance measurements, biologically effective UV-doses, total ozone amount and cloud transmission can be determined.

Modelling.

Road network emission and dispersion model. The effect of road traffic pollution on urban populations is expected to increase during the next few years. Traffic planners are often in need of practical tools for studying the effect of such measures on the environment. Quite a few air dispersion models exist and can be used for this purpose.

NILU has developed a personal computer-based model *RoadAir*, for quantitative descriptions of air pollution along road networks. RoadAir calculates total emissions, concentrations along each road segment and the air pollution exposure of the population and *buildings* along each road. Calculations can be carried out for road networks, defined by road and traffic data. The model was primarily developed for conditions in Scandinavia, but can easily be adapted to conditions in other parts of the world. RoadAir is incorporated into the AirQUIS system.

Conclusions

Concerning the characterisation of environmental degradation factors the following R&D and corresponding standardisation needs are listed:

1. International Standardisation of PSL Methodology for Materials is well underway, based on Damage Functions Approach, and extensive Characterisation of the Exposure Environment on macro-, meso- and microscale.
2. Damage functions research in field and laboratory has to be carried out to define the degradation mechanism and type and form of the degradation factors.
3. In the context of assessing building performance, a huge bulk of data on global, continental (macro) and national (meso) levels are available for exploitation from the Environmental Research Area.
4. *Some methods* for automatic and continuous monitoring of important degradation factors in the *micro environment* on buildings exist, but further development is needed.

5. Quite a *few dose response functions* have emerged from the environmental research area. However, these functions have to be tested and validated in the *micro environment* on buildings.
6. Integrated Information and Management systems for *Air Quality* exist, with *effect modules* allowing for assessment of damages to population and building *exposure* and for abatement strategies.
7. Interdisciplinary co-operation between the building and environmental research community is a must.

References

(1) Haagenrud, S.E. and Henriksen, J.F. (1996) Degradation of built environment – Review of cost assessment model and dose response functions. In: *7th International Conference on the durability of building materials and components. Stockholm 1996*. Proceedings. Ed. by C. Sjöström. London, E & FN Spon. pp. 85-96.

(2) Architectural Institute of Japan (1993) Principal guide for service life planning of buildings. Tokyo.

(3) CSA (1994) Guideline on durability in buildings – Draft 9, September 1994 (CSA-S478-1994).

(4) International Organisation for Standardisation (1995) Design life of buildings, draft 2. Geneve (ISO TC 59/SC 3/WG9).

(5) Masters, L.W. and Brandt, E. (1984) Systematic Methodology to Service Life Tradiction of Building Materials and Components. *Materials and Structures, 22,* 385-392.

(6) Caluwaerts, P., Sjöström, C., Haagenrud, S.E. (1996) Service Life Standards - Background and Relation to the European Construction Products Directive. In: *7th International Conference on the durability of building materials and components. Stockholm 1996*. Proceedings. Ed. by C. Sjöström. London, E & FN Spon. pp. 1353-1363.

(7) Haagenrud, S.E. (1997) Environmental Characterisation including equipment for monitoring. CIB W80/RILEM 140PSL. Subgroup 2 Report. Kjeller (NILU OR 27/97).

(8) International Organisation for Standardisation (1984) Performance standards in buildings – Principles for their preparation and factors to be considered. Geneve (ISO 6241-1984).

(9) Butlin, R.N. et al. (1994) Effects of pollutants on buildings. BRE final report. May 1994 (DOE report no. DOE/HMIP/RR/94/030).

(10) Tomiita, T. (1993) Service life prediction system of polymeric materials exposed outdoors. Personal communication.

(11) Martin, J.W., Saunders, S.C., Floyd, F.L. and Weinburg, J.P. (1994) Methodologies for predicting the service lives of coating systems. Gaithersburg, MD., U.S. Department of Commerce, Technology Administration, National Institute of Standards and Technology (NIST building science series, 172).

(12) Kucera, V., Tidblad, J., Henriksen, J.H., Bartonova, A. and Mikhailov, A.A. (1995) Statistical analysis of 4-year materials exposure and acceptable deterioration and pollution levels. Convention on long-range transboundary air pollution. Prepared by the main research centre, Swedish Corrosion Institute, Stockholm. (UN ECE ICP on effects on materials including historic and cultural monuments, Report No. 18).

(13) Tomiita, T. (1992) Solar UV, wetness and thermal degradation maps of Japan. *Construction and building materials, 6,* 195-199.

(14) Keeble, E.J. (1986) Microclimate data and its interpretation for problems of building deteriology. Paper to SCI/BBA symposium on building deteriology.

(15) Cole, I.S. (1994) The implications of the building envelop microclimate to the durability and serviceability of wood and wood products. *24th Forest Products Research Conference, Melbourne, Australia, Nov. 1994.*

(16) Lacy, R.E. (1976) Driving Rain Index. HMSO (Building Research Establishment, BRE Report).

(17) Eckermann and Veit, 1996

(18) Janson, J. and Sjöström, C. (1979) Fabrikslackerad plåt – åldrande och provningsmetoder. Gävle, Statens institut för byggnadsforskning (Meddelande M79:10) (in Swedish).

(19) Dahlback, A. (1996) Measurements of biological effective UV doses – Total ozone abundance and cloud effects with multi-channel moderate band-width filter instruments. *Applied Optics, 35,* 6514-6521.

(20) Yamasaki, R.S. (1983) Solar Ultraviolet Radiation on Horizontal South/46° angle and south/vertical surfaces. *Durability of Building Materials, 2,* 17-26.

(21) Haagenrud, S.E., Henriksen, J.F. and Gram, F. (1985) Dose-response functions and corrosion mapping of a small geographical area. Electrochemical Society. Symposium on corrosion effects of acid deposition. Las Vegas 14-15 October 1985. Lillestrøm, Norwegian Institute for Air Research (NILU F 53/85).

(22) Morcillo, M. and Feliu, S. (1993) Mapas de España de Corrosividad Atmosferica. Madrid, Spain, Gráficas Salué.

(23) Tomiita, T. and Kashino, N. (1989) Temperature modified wetness time and wet-dry cycle map in Japan. *Transactions of AIJ, 405,* 1-7.

(24) S Toll, F.K. (1977) Untersuchungen zur Korrosions- und Witterungsbeständigkeit von Coil-Coating Verbundsystemen. Aachen, Germany, Institut für Kunststoffverarbeitung, RWTH (in German).

(25) Mattson, J. (1995) Damages from rottening and insects. Oslo, Norges forskningsråd (Fuktprogrammets skriftserie nr. 23) (in Norwegian).

(26) Svennerstedt, B. (1989) Ytfukt på fasadmaterial. Gävle, Sweden, Statens institut för byggnadsforskning (TN:16) (in Swedish).

(27) Cole, I.S. and Ganther G. (1996b) A preliminary investigation into airborne salinity adjacent to and within the envelope of Australian houses. *Constr. Building Materials, 10, No. 3, 203-207.*

(28) Sjöström, C. and Henriksen, J.F., 1987

(29) Scheffer, T.C. (1970) A climate index for estimating potential for decay in wood structures above ground. *Forest products journal, 21,* 10-25.

(30) Dovland, H. (1993) EMEP – The European Monitoring and Evaluation Programme. Presented at the Expert Meeting on Acid Precipitation Monitoring Network in East Asia. Toyama, Japan, 26-28 October 1993. Lillestrøm, Norwegian Institute for Air Research (NILU F 30/93).

(31) Sivertsen, B. and Haagenrud, S.E. (1994) EU 833 ENSIS '94 – An environmental surveillance system for the 1994 Winter Olympic Games. Presented at Vision Eureka Item Conference, Lillehammer, 14–15 June 1994. Lillestrøm, Norwegian Institute for Air Research (NILU F 10/94).

(32) Glomsrød, S., Godal, O., Henriksen, J.F., Haagenrud, S.E. and Skancke, T. (1996) Corrosion costs in Norway. *Paper presented at the UN ECE Convention on Long-Range Transboundary Air Pollution workshop on Economic evaluation of air pollution abatement and damage to buildings including cultural heritage, Stockholm, January 1996.*

(33) Haagenrud, S.E. and Henriksen, J.F. (1994) Materialkorrosjon forårsaket av luftforurensninger - med vekt på dose-respons-sammenhenger. Kjeller, Norwegian Institute for Air Research (NILU OR 74/94) (in Norwegian).

Chapter 9

Accelerated Weathering: Science, Pseudo-Science, or Superstition?

J. A. Chess, D. A. Cocuzzi, G. R. Pilcher, and G. N. Van de Streek

Akzo Nobel Coatings Inc.,
1313 Windsor Avenue, Columbus, OH 43211–2898

Undoubtedly without fully appreciating the full force of what they are assenting to, it is common for a room full of people to nod their heads vigorously when someone says, "the more things change, the more they stay the same." In a literal sense this is, of course, seldom true. But in a figurative sense--and, more importantly, in a *practical* sense--this is quite often true. While the details may change with the passing of time, the essence of a given situation may change but little. This paper, entitled "Accelerated Weathering: Science, Pseudo-science or Superstition," was written in 1997, yet it could easily have been presented at the symposium entitled "Accelerated Weathering: Myth vs. Reality," which was organized by the Cleveland Society for Coatings Technology two decades ago, in 1977. In fact, one of the co-authors of this current paper, J. A. Chess, was also a co-author of a paper presented at the Cleveland symposium. We've certainly generated a lot of data in the twenty years between that meeting and this meeting, but do we know any more about accelerated weathering now than we did then? Perhaps.... but we have only moved forward incrementally; no "silver bullets" have been discovered during the past twenty years, nor has anyone postulated a "unified weathering theory" pulling together all known information on the chemical and physical mechanisms by which diverse materials degrade, and how such degradation processes may be *accelerated and correlated* with "real world" results.

This paper will provide no solutions to a decades' old dilemma, but it will provide information and opinions, based upon extensive empirical data drawn from our experience with coatings for building products, which may be of value to a selected audience. First of all, the authors would like to declare "what we believe," so that there can be no doubt about where this paper is headed:

- *The best prediction of durability is real-time exposure in the location where the coating will be installed.*

© 1999 American Chemical Society

- *The second-best predictor is real-time exposure in Florida at 45° facing south (for roofs) and 90° facing south (for sidewalls.)*
- *The (distant) third-best predictor is one of several available accelerated weathering devices.*
- *Such devices, however, are generally paint system-specific in their predictive ability, e.g., weathering device "T" may successfully predict the real-time performance of System X, but cannot successfully predict the performance of System Y. Of all such accelerated weathering devices currently available, overwhelming evidence suggests that a UV-A tester is more likely to correctly predict the results of real-time, exterior exposure, in a variety of chemical systems.*
- *Like any accelerated data, UV-A data is only valuable when it is placed into the proper context by a skilled coatings scientist, who uses it as one of several tools at his/her disposal to predict long-term weathering effects.*
- *No accelerated weathering device, when used in a "stand alone" fashion, can predict "real world" weathering with any level of accuracy.*
- *Even the "real world" is a kind of myth--ever changing climatic, atmospheric, environmental, chemical, thermal, humidity and other considerations render the "real world" a dynamic exposure site.*
  (See Figure 1.)

This is our creed, based upon over a quarter of a century's worth of attempts to correlate accelerated weathering tests with the "real world," whatever that is. It applies to coatings for metal building product components which are applied by the coil coating process and thermally set or cured. This is our area of expertise and interest. Our work has benefited from the scientific inquiries and different practical approaches of several companies, since Akzo Nobel's current worldwide Coil Coatings organization is the product of a series of mergers, beginning in the late 1970's, which eventually involved the Wyandotte Paint Co., Pontiac Paint, Hanna Chemical Coatings Corp., Celanese, Reliance Universal, Akzo Coatings, Midland Dexter, Svensk Färgindustri, and Nobel. Each organization brought its own theoretical and experimental approach to the table, and each brought the results of its empirical testing, as well. Our current work is able to draw upon information generated by tens of thousands of exterior exposure panels exposed in over two dozen "UV," "aggressive chemical," and "corrosion" sites worldwide. It is upon these panels that we have based our current beliefs.

Before proceeding, it is important to define terms, because a term like "durability" can be not only qualitative--it can be downright slippery. "Durability" is often in the eyes of the beholder. There are parts of the world where rapid gloss loss of exterior coatings is valued because the coating has reached its "final" color very early in its life cycle and its appearance can be expected to remain relatively unchanged with time. In other areas of the world, however, rapid, precipitous gloss loss might cause the local denizens to clutch their chests because they prize the appearance of a glossy building and their concept of durability transcends mere

132

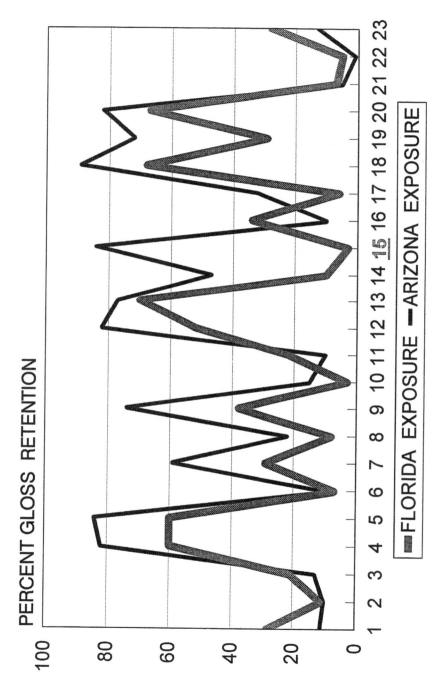

Figure 1. 23 Different Polyesters, High Gloss Brown.

color stability. Clearly, *appearance* depends upon far more than just color, and here's where the complications begin-- "durability" is a concept which involves freedom from some combination of color change, gloss change, chalking, cracking, crazing, blistering, peeling, etc., etc., etc., but that "special combination" may differ from observer to observer, or market to market. Since no two observers can be reasonably counted upon to agree on the exact weighting of these various factors, they must all be tested for, and it is desirable that they all be "predictable."

Over the years, in pursuit of this "predictability," our various predecessor organizations have worked with just about every predictive methodology--and every piece of predictive equipment--that has come along. To place our work in a context, following is a brief summary of the evaluation of accelerated testing methods from 1906 to the present:

- *1906--North Dakota Agricultural Experiment Station*

  This was the first test site for comparison of coatings in this country. Over time, this exterior test method moved to Florida and became ASTM G-7. Today, 5, 45, and 90 degree testing are common, along with black box, direct, and heated methods of testing.

- *1918--Fade-Ometer*

  First used for textiles, this was a dry test method and is no longer used for coatings. The carbon arc used did not achieve sufficiently short wavelengths for correlation with exterior exposure.

- *1927--Weather-Ometer*

  Water spray was added to the Fade-Ometer for improved results. However, the light distribution was still very poor.

- *1933--Sunshine Carbon ARC Lamp*

  This is the "Weather-Ometer" or "Atlas Weather-Ometer" as it is known today. The spectral light distribution from this equipment contains considerably more energy than was obtainable before. The carbons used are so strong that they give a poor simulation of natural sunlight in the UV region. Light filters are used to help improve the correlation. Two different types of filters are currently in use--the 2.5mm Corex 7058 and the newer 3mm Pyrex 7740 type. All types of carbon arc accelerated weathering are now covered by ASTM G-23.

- *1960--Xenon Arc Lamp*

With filters, the "Xenon Arc" spectral light distribution places it in the position of being one of the best light sources for accelerated weathering. This type of equipment can reasonably approximate sunlight in both the UV and visible light range. The equipment, however, is both expensive to purchase and expensive to operate. **(See ASTM G-26.)**

- *1960--Equatorial Mounts with Mirrors*

EMMAQUA testing is a method of employing natural sunlight as the source of light but concentrating it with mirrors. With the addition of air flow for cooling and water spray for moisture, quite reasonable correlation to natural weathering has been reported. The most important factor for correlation has been to record only UV light exposure as UV-MJ/M^2, not total light exposure [Langley]. ASTM G-90 covers this test method. Both "night time" and "day time" wetting methods are used, with "night time" being closer to Florida exposure, in our opinion. **(See Figures 2 & 3.)**

- *1965--Dew Cycle Weather-Ometer*

The development of coatings with greatly improved durability made normal carbon arc exposure methods too time consuming. The "Dew Cycle" method, ASTM 3361, is based on the assumption that, when both moisture and radiant energy are present at the same time, the most rapid film degradation takes place. To maintain humidity during the dark period, cold water is sprayed on the back of the test panels. Also, to speed results, the filters are removed to increase the intensity of the light.

By 1970, reports in the ***Journal of Coatings Technology*** were showing a lack of positive correlation to Florida exposure with this method. This became a common observation, and--by 1987--the ASTM test method included the following statement: "Failure caused by this light may bear no relationship to failure in natural sunlight." It has been reported that this test is 10-30 times more severe than the Atlas Weather-Ometer.

- *1970--Fluorescent UV Condensation*

This relatively inexpensive method of accelerated weathering uses fluorescent UV lights in a humidity condensation apparatus. Until 1987, it had the same excess UV light drawback as an unfiltered Atlas Weather-Ometer. Only three different fluorescent light bulbs were sold at the

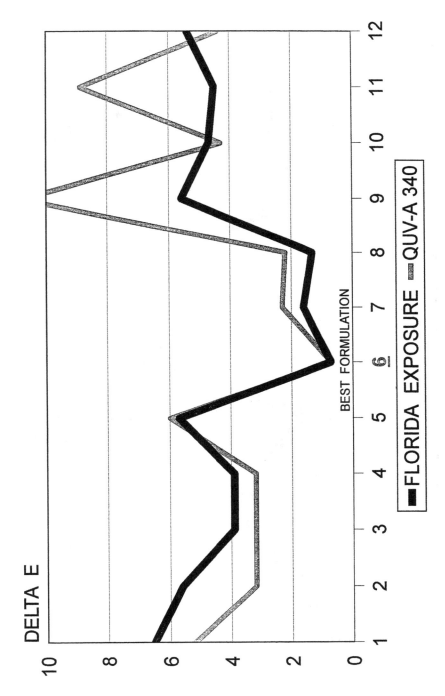

Figure 2. 12 Different Polyesters, Mist Green Color.

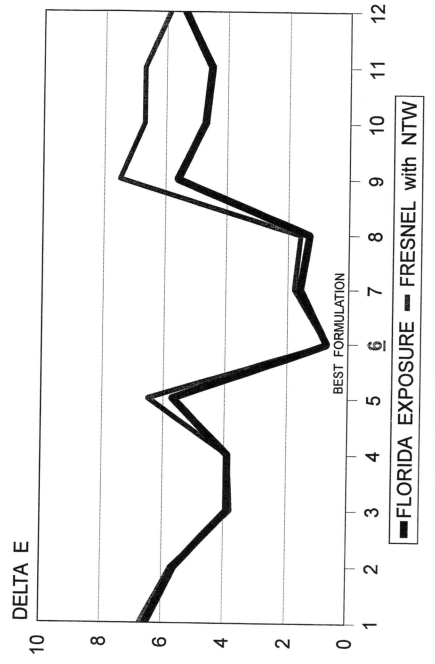

Figure 3. 12 Different Polyesters, Mist Green Color.

time: The first two were FS-40 and QUV-B 313, both of which give off shorter-wavelength UV radiation than is found on earth. The third bulb was QUV-A 351. This bulb is a good match for exposure under glass but cuts off the shorter wavelengths needed for film breakdown. In 1987, a new bulb, QUV-A 340, was introduced which gives a good match to the short wave bands of UV sunlight. This bulb's greatest strength lies in the evaluation of polymers, rather than pigments, since it concentrates all its energy within a narrow wave band. (See ASTM G-53 for operating principles.) **(See Figures 4 & 5.)**

- *1985--Predicting Coating Durability with Analytical Methods*

Several papers have appeared, based on short-term exposure, with infrared spectroscopy and other analytical methods used to follow chemical changes in the coating. The goal has been to predict the service life of the coating without using harsh or misleading acceleration factors. To date, the methods have worked best with basecoat/clearcoat systems. It is not yet certain how effective they will be with low gloss pigmented systems, but this work looks very promising.

- *Future*

  ◊ Reliability Theory?
  ◊ More advanced analytical methods?
  ◊ Computer modeling?
  ◊ Other?

At various points during the past four decades, our researchers explored, to a greater or lesser extent, just about all of these approaches. Each was greeted with the naive hope that "this one will finally provide the accelerated weathering test that we can take to the bank," and each was decried with great wailing and gnashing of teeth when this proved not to be the case. Like so many other companies, we "worked our way" through various accelerated devices--filtered carbon arc, unfiltered carbon arc, dew cycle, xenon arc, UV-B with several different "B" bulbs, and a few others as well. This work, of course, was done sequentially as the methods and testing devices were developed. We have no "master panel set" in which identical panels were weathered simultaneously in all known accelerated devices, then compared to Florida (or other "real world") exposure panels.

We do, however, have a very large number of panels which we have exposed in certain accelerated testing devices and have compared not only to actual exterior results, but to each other as well. When we have done this, one "truth" has made itself evident throughout our testing in the past ten years and has significantly influenced our thinking with regard to accelerated weathering: The truth is that there is no single, infallible predictor of outdoor durability, but there are some weathering devices which--when placed into the proper context and interpreted by

138

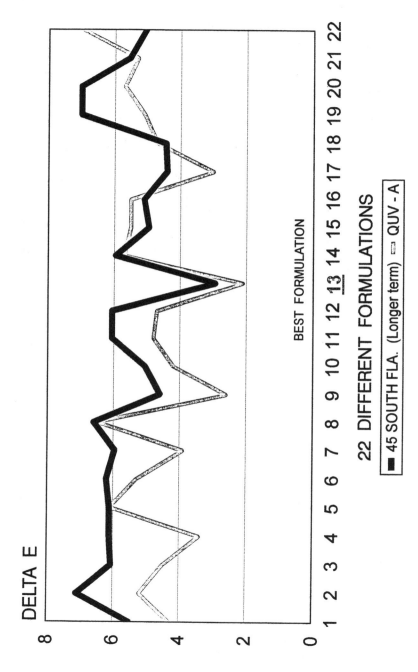

Figure 4. 22 Green Plastisols, Longer Term Fla. vs QUV-A 340.

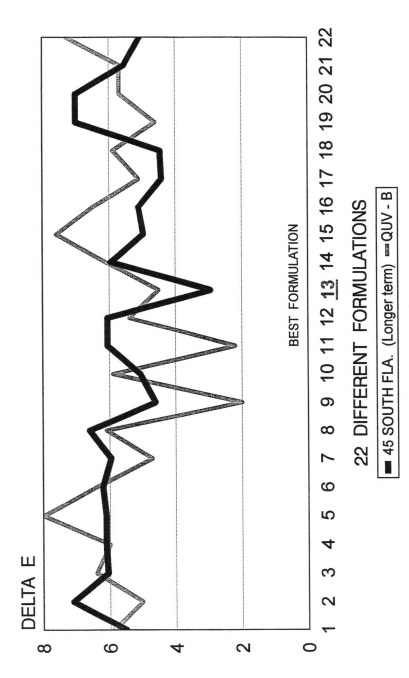

Figure 5. 22 Green Plastisols, Longer Term Fla. vs QUV-B 313.

an expert coatings specialist--greatly increase the confidence level with which out-door durability can be predicted. One such device is the QUV-A Weathering Tester, originally engineered and marketed by Q-Panel Lab Products, a leading company which has traditionally been on the cutting edge of accelerated testing methodology, whether radiation-related (UV-A and UV-B), cyclic corrosion, or condensing humidity. Q-Panel engineered both of the fluorescent-UV/condensation testers which the coatings industry commonly refers to as QUV-A and QUV-B. Here is what Q-Panel has to say about these two devices:

> *UV-A's are especially useful for tests comparing generically differ-ent types of polymers. Because UV-A lamps have no UV output below the normal solar cutoff of 295 mm, they usually do not de-grade materials as fast as UV-B lamps. However, they usually give better correlation with actual outdoor weathering.* (1)

At the time that QUV-A was introduced, "QUV" testing was being done by nearly everyone, and it wasn't designated as "UV-B" testing. A lot of people felt that it **was** the silver bullet, and--for certain systems, under certain conditions--perhaps it was. Our team approached QUV-A, nearly ten years ago, with some skepticism; the test is slow and considerably dilutes the phrase "accelerated." Nonetheless, the opinion that the ubiquitous QUV-B tester was not perfect and might be improved upon was not an opinion unique to its creators; in the late 1980's, it was being echoed throughout the industry. After comparing the weather-ing of automotive clearcoats in three different accelerated weathering instruments (carbon arc, xenon arc, and UV-B), Dr. David R. Bauer, Ford Motor Company's internationally eminent coatings scientist, and his colleagues found that "....since the degradation chemistry that occurs in these tests is unnatural," none of these devices were acceptable for the coatings that Ford was testing. Dr. Bauer con-cluded that "although acceleration factors can be calculated (based on amide II signal loss rates) they cannot be used reliably to predict service life." (2) This was recently reinforced, nearly a decade later, at the Spring, 1996, European Coil Coating Meeting, by Dr. G. C. Simmons (Becker Industrial Coatings) in his paper, "New Developments in Coil Coatings Paints": "It is now established fact that they [ASTM B117 salt spray and QUV-B] do not correlate well to natural exposures, and in some specific cases can lead to totally wrong conclusions being made."

Nothing has occurred, in the ten years since Dr. Bauer's comments, to change our thinking with regard to our dissatisfaction with broad use of the weath-ering devices utilizing UV-B bulbs. In fact, it was reinforced by Ford's Dr. John L Gerlock at a major scientific gathering in 1997 when he indicated that an FS40 bulb (UV-B) might be "good for studying the aging of the Taurus in low earth orbit," but not on the earth's surface. (3) Nor have such observations been limited to the automotive and coatings industries. In an important study by 3M Company, in which an attempt was made to develop an accelerated weathering test for films using two types each of carbon arc and UV-B bulbs, the researchers concluded that the results indicated "poor predictive ability using any of the laboratory devices."

(*4*) They further noted that Spearman ranking of the results from the UV-B samples ranged from "perfect correlation (1.0) to an almost complete reversal (-0.8) with essentially random scatter in between." (*5*) 3M's conclusion was that "commonly used cycles in carbon arc and fluorescent UV-condensation [UV-B] test equipment exhibited generally unacceptable correlation levels for these materials." (*6*) A major study by Dr. Carl J. Sullivan of ARCO Chemical Company (UV-A, UV-B, carbon arc and xenon arc) seconded this opinion: "These four accelerated test procedures yield contradicting conclusions on the weatherability of these four resin systems." (*7*) Dr. Sullivan notes that "the Florida exposure data clearly corroborate conclusions drawn from A-340, Xenon, and EMMAQUA studies and contradict B-313 results." (*8*)

While it is true that UV-B testing certainly accelerates the aging and degradation processes, it may--depending upon the coatings system--be accelerating the wrong chemistry, thereby vitiating any value that the information might provide and discrediting the test. Our own work has repeatedly shown that UV-B testing is so riddled with anomalies that--even when its predictions are in the right church--they are only rarely (and possibly coincidentally) in the right pew. It was this general lack of correlation with real chemical reactions and authentic exterior weathering results that led to the development of the UV-A lamp testing devices, which correlate more closely with sunlight. This was a very important advance because many of the most durable coatings in the "real world" are unnaturally damaged by the more destructive short wavelength of UV radiation below 295nm that is emitted by UV-B--radiation which **does not occur** in natural sunlight.

Even the American Society for Testing Materials (ASTM), which usually maintains a discreet silence on the appropriateness of its testing methods, allowed the inclusion of the following comments under the "non-mandatory information" section of Standard Method G53, "Standard Practice for Operating Light and Water Exposure Apparatus (Fluorescent UV-Condensation Type) for Exposure of Nonmetallic Materials," which is followed by laboratories around the world for running UV-A and UV-B accelerated testing:

> *All UV-B lamps emit UV below the normal sunlight cut-on. [sic]*
> *This short wavelength UV can produce rapid polymer degradation*
> *and **often causes degradation mechanisms that do not occur when***
> ***materials are exposed to sunlight. This may lead to anomalous***
> ***results....** For certain applications, the longer wavelength spectrum*
> *emitted by UV-A lamps is useful. Because UV-A lamps typically*
> *have little or no UV output below 300 nm, they usually do not degrade materials as rapidly as UV-B lamps, but **they may allow enhanced correlation with actual outdoor weathering.** (9) (**See Figures 6 & 7.**)*

In spite of our skepticism--in spite of our concerns that no accelerated weathering device can even hope to exactly predict the appearance that will result when the ravages of time, working hand-in-hand with humidity, heat, UV radiation

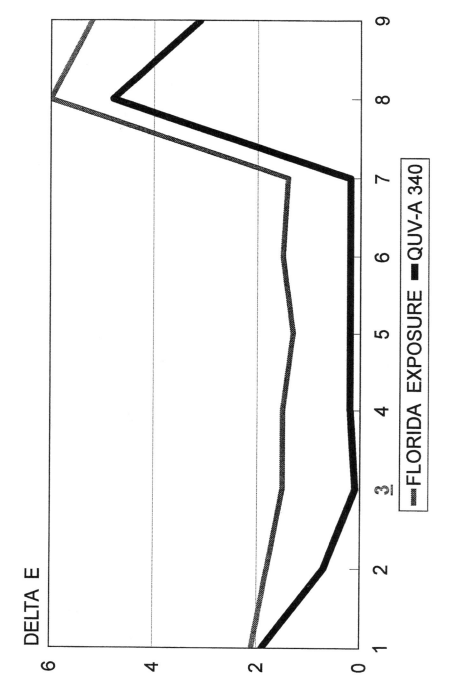

Figure 6. 9 Different Brown Polyesters.

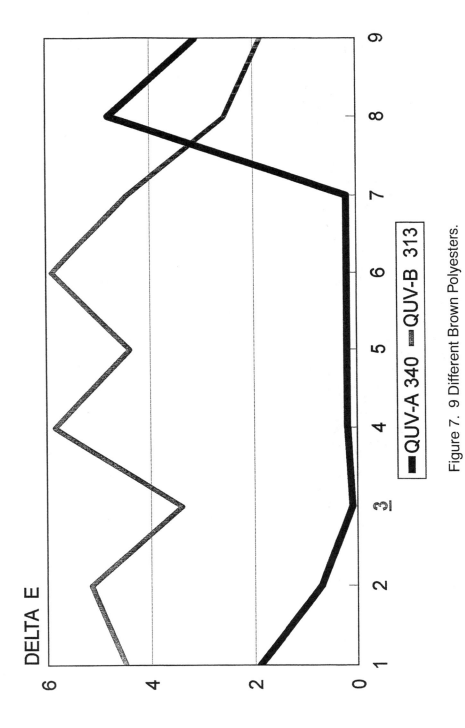

Figure 7. 9 Different Brown Polyesters.

and chemical agents such as SO_2 and NO_x, wreak their worst on a coated steel or aluminum building panel--we embarked upon a course to explore the new UV-A weathering device....

...A course which we have been on for a decade, and which has taken us in an enlightening and profitable direction. We certainly haven't found the mythical "silver bullet," but, in UV-A testing, we have definitely found an accelerated device which has enabled us to make surprisingly accurate predictions of the "real world" behavior of experimental polymer systems, and--especially--of modifications to existing polymers.

We have not, however, found UV-A accelerated exposure to be especially valuable for the prediction of the service performance of pigments, but we are not disappointed or disillusioned by this fact either. Although little generalization can be made about the huge group of different chemistries, both inorganic and organic, that make up the group of coatings components which are collectively referred to as "pigments," it seems clear that their performance can be so dramatically affected by chemical environments which are present at various places on the surface of the earth that it cannot be predicted with a high level of confidence by testing devices which lack these specific chemical atmospheres. (Xenon arc testing, which adds infrared and visible light components absent from UV-A testing, may prove helpful in evaluating the service life of pigments, albeit at a significantly higher operating cost--and still without benefit of the localized chemical environments at work in the "real world".) It would be nice to compare "brown coatings X, Y, and Z" from different sources based on different polymeric systems in a simple weathering device, but obtaining meaningful, valid data from such attempts is more wishful thinking than sound science.

Perhaps the most attractive aspect of UV-A testing, at least in our own program, has been the consistent absence--not of anomalous data, since this can always occur--but of *truly damaging, seriously misleading* data. This was the most treacherous aspect of our past work with such devices as the dew cycle carbon arc tester and the UV-B tester. While not ideal, we can all live with systems that look bad in an accelerated test but good in "real life." The worst that can happen is that we fail to sell a good product. If one were to believe anomalous testing results, however, that predicted *good* weatherability and went to the market with what turned out to be a *bad* product, this could spell disaster. *This does happen, however.* We found many examples of coatings that looked good in an accelerated test but bad under actual exterior service life conditions. This, of course, is the worst possible scenario. We have seen cases where dark brown plastisol films, based on poly(vinyl)chloride, have compared favorably in UV-B exposure to similarly-pigmented dark brown fluoropolymer films, based on poly(vinylidene)fluoride (usually abbreviated PVDF or PVF_2). Plastisol coatings have a very definite specialized niche in the building product marketplace, which they serve admirably, but no one expects them to perform over time at the level of a fluoropolymer--nor do they. In another recent, dramatic case involving a new, experimental polymer, we ran multiple accelerated tests which we are in the process of correlating with Florida. As can be seen from the data in Figures 6 and 7, if we had taken this product to the marketplace based upon UV-B results, we would have been making a grave mistake. Fortunately, our research policy prevents such mistakes from occurring, which is good, because--in the case of the unusual chemistry on which this polymer was based--even the UV-A accelerated data looked "OK," although nowhere nearly as misleading as the UV-B data. These disastrous results occurred because the new polymer was partially based on a relatively new, low-use cycloaliphatic monomer which does not absorb UV radiation in the 280-295nm range, where UV-B does its greatest damage. For this reason, it looks wonderful in the UV-B test, even though it does not look acceptable under actual service conditions, where other factors--most notably humidity and attendant hydrolytic instability--were apparently at work. The manufacturer of this monomer now includes this *caveat* in their product literature: "The use of QUV-B 313 is not suggested as a screening tool because the low-wavelength portion of the exposure spectrum can lead to anomalous results." (*10*) The fact that even the UV-A failed to predict the full extent of the poor field performance of this polymer, based on new, unusual chemistry, leads us to another important aspect of "what we believe":

Accelerated weathering devices only have proven value--

- *when testing materials which are very similar to other materials for which correlation with real-time outdoor exposure at a variety of test sites has already been established;*

- *when they are used in conjunction with other "real-time, real conditions" test data; and*

- *when the data which they yield are analyzed by an expert and compared against real-time, long-term exterior exposure data.*

We cannot stress these points too much or too often because a distressing new trend is emerging in the market place: Tools and data intended strictly for use in the scientific community, in the hands of skilled and experienced coatings scientists, are being moved into the marketplace where they are being misused as marketing tools. This is clearly exemplified by cases where sales or marketing representatives are attempting to sell coatings, based on new chemistry, solely on the basis of their performance in an accelerated weathering device. The device of choice is often UV-B, probably because of its ubiquitous nature and "known destructiveness." Since almost any coating looks good in *some* accelerated test, there are those who arm themselves with such testing data, then imprudently and improperly enter the marketplace crying "Eureka" from the housetops. Potential customers then attempt to require competitive products--based upon completely different chemistry--to match the same set of test results, and the "accelerated testing wars" have begun. This is a dangerous development in the marketplace because it places the emphasis not on actual field performance but on accelerated testing data, which--in this context--is only so much *hocus-pocus*. Which brings us to the final major aspect of "what we believe":

> **Running unknowns, such as competitive products, in an accelerated testing device for the purpose of predicting actual field performance is extremely risky business and becomes "pseudoscience" in the hands of anyone other than an expert coatings scientist.**

Strong words? Certainly--but impropriety must always be called on the carpet, lest it become the norm of our industry. We don't doubt that those who are misusing the results of accelerated testing may be doing so in good faith--our entire industry, after all, routinely runs many accelerated tests which we all acknowledge have no value (the B-117 salt spray test leaps to mind, but there are others), yet we continue to use the results from these tests as marketing tools. This is a form of the "big lie" technique that George Orwell described so chillingly in his novel, *1984, i.e.,* "if you tell a lie often enough, even though everyone knows that it is a lie, people will eventually come to believe it." By analogy, "if you present certain types of accelerated testing data to the marketplace in a consistent, insistent manner, even if it is demonstrably lacking a sound scientific basis, people will eventually come to believe that it must have some validity." Someone has to stand up and say, "enough." Accelerated testing data has its place--but that place is in the laboratory, in the hands of skilled coatings scientists, and in the company of related data from other complementary testing regimes, which place the data in a proper context.

Prof. William F. Kieffer, former editor of the *Journal of Chemical Education,* was internationally known as one of the foremost educators in the field of chemistry during the years from the 1940's through the 1970's, and his students fondly remember the way in which he could reduce even the most complex concepts to simple, easily-comprehensible terms. For instance, many generations of his chemistry students can still recite the laws of thermodynamics, courtesy of the "Kieffer Reduction":

Law #1   • The energy of the universe is constant.

Kieffer's Reduction   ♦ "You can't win."

Law #2   • The entropy of the universe strives toward a maximum.

Kieffer's Reduction   ♦ "You can't break even."

Law #3   • All systems have a specific entropy.

Kieffer's Reduction   ♦ "You can't even get out of the game."

In homage to Prof. Kieffer--and to all great educators who triumph in their craft because of their interest in conveying the *essence of what is important* in clear, concise terms that students will remember--we offer our own attempt to render the concept of "accelerated weathering" in equally clear, memorable terms:

Science:	• There are no silver bullets -- it takes time, many experiments, and a lot of testing to arrive at proper products for field performance.
Pilcher's Reduction	♦ You get what you pay for.
Pseudo-science:	• Anyone can compare system "A" from one coatings producer to system "B" from another in an accelerated testing regime and draw appropriate conclusions regarding which one will be "better" under actual service conditions.
Pilcher's Reduction	♦ Flip a coin; its faster, cheaper, and generally just as accurate.
Superstition:	• That "somewhere out there" is a weathering device that will predict long-term durability of all new coatings systems under all service life conditions.
Pilcher's Reduction	♦ There is no free lunch.

This is not a simple subject, and we do not feel that we have all the answers. Certainly UV-A, EMMAQUA (NTW)--and possibly xenon arc, as well-- offer improved tools for the coatings scientist of the 1990's, but *they are only tools.* As we look ahead, the work being done by both Ford Motor Company and Akzo Nobel (possibly other laboratories, as well) to relate early chemical changes of "real world" samples to long-term durability should greatly enhance our ability to run--and trust--predictive testing regimes. Ongoing work at NIST, under the direction of Dr. Jonathan L. Martin, on Reliability Theory--and work being undertaken (also at NIST), under the expert guidance of Dr. Mary E. McKnight, to relate durability as one aspect of overall *appearance prediction,* may revolutionize the way in which we think of the entire subject of "appearance," both initial and weathered. But the results of this work will be a long way off. For now, our best suggestion is this: "Thinking ahead is the best form of accelerated weathering."

Literature Cited

1. Q-Panel Lab Products, Product Bulletin LU-8160, **1994**.
2. Bauer, D. R.; Paputa Peck, M.C.; Carter, R. O. *J. Coatings Tech.,* **1987**, 59, p.103.

3. Gerlock, J. L., Ford Motor Company, Personal Communication, 1987.
4. Fischer, R. M. *SAE Technical Paper Series, #841022,* **1984**, p.2.
5. *Ibid.*
6. *Ibid.*, p. 1.
7. Sullivan, C. J. *J. Coatings Tech.,* **1995**, 67, p. 55.
8. *Ibid,* p. 56.
9. *ASTM Standard Method G53-95,* **1995**, p. 7.
10. Eastman Chemical Company, Publication N-335A, **1996**.

Chapter 10

Accelerated Life Tests: Concepts and Data Analysis

William Q. Meeker[1] and Luis A. Escobar[2]

[1]Department of Statistics, Snacedor Hall, Iowa State University,
Ames, Iowa 50011–1210
[2]Department of Experimental Statistics, Louisiana State University,
Baton Rouge, LA 70803

Abstract

Today's manufacturers face strong pressure to develop new, higher technology products in record time, while improving productivity, product field reliability, and overall quality. Estimating the failure-time distribution or long-term performance of high-reliability products is particularly difficult. Most modern products are designed to operate without failure for years, decades, or longer. Thus few units will fail or degrade appreciably in a test of practical length at normal use conditions. Thus the requirements for rapid product development and higher reliability have increased the need for accelerated testing of materials, components, and systems.

This paper describes methods for analyzing and planning accelerated life tests with an application in the area of microelectronics. In this application, the purpose of the accelerated life test was to study the dominant failure mechanism of an integrated circuit. It was believed that a first-order chemical reaction would provide an adequate description of this failure mechanism. Similar methods can be used to do laboratory evaluations of the life of products like paints and coatings. We have also included a section describing methods for using laboratory tests to predict life in the field with variability in use rates as well a both spatial and temporal variability in environmental conditions.

© 1999 American Chemical Society

Introduction

Motivation. Estimating the failure-time distribution or long-term performance of components of *high reliability* products is particularly difficult. For example, the design and construction of a communications satellite, may allow only 8 months to test components that are expected to be in service for 10 or 15 years. For such applications, Accelerated Tests (ATs) are used widely in manufacturing industries, particularly to obtain timely information on the reliability of simple components and materials. There are difficult practical and statistical issues involved in accelerating the life of a complicated product that can fail in different ways. Generally, information from tests at high levels of one or more accelerating variables (e.g., use rate, temperature, voltage, or pressure) is extrapolated, through a physically reasonable statistical model, to obtain estimates of life or long-term performance at lower, normal levels of the accelerating variable(s). In some cases, the level of an accelerating variable is increased or otherwise changed during the course of a test (step-stress and progressive-stress ATs). AT results are used in the reliability-design process to assess or demonstrate component and subsystem reliability, to certify components, to detect failure modes so that they can be corrected, compare different manufacturers, and so forth. ATs have become increasingly important because of rapidly changing technologies, more complicated products with more components, higher customer expectations for better reliability, and the need for rapid product development.

Different types of acceleration. The term "acceleration" has many different meanings within the field of reliability, but the term generally implies making "time" (on what ever scale is used to measure device or component life) go more quickly, so that reliability information can be obtained more rapidly.

There are several different methods of accelerating a reliability test:

- Increase the use-rate of the product. Consider the reliability of a toaster, which is designed for a median lifetime of 20 years, assuming a usage rate of twice each day. If, instead, we test the toaster 365 times each day, we could reduce the median lifetime to about 40 days. Also, because it is not necessary to have all units fail in a life test, useful reliability information could be obtained in a matter of days instead of months.

- Increase the aging-rate of the product. For example, increasing the level of experimental variables like temperature or humidity can accelerate the chemical processes of certain failure mechanism such as chemical degradation (resulting in eventual weakening and failure) of an adhesive mechanical bond or the growth of a conducting filament across an insulator (eventually causing a short circuit).

- Increase the level of stress (e.g., temperature cycling, voltage, or pressure) under which test units operate. A unit will fail when its *strength* drops below applied stress. Thus a unit at a high stress will generally fail more rapidly than it would have failed at low stress.

Combinations of these methods of acceleration are also employed. Variables like voltage and temperature cycling can both increase the rate of an electro-chemical reaction (thus accelerating the aging rate) and increase stress relative to strength. In such situations, when the effect of an accelerating variable is complicated, there may not be enough physical knowledge to provide an adequate physical model for acceleration (and extrapolation). Empirical models may or may not be useful for extrapolation to use conditions.

Acceleration Models. Interpretation of accelerated test data requires models that relate accelerating variables like temperature, voltage, pressure, size, etc. to time acceleration. For testing over some range of accelerating variables, one can fit a model to the data to describe the effect that the variables have on the failure-causing processes. The general idea is to test at high levels of the accelerating variable(s) to speed up failure processes and then to extrapolate to lower levels of the accelerating variable(s). For some situations, a physically reasonable statistical model may allow such extrapolation.

Physical Acceleration Models. For well-understood failure mechanisms, one may have a model based on physical/chemical theory that describes the failure-causing process over the range of the data and provides extrapolation to use conditions. The relationship between accelerating variables and the actual failure mechanism is usually extremely complicated. Often, however, one has a simple model that adequately describes the process. For example, failure may result from a complicated chemical process with many steps, but there may be one rate-limiting (or dominant) step and a good understanding of this part of the process may provide a model that is adequate for extrapolation.

Empirical Acceleration Models. When there is little understanding of the chemical or physical processes leading to failure, it may be impossible to develop a model based on physical/chemical theory. An empirical model may be the only alternative. An empirical model may provide an excellent fit to the available data, but provide nonsense extrapolations. In some situations there may be extensive empirical experience with particular combinations of variables and failure mechanisms and this experience may provide the needed justification for extrapolation to use conditions. The next section describes some simple acceleration models that have been useful in specific applications.

Temperature Acceleration

It is sometimes said that high temperature is the enemy of reliability. Increasing temperature is one of the most commonly used methods to accelerate a failure mechanism.

152

Arrhenius Relationship Time-Acceleration Factor. The Arrhenius relationship is a widely-used model describing the effect that temperature has on the rate of a simple chemical reaction. This relationship can be written as

$$\mathcal{R}(\text{temp}) = \gamma_0 \exp\left(\frac{-E_a}{k_B \times \text{temp K}}\right) = \gamma_0 \exp\left(\frac{-E_a \times 11605}{\text{temp K}}\right) \quad (1)$$

where $\mathcal{R}$ is the reaction rate and $\text{temp K} = \text{temp}\,°C + 273.15$ is temperature in the absolute Kelvin scale, $k_B = 8.6171 \times 10^{-5} = 1/11605$ is Boltzmann's constant in units of electron volts per $°C$, and E_a is the activation energy in units of electron volts (eV). The parameters E_a and γ_0 are product or material characteristics. The Arrhenius acceleration factor is

$$\mathcal{AF}(\text{temp}, \text{temp}_U, E_a) = \frac{\mathcal{R}(\text{temp})}{\mathcal{R}(\text{temp}_U)} = \exp\left[E_a\left(\frac{11605}{\text{temp}_U\,K} - \frac{11605}{\text{temp K}}\right)\right].$$
$$(2)$$

When $\text{temp} > \text{temp}_U$, $\mathcal{AF}(\text{temp}, \text{temp}_U, E_a) > 1$. When temp_U and E_a are understood to be, respectively, product use temperature and reaction-specific activation energy, $\mathcal{AF}(\text{temp}) = \mathcal{AF}(\text{temp}, \text{temp}_U, E_a)$ will be used to denote a time-acceleration factor. The Arrhenius relationship does not apply to all temperature acceleration problems and will be adequate over only a limited temperature range (depending on the particular application). Yet it is satisfactorily and widely used in many applications. Nelson [6], page 76) comments that " ... in certain applications (e.g., motor insulation), if the Arrhenius relationship ... does not fit the data, the data are suspect rather than the relationship."

Accelerated Life Test Models. Most parametric ALT models have the following two components:

1. A parametric distribution for the life of a population of units at a particular level(s) of an experimental variable or variables. It might be possible to avoid this parametric assumption for some applications, but when appropriate, parametric models (e.g., Weibull and lognormal) provide important practical advantages for most applications.

2. A relationship between one (or more) of the distribution parameters and the acceleration or other experimental variables. Such a relationship models the effect that variables like temperature, voltage, humidity, and specimen or unit size will have on the failure-time distribution. This part of the accelerated life model should be based on a physical model such as one relating the accelerating variable to degradation, on a well-established empirical relationship, or some combination.

The example in this paper uses the log-location-scale regression model described in Chapters 17 and 19 of [3] and the relationship between life and temperature implied by the Arrhenius relationship in (1) and (2).

Strategy for Analyzing ALT Data. The following strategy is useful for analyzing ALT data consisting of a number of groups of specimens, each having been run at a particular set of test conditions (e.g., three levels of temperature). The basic idea is to start by examining the data graphically. Use probability plots to analyze each group separately and explore the adequacy of candidate distributions. Then fit a model that describes the relationship between life and the accelerating variable(s). Briefly, the strategy is to

1. Examine a scatter-plot of failure time versus the accelerating variable.

2. Fit distributions individually to the data at separate levels of the accelerating variable. Plot the fitted ML lines on a multiple probability plot along with the individual nonparametric estimates at each level of the accelerating variable. Use the plotted points and fitted lines to assess the reasonableness of the corresponding life distribution and the constant-σ assumption. Repeat with probability plots for different assumed failure-time distributions.

3. Fit an overall model with the proposed relationship between life and the accelerating variable.

4. Compare the combined model from Step 3 with the individual analyses in Step 2 to check for evidence of lack of fit for the overall model.

5. Perform residual analyses and other diagnostic checks to assess the adequacy of the model assumptions.

6. Assess the reasonableness of the ALT data and model to make the desired inferences.

The example in this paper has just one accelerating variable (the simplest and most common type of ALT in electronic applications). Chapter 4 of [6] and Chapter 19 of [3] show how to apply the same general strategy to ALTs with two or more accelerating variables.

Analysis of Single-Variable ALT Data

This section describes methods for analyzing ALT data with a single accelerating variable. The subsections illustrate, in sequence, the steps in the strategy described above.

Example 1 Data from an ALT on a New-Technology IC Device. Table 1 gives data from an accelerated life test on a new-technology integrated circuit (IC) device. The device inspection involved an expensive electrical diagnostic test. Thus only a few inspections could be conducted on each device. One common method of planning the times for such inspections is to choose a first inspection time and then space the inspections such that

Table 1: Failure interval or censoring time and testing temperature from an ALT experiment on an integrated circuit device.

Hours			Number of	Temperature
Lower	Upper	Status	Devices	°C
—	1536	Censored	50	150
—	1536	Censored	50	175
—	96	Censored	50	200
384	788	Failed	1	250
788	1536	Failed	3	250
1536	2304	Failed	5	250
—	2304	Censored	41	250
192	384	Failed	4	300
384	788	Failed	27	300
788	1536	Failed	16	300
—	1536	Censored	3	300

they are equally spaced on a log axis. In this case, the first inspection was after one day with subsequent inspections at two days, four days, and so on (except for one day when the person doing the inspection had to leave early). Tests were run at 150, 175, 200, 250, and 300°C junction temperature. Failures had been found only at the two higher temperatures. After an initial analysis based on early failures at 250°C and 300°C, there was concern that no failures would be observed at 175°C before the time at which decisions would have to be made. Thus the 200°C test was started later than the others to assure some failures and only limited running time on these units had been accumulated by the time of the analysis.

The developers were interested in estimating the activation energy of the failure mode and the long-life reliability of the ICs. Initially engineers asked about "MTTF" at use conditions of 100°C junction temperature. After recognizing that the estimate of the mean would be on the order of 6 million hours (more than 700 years) they decided that this would not be a useful reliability metric. Subsequently they decided that the average hazard rate or the proportion that would fail by 100 thousand hours (about 11 years) would be more useful for decision-making purposes. ∎

Scatter-Plot of ALT Data. Start by examining a scatter-plot of failure-time data versus the accelerating-variable data. A different symbol should be used to indicate censored observations.

Example 2 Scatter-plot of the IC device data. Figure 1 is a scatter-plot of the C device data introduced in Example 1. As expected, units fail sooner at higher levels of temperature. The heavy censoring (note for example that there were no failures at 150°C or 175°C) makes it difficult to see the form of the life/accelerating variable relationship from this plot. ∎

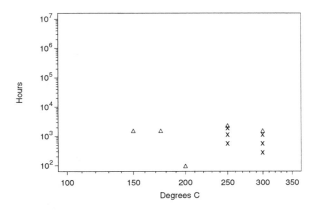

Figure 1: Scatter-plot showing inspection times where failures and right-censored observations were observed versus °C junction temperature for the IC device data. Right-censored observations are indicated by a △

Multiple probability plot of nonparametric cdf estimates at individual levels of the accelerating variable.

To make a multiple probability plot, first compute nonparametric estimates of the failure-time distribution for each group of specimens tested at the same level of the accelerating variable. Then plot these on probability paper corresponding to the particular distribution being entertained as a possible model for the data. Such plots provides a powerful tool for assessing the distributional model for the different levels of the accelerating variable (or variable-level combinations). One can make and compare plots for different distributions (e.g., Weibull and lognormal).

It is useful to plot, in addition, the individual fitted distribution at each level of the accelerating variable. Particularly if a suitable parametric distribution can be found, then ML estimates of the cdf at each level of the accelerating variable should be computed and put on the probability plot along with the corresponding nonparametric cdf estimates. This plot is useful for assessing the commonly used assumptions that distribution shape does not depend on the level of the accelerating variable and that the accelerating variable only affects the distribution scale parameter. The slopes of the lines are related to the distribution shape parameter values. Thus we can assess graphically the assumption that temperature has no effect on distribution shape.

Example 3 Multiple probability plot and ML estimates for the new-technology IC device life at 250°C and 300°C. Figure 2 is a log-

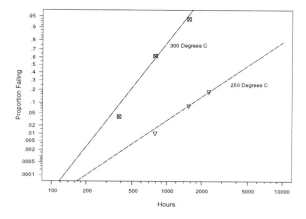

Figure 2: Lognormal probability plot of the proportion failing at 250°C and 300°C for the new-technology integrated circuit device ALT experiment.

normal probability plot of the failures at 250°C and 300°C along fitted lines that will be described below. This plot shows that the lognormal distribution provides a reasonable description of the observed proportion failing at 250°C and 300°C. Table 2 summarizes the individual lognormal ML estimates.

The different slopes in the plot suggests that the lognormal shape parameter σ changes from 250 to 300°C. Such a change could be caused by the occurrence of a different failure mode at high temperatures, casting doubt on the simple first-order Arrhenius model. Failure modes with a higher activation energy, that might never be seen at low levels of temperature, can appear at higher temperatures (or other accelerating variables). A 95% confidence

Table 2: Individual lognormal ML estimation results for the new-technology IC device.

	Parameter	ML Estimate	Standard Error	95% Approximate Confidence Intervals Lower	Upper
250°C	μ	8.54	.33	7.9	9.2
	σ	.87	.26	.48	1.57
300°C	μ	6.56	.07	6.4	6.7
	σ	.46	.05	.36	.58

The log-likelihood values were $\mathcal{L}_{250} = -32.16$ and $\mathcal{L}_{300} = -53.85$. The confidence intervals are based on the normal-approximation method.

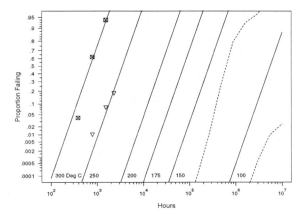

Figure 3: Lognormal probability plot showing the ML fit of the Arrhenius-lognormal model for the new-technology IC device. The dotted lines are a set of approximate 95% confidence intervals for the failure time distribution at 100°C junction temperature.

interval for the ratio $\sigma_{250}/\sigma_{300}$ is $[1.01, \quad 3.53]$. This suggests that there could be a real difference between σ_{250} and σ_{300}. These results also suggested that detailed physical failure mode analysis should be done for at least some of the failed units and that the accelerated test should be extended until some failures are observed at lower levels of temperature. Even if the Arrhenius model is questionable at 300°C, it could be adequate below 250°C. ∎

Multiple Probability Plot of ML Estimates with a Fitted Acceleration Relationship. In order to draw conclusions about life at low levels of accelerating variables, one needs to use a life/accelerating variable relationship to tie together results at the different levels of the accelerating variable. The cdfs estimated from the model fit can also be plotted on a probability plot along with the data to assess how well the life/accelerating variable model fits the data. Extrapolations to other levels of the accelerating variable can also be plotted.

Example 4 ML estimates of the Arrhenius-lognormal model for the new-technology IC device data. The Arrhenius-lognormal failure-time regression model follows from (2) and can be expressed as

$$\Pr[T \leq t; \texttt{temp}] = \Phi_{\text{nor}} \left[\frac{\log(t) - \mu}{\sigma} \right]$$

Table 3: Arrhenius-lognormal model ML estimation results for the new-technology IC device.

Parameter	ML Estimate	Standard Error	95% Approximate Confidence Intervals	
			Lower	Upper
β_0	-10.2	1.5	-13.2	-7.2
β_1	.83	.07	.68	.97
σ	.52	.06	.42	.64

The log likelihood is $\mathcal{L} = -88.36$. The confidence intervals are based on the normal-approximation method.

where $\mu = \beta_0 + \beta_1 x$, $x = 11605/(\text{temp K})$, and $\beta_1 = E_a$ is the activation energy. Table 3 gives Arrhenius-lognormal model ML estimates for the new-technology IC device.

Figure 3 is a lognormal probability plot showing the Arrhenius-lognormal model fit to the new-technology IC device ALT data. This figure shows lognormal cdf estimates for all of the test levels of temperature as well at the use-condition of 100°C junction temperature. The dotted curves are a set of pointwise 95% normal-approximation confidence intervals. They reflect the random "sampling uncertainty" arising from the limited sample data. This plot shows the rather extreme extrapolation needed to estimate the failure-time distribution at the use conditions of 100°C. If the projections are close to the truth, it appears unlikely that there will be any failures below 200°C during the remaining 3000 hours of testing and, as mentioned before, this was the reason for starting some units at 200°C. It is important to note that these intervals do *not* reflect model-specification and other errors (and we know that the model is only an approximation for the exact relationship). Figure 4 shows directly the fitted life/accelerating variable relationship and the estimated densities at each level of temperature, and lines indicating ML estimates of percent failing as a function of temperature. The density estimates are normal densities because time is plotted on a log scale. ∎

Comparing Individual and Model Analyses. It is useful to compare individual analyses with model analyses. This can be done both graphically and analytically. A likelihood ratio test provides an analytical assessment about whether observed deviations between the individual model fit and the overall life/accelerating variable relationship can be explained by random variability or not. A statistical test (details given in Example 19.3.2 of [3]) indicates that there is some lack of fit in the constant-σ Arrhenius-lognormal model. This suggests that the underlying failure mode and or the underlying chemical reaction might have changed at 300°C.

ALT Predictions with a Known Acceleration Factor. In some applications, temperature-accelerated life tests are run with only one level of

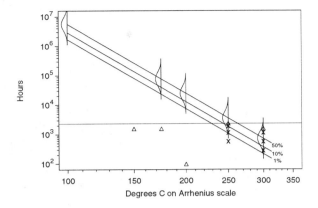

Figure 4: Arrhenius plot showing the new-technology IC device data and the Arrhenius-lognormal model ML estimates. Censored observations are indicated by $\triangle$.

temperature. Then a given value of activation energy is used to compute an acceleration factor to estimate life at use temperature. Resulting confidence intervals are generally unreasonably precise because activation energy is generally not know exactly. For example, MIL-STD-883 provides reliability demonstration tests based on a given value of E_a.

Example 5 Analysis with activation energy E_a given. Figure 5, similar to Figure 2 shows the effect of assuming that $E_a = .8eV$ and having to estimate only β_0 and σ from the limited data. Using a given E_a results in a set of approximate 95% confidence intervals for $F(t)$ at 100°C that are unrealistically narrow. ∎

Using Prior Information in Accelerated Testing

This section uses an extension of the Bayesian methods presented in Chapter 14 of [3], to reanalyze the data from Example 1. The computational methods used here follow those used in [2] and [3].

Example 1 illustrated the analysis of accelerated life test data on a new-technology IC device. As a contrast, Example 5 showed how much smaller the confidence intervals on $F(t)$ would be if the Arrhenius activation energy were known. Generally it is unreasonable to assume that a parameter like

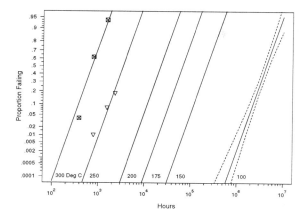

Figure 5: Lognormal probability plot showing the Arrhenius-lognormal model ML estimates and 95% confidence intervals for $F(t)$ at $100°C$ for the new-technology IC device with given $E_a = .8$.

activation energy is known exactly. For some applications, however, it may be useful or even important to bring outside knowledge into the analysis. Otherwise it would be necessary to spend scarce resources to conduct experiments to learn what is already known. In some applications, knowledgeable reliability engineers can, for example, specify the approximate activation energy for different expected failure modes. Translating the information about activation energy into a prior distribution will allow the use of Bayesian methods like those described in Chapter 14 of [3]. This section shows how to incorporate prior information on the activation energy for a failure mode into an analysis of the new-technology IC device data.

Prior Distributions. The most important motivation for using prior information to supplement data in an analysis is to combine it with data to provide more and better information about model parameters of interest. It is convenient to divide available prior information about a parameter into three different categories:

1. Parameters that are given as known, leading to a degenerate prior distribution.

2. Parameters with a diffuse or approximately noninformative prior distribution.

3. An informative, nondegenerate prior distribution.

In general, there are two possible sources of prior information: a) expert or other subjective opinion or b) past data. The prior pdf $f(\theta)$ may be either informative or not. Loosely speaking, a noninformative (there is a particular technical definition for a noninformative prior distribution, but we use the term loosely to indicate a prior distribution that carries little or no weight in estimation relative to the information in the available data) prior distribution is one that provides little or no information about any of the parameters in θ. Such a prior distribution is useful when it is desired to let the data speak for themselves without being influenced by previous data, expert opinion, or other available prior information.

This section reanalyzes the new-technology IC device ALT data in order to compare

- A diffuse (wide uniform) prior distribution for E_a.

- A given value (degenerate prior distribution) for E_a.

- The engineers' prior information, converted into an informative prior distribution for E_a.

On the basis of previous experience with a similar failure mode, the engineers responsible for this device felt that it would be safe to presume that, with a "high degree of certainty," the activation energy E_a is somewhere in the interval .80 to .95. They also felt that a normal distribution could be used to describe the uncertainty in E_a. We use normal distribution 3-SD limits (i.e., mean $\pm$ three standard deviations) to correspond to an interval with a high degree of certainty, corresponding to about 99.7% probability. This is an informative prior distribution for E_a. The engineers did not have any firm information about the other parameters of the model. To specify prior distributions for the other parameters, it is then appropriate to choose a diffuse prior. A convenient choice is a uniform distribution that extends far beyond the range of the data and physical possibility. We use UNIF(a, b) to denote a distribution with probability distributed uniform between the limits a and b. As in described in Chapter 14 of [3], the parameters used to specify the joint posterior distributions should be given in terms of parameters that can be specified somewhat independently and conveniently. For this example, the prior distributions for σ was specified as UNIF$(.2, .9)$ and the prior distribution for $t_{.1}$ at 250°C was specified to be UNIF$(500, 7000)$ hours. Comparison with the ML estimates from Example 1 shows that the corresponding joint uniform distribution is relatively diffuse.

Figure 6 compares a NOR prior distribution with a 3-SD range of $(.80, .95)$ and a UNIF$(.4, 1.4)$ (diffuse) prior for E_a. The corresponding marginal posterior distributions for E_a are also shown. The center of the marginal posterior distribution for E_a corresponding to the informative NOR prior is very close to that of the prior itself. This is mostly because the prior is strong relative to the information in the data. Figure 6 also shows a posterior distribution corresponding to the uniform (diffuse) prior. The corresponding joint posterior is approximately proportional to the profile likelihood for E_a. The uniform

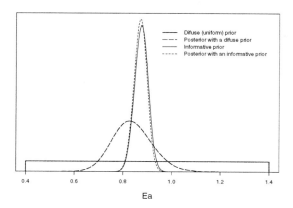

Figure 6: Plot of diffuse and informative prior distributions for the new technology device activation energy E_a along with corresponding posterior distributions.

prior has had little effect on the posterior and therefore is approximately noninformative.

In addition to activation energy E_a, the reliability engineers also wanted to estimate life at 100°C. Figure 7 shows different posterior distributions for $t_{.01}$ at 100°C. The plots on the top row of Figure 7 compare posteriors computed under the informative and diffuse prior distributions for E_a. This comparison shows the strong effect of using the prior information in this application.

Recall from Example 1 that there was some concern (because of the different slopes in Figure 2) about the possibility of a new failure mode occurring at 300°C. Sometimes physical failure mode analysis is useful for assessing such uncertainties. In this application the information was inconclusive.

When using ML estimation or when using Bayesian methods with a diffuse prior for E_a, it is necessary to have failures at two of more levels of temperature in order to be able to extrapolate to 100°C. With a given value of E_a or an informative prior distribution on E_a, however, it is possible to use Bayes methods to estimate $t_{.01}$ at 100°C with failures at only one level of temperature. The posterior distributions in the bottom row of Figure 7 assess the effect of dropping the 300°C data, leaving failures only at 250°C. Comparing the graphs in the NW and SW corners, the effect of dropping the 300°C data results in a small leftward shift in the posterior. Relative to the confidence intervals, however, the shift is small. Comparing the two plots in the bottom row suggests that using a given value of $E_a = .8$ results in an interval that is probably unreasonably narrow and potentially misleading.

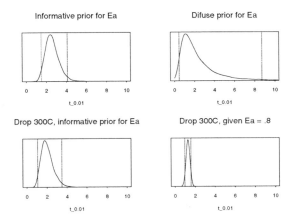

Figure 7: Plot of the marginal posterior distribution of $t_{.01}$ at $100°$C for the new technology device, based on different assumptions. NW corner: all data and an informative prior for E_a. NE corner: all data and a diffuse prior for E_a. SW corner: drop $300°$C data and an informative prior for E_a. SE corner: drop $300°$C data and given $E_a = .8$. The vertical lines are two-sided 95% Bayes confidence intervals for $t_{.01}$ at $100°$C.

If the engineering information and previous experience used to specify the informative prior on E_a is credible for the new device, then the SW analysis provides an appropriate compromise between the commonly used extremes of assuming nothing about E_a and assuming that E_a is known.

If one tried to compute the posterior after dropping the $300°$C, using a uniform prior distribution on E_a, the posterior distribution would be strongly dependent on the range of the uniform distribution. This is because with failures only at $250°$C, there is no information on how large E_a might be. In this case there would be no approximately uninformative prior distribution.

To put the meaning of the results in perspective, the analysis based on the informative prior distribution for E_a after dropping the suspect data at $300°$C would be more credible than the alternatives. The 95% Bayesian confidence intervals for $t_{.01}$ at $100°$C for this analysis are [.6913, 2.192] million hours or [79, 250] years. This does not imply that the devices will last this long (we are quite sure that they will not!). Instead, the results of the analysis suggest that, if the Arrhenius model is correct, this particular failure mode is unlikely to occur until far beyond the technological life of the system into which the IC would be used. It is likely, however, that there are other failure modes (perhaps with smaller E_a) that will be observed, particularly at lower levels of temperature (see also Pitfall 3, discussed below).

Cautions on the Use of Prior Information. In many applications, engineers really have useful, indisputable prior information (e.g., information from physical theory or past experience deemed relevant through engineering or scientific knowledge). In such cases, the information should be integrated into the analysis. Analysts and decision makers must, however, beware of and avoid the the use of "wishful thinking" as prior information. The potential for generating seriously misleading conclusions is especially high when experimental data will be limited and the prior distribution will dominate in the final answers (common in engineering applications). [1] describes such concerns from an engineering point of view.

As with other analytical methods, when using Bayesian statistics, it is important to do sensitivity analyses with respect to uncertain inputs to ones model. For some model/data combinations, Bayes' estimates and confidence bounds can depend entirely on prior assumptions. This possibility can be explored by changing prior distribution assumptions and checking the effect that the changes have on final answers of interest.

Pitfalls and Suggestions

ALTs can be a useful tool for obtaining timely information about materials and products. There are, however, a number of important potential pitfalls that could cause an ALT to lead to seriously incorrect conclusions. Users of ALTs should be careful to avoid these pitfalls.

One of the most important assumptions of accelerated life testing is that increases in the acceleration variables do not change the underlying failure mechanism. In some cases new failure modes result from a fundamental change in the way that the material or component degrades or fails at high levels of the accelerating variable(s). For example, instead of simply accelerating a failure-causing chemical process, increased temperature may actually change certain material properties (e.g., cause melting). In less extreme cases, high levels of an accelerating variable will change the relationship between life and the accelerating variable (e.g., life at high temperatures may not be approximate linear in inverse absolute temperature, as predicted by the Arrhenius relationship).

If other failure modes are caused at high levels of the accelerating variables and this is recognized, it can be accounted for in the data analysis by treating the failure for the new failure modes as a censored observation (as long as the new failure mode does not completely dominate the failure mode(s) of interest). Chapter 7 of [6] gives several examples. In this case, however, such censoring can severely limit the information available on the failure mode of interest. If other failure modes are present but not recognized in data analysis, seriously incorrect conclusions are likely.

In general, ALT experiments should be planned and executed with a great deal of care. Inferences and predictions should be made with a great deal of caution. Some particular suggestions for doing this are

- Use previous experience with similar products and materials.

- Conduct initial studies (pilot experiments) to evaluate the effect that the accelerating variable or variables will have on degradation and the effect that degradation will have on life or performance. Information from preliminary tests provide useful input for planning ALTs (as described in Chapter 20 of [3]).

- Use failure mode analysis and physical/chemical theory to improve or develop physical understanding to provide a better physical basis for ALT models.

- Limit, as much as possible, the amount of extrapolation (in time and in the accelerating variable). Methods for doing this are described in Chapter 20 of [3].

Issues Relating to Outdoor Service Life Prediction

This section describes some particular issues that will be important in the application of accelerated test methods to the prediction of outdoor service life .

Relating AT Results with Field Performance. Using the results of laboratory accelerated test to predict the life of product in the field is difficult. In the simplest cases such predictions involve extrapolation from high levels of accelerating variables to use-levels of those variables. With an appropriate model describing the relationship between life and the accelerating variable, accelerated tests like those described in the earlier sections of this paper can provide a useful description of the failure behavior of a product under homogeneous well-controlled conditions.

When, however, the product is exposed to harsh variable environments that differ from the controlled environment of the laboratory and when different products are used in different environments, the prediction problem is more complicated. It is possible, however, to make such predictions if the following information is available:

- An appropriate model for relating cumulative damage rates to cumulative damage resulting from well-understood failure mechanisms, and

- A model describing the temporal variability in environmental conditions for a particular unit, and

- A model or other population data to describe the unit-to-unit differences (e.g., a spatial description of weather variables).

Laboratory Experiments. Important environmental variables need to be studied experimentally. Importance depends on failure mechanism. Potentially important variables for different failure modes include temperature, humidity, UV radiation, acid rain, etc. Estimation of interactions between environmental variables (e.g., if the effect of a change in UV radiation depends on the level of temperature) requires experiments that vary more than one variable (or factor) at a time (e.g., factorial or fractional factorial experiments). When laboratory environmental conditions and use-rates (i.e., test conditions) cannot be carefully controlled, the test conditions should, at least, be carefully measured and recorded. Laboratory experiments are (or should be) run under carefully controlled environmental conditions and use-rates.

Field Data. Consideration of environmental conditions is particularly important for products that are used out doors. Field data arise from a complicated mixture of environmental conditions and use-rates. Environmental conditions vary geographically (spatially) and over time (temporally). Different failure mechanisms operate on different time scales (e.g., real time, number of use cycles, amount of time exposed to sunlight, etc.) Use rate varies from user to user and according to application. Use rate may also vary over temporally.

Modeling the Relationship between Service Life and Environmental Conditions. We use e to denote or describe the environmental conditions to which a *particular* unit of product will be exposed (e.g., UV radiation, temperature, humidity, acid rain pH might be the important variables). Also, r is a corresponding scalar that is used to denote or describe the product use rate (e.g., the number of hour/day of use or exposure to the sun). For service life prediction there is need to use information from the accelerated testing and other available knowledge about the failure mechanism to predict service life at specified conditions (e, r).

Modeling the Effect of Temporal Variability in Environmental Conditions. Many environmental variables (e.g., UV radiation and temperature) have both daily and seasonal periodicities as well as other random variabilities. In some situations, for purposes of prediction, it may be satisfactory to suppose that the (e, r) represents environmental conditions and use-rates that are approximately constant over time (e.g., when use temperature can be described adequately by average temperature and humidity). In other applications, however, especially when temporal variability in environmental conditions is large, such variability might have to be described by a stochastic process model and the use of a cumulative damage (or degradation) model to model the relationship between (e, r) and the failure-time distribution. Then (e, r) would be used to denote the parameters of such a model.

Time Transformation Model for Environment and Use Rate Differences. A time transformation function provides a general approach to relating the failure-time distributions at different conditions. Using this approach, the failure time at conditions (e, r) is related to the failure time at specified baseline conditions (e_0, r_0) by

$$T(e, r) = \Upsilon[T(e_0, r_0); e, r]$$

To be a time transformation, the function $\Upsilon(t; e, r)$ must have the following properties:

- For any e, r, $\Upsilon(0; e, r) = 0$.

- $\Upsilon(t, e, r)$ is non-negative for all t and e, r.

- For fixed e, r, $\Upsilon(t; e, r)$ is monotone in t.

- When evaluated at e_0, r_0, the transformation is the identity transformation [i.e., $\Upsilon(t; e_0, r_0) = t$ for all t].

Quantiles and Failure Probabilities as a Function of Environment and Use Rate. When $\Upsilon(t; e, r)$ is monotone increasing in t, the quantile of the failure-time distributions at (e, r) and (e_0, r_0) are related by

$$t_p(e, r) = \Upsilon[t_p(e_0, r_0); e, r], \quad 0 < p < 1.$$

The cdfs of the failure-time distributions at (e, r) and (e_0, r_0) are related by

$$\begin{aligned}
\Pr[T \leq t; e, r] &= \Pr[\Upsilon(T; e_0, r_0) \leq t; e, r] \\
&= \Pr[T \leq \Upsilon^{-1}(t; e, r); e_0, r_0]
\end{aligned}$$

where $\Upsilon^{-1}(t; e, r)$ is the inverse of the Υ function.

Scale Accelerated Failure Time Model for Environment and Use Rate Differences. A particularly simple time transformation function is the scale accelerated failure time (SAFT) model. In this model the time to failure $T(e, r)$ at environment and use rate conditions (e, r) is related to the time to failure time $T(e_0, r_0)$ at environment and use rate conditions (e_0, r_0) through the relationship

$$T(e, r) = T(e_0, r_0)/\mathcal{AF}(e, r)$$

where $\mathcal{AF}(e, r) > 0$ is a time-invariant scale factor that depends on (e, r) and (e_0, r_0). Under the SAFT model

$$t_p(e, r) = t_p(e_0, r_0)/\mathcal{AF}(e, r)$$

and

$$\Pr[T \leq t; e, r] = \Pr[T \leq \mathcal{AF}(e, r) \times t; e_0, r_0]$$

Modeling Unit-to-Unit or Spatial Variability in Environmental Conditions and Use-Rates. When product is installed and operated in different environments, the variability in environments will lead to additional variability in reported product service life. With an adequate description of the subpopulations involved (i.e., size and corresponding environmental conditions, it is possible to develop a model for the overall failure-time distribution or the the distribution of the number of future failures. To simplify presentation, we consider a model for a population of units that was placed into service over a relatively short period of time (perhaps one month).

Mixture of Environmental Conditions and Use-Rates. Suppose that the population of units in the field can be subdivided into k subpopulations of units, according to the environmental conditions and use-rates to which they are exposed. There are $n_i = n(e_i, r_i)$ units in subpopulation i having environmental conditions and use-rates denoted by (e_i, r_i), $i = 1, 2, \ldots, k$. The total number of units in the field is $n = \sum_{i=1}^{k} n_i$. The relative frequency (or proportion) of units at conditions (e_i, r_i) will be denoted by $f(e_i, r_i) = f_i = n_i/n$, $i = 1, \ldots, k$.

Mixture Population Failure-Time Distribution. For subpopulation i with environment and use rates (e_i, r_i) occurring with relative frequency $f_i, i = 1, \ldots, k$

$$\Pr(T \leq t) = \sum_{i=1}^{k} \Pr[T \leq t; e_i, r_i] \times f_i$$

For the SAFT model

$$\Pr(T \leq t) = \sum_{i=1}^{k} \Pr[T \leq \mathcal{AF}(e_i, r_i) \times t; e_0, r_0] \times f_i$$

Prediction for the Number of Units Failing. Suppose that n is the total number of exposed units. For the SAFT model, the expected number of units failed by time t is

$$\begin{aligned}
\mathrm{E}[N(t)] &= n \times \Pr(T \leq t) \\
&= \sum_{i=1}^{k} n \times \Pr[T \leq \mathcal{AF}(e_i, r_i) \times t; e_0, r_0] \times f_i \\
&= \sum_{i=1}^{k} \Pr[T \leq \mathcal{AF}(e_i, r_i) \times t; e_0, r_0] \times n_i
\end{aligned}$$

where $n_i = n \times f_i$ is the number of units on test at environment and rate conditions (e_i, r_i).

Extension to Populations of Units Entering Service at Different Times. Most applications consist of a population of units that was placed into service over long periods of time (many months or years). Often it is necessary to predict the number of failures over some future interval of time (e.g., over the next year) from all of the units in the field (or all that are still under warranty). In this case, one can do separate analyses, as described above, for each cohort of units and aggregate the results into an overall prediction. In some areas of application, this is called a risk analysis. See, for example, Chapter 12 of Meeker and Escobar [3] for an example and description of methods for constructing prediction bounds in a somewhat simpler situation.

Further Reading

Meeker and Escobar [3] describe the basic ideas of reliability data analysis methods and methods for accelerated testing. Nelson [6] is an extensive and comprehensive source for further material, practical methodology, basic theory, and examples for accelerated testing while [4] describes important areas of statistical research in accelerated testing. Chapter 7 of [6] describes graphical and ML methods for analyzing ALT data with competing failure modes.

Acknowledgments

Parts of this paper were taken from Meeker and Escobar [3] with permission from John Wiley & Sons, Inc. We would like to thank Jonathan Martin for helpful discussions about some of the important issues involved in applying the methods of accelerated testing to the prediction of service life. Computing was done, in part, using equipment purchased with funds provided by an NSF SCREMS grant award DMS 9707740 to the Department of Statistics at Iowa State University.

References

[1] Evans, R. A., *IEEE Transactions on Reliability,* 1989, vol. 38, pp.401.

[2] Gelfand, A. E., and Smith, A. F. M., *Journal of the American Statistical Association,* 1990, vol. 85, pp. 398-409.

[3] Meeker, W. Q., and Escobar, L. A. (1998), *Statistical Methods for Reliability Data.* John Wiley & Sons: New York, 1998.

[4] Meeker, W. Q., and Escobar, L. A., *International Statistical Review,* 1993, 61, 147-168.

[5] MIL-STD-883 (1985), *Test Methods and Procedures for Microelectronics,* available from Naval Publications and Forms Center, 5801 Tabor Ave, Philadelphia, PA 19120.

[6] Nelson, W., *Accelerated Testing: Statistical Models, Test Plans, and Data Analyses,* John Wiley & Sons: New York, Inc, 1990.

Chapter 11

A Unique Facility for Ultra-Accelerated Natural Sunlight Exposure Testing of Materials

G. Jorgensen, C. Bingham, J. Netter, R. Goggin, and A. Lewandowski

National Renewable Energy Laboratory, 1617 Cole Boulevard,
Golden, CO 80401

An ultra-accelerated material exposure test facility that uses highly concentrated natural sunlight has been developed at the National Renewable Energy Laboratory (NREL). By adequately controlling sample temperatures and demonstrating that reciprocity relationships are obeyed (i.e., level of applied accelerated stresses does not change failure/degradation mechanisms from those experienced in real-world use), this unique facility allows materials to be subjected to accelerated irradiance exposure factors of 50-100X. In terms of natural sunlight exposure, one year's equivalent of representative weathering can be accumulated in just 3-10 days.

To enhance the commercialization of renewable energy devices, a critical need exists to predict accurately their service lifetimes. New advanced materials and devices, many associated with a broad range of renewable energy technologies, are being developed to meet increasingly demanding service lifetime requirements (*1*). Potential industrial manufacturers of solar thermal electric and solar industrial systems require silvered polymer reflector materials capable of maintaining high performance during ten years of exposure in outdoor environments. Users of PV modules desire units that can operate for 30 years at defined efficiency levels. Electrochromic switching devices for energy efficient glazing applications must last 20 years when incorporated into windows for buildings (*2*).

Other important technologies have similarly stringent service lifetime constraints. A service lifetime requirement for polymeric materials of up to 10-30 years, which is a significant increase, is anticipated. There is an urgent need to extend the service lifetime requirements of decorative and protective coatings for the automotive industry beyond the present 5-10 years. The coatings industry in general has a significant need for facilities that provide real-world exposure testing of their products (e.g., paints, clearcoats). Businesses will simply be unable to afford to wait for extended periods of time to directly measure

© 1999 American Chemical Society

product lifetimes or to risk, without substantiating data, providing warranties demanded by consumers.

To predict service lifetime, the usual approach is to correlate real-world test results with accelerated lifetime test (ALT) results. ALT usually makes use of laboratory controlled elevated intensity levels associated with artificial light sources. Two limitations are generally inherent in this strategy. First, spectral differences exist between natural sunlight and artificial light sources; great care must be exercised to assure unrealistic wavebands are not used in ALT. Second, the outdoor exposure environment is variable (in terms of spectral intensity as well as other important stresses) both spatially (geographic location) and temporally (seasonally and annually), thereby making correlations with non-reproducible conditions difficult at best. Because of these concerns, various devices have been developed to provide exposure to natural sunlight for accelerated durability testing. The most successful commercial concept, which uses Fresnel reflector elements, allows terrestrial sunlight acceleration factors of only 5-6X (3).

To address these two major limitations, an ultra-accelerated material exposure test facility that uses highly concentrated natural sunlight has been developed at NREL. It makes use of NREL's existing High-Flux Solar Furnace (HFSF) (4), and an innovative irradiance redistribution guide (IRG) (5), to provide the unique capability of being able to modify (redistribute) the Gaussian-shaped beam from the HFSF to a more uniform profile on a sample exposure plane. A highly functional and flexible sample exposure chamber has been fabricated and integrated with the HFSF/IRG. This chamber provides concurrent control of multiple levels of other important stress factors during light exposure to allow materials whose degradation may be sensitive to multi-stress damage functions to be tested. In situ monitoring of solar irradiance at the sample exposure plane within the chamber can also be performed using a fiber optic-based spectral radiometer system. The NREL facility has been used to expose a number of solar reflector materials to an equivalent of 1 year of solar exposure at 50, 75, and 100 suns. Measurement and analysis of degradation data from these experiments verify reciprocity relationships for the materials tested.

Experimental Procedure

System Description. A schematic diagram of the system design is shown in Figure 1. Sunlight is continuously directed onto the primary concentrator array of the HFSF by a tracking heliostat. Concentrated sunlight passes through an attenuator to a novel secondary concentrator (the IRG). The existing IRG was designed specifically to provide a uniform concentration of 400X (400 kW/m² flux at nominal solar irradiance levels) within a 10 cm diameter spot at the optimal sample plane. For normal materials, lower concentration levels are needed to provide accelerated exposure conditions without destroying the samples. An attenuator is used to moderate and control the level of concentrated sunlight input to the IRG. The spatial uniformity of the flux incident upon the sample plane was documented by Lewandowski et al. (6); the measured results are shown in Figure 2.

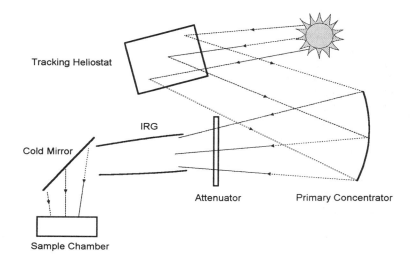

Figure 1. System Schematic

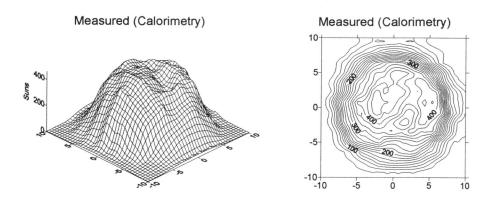

Figure 2. Measured Beam Profile at the Nominal Target Plane (0 cm; Distance units are centimeters, intensity units are concentration in suns)

The IRG is orientated along the optical axis of the HFSF just behind the nominal focal point and a "cold" mirror is placed between the IRG and the sample chamber to allow the sample plane to remain horizontal. The window of the sample chamber is a quartz plate. The cold mirror reflects ultraviolet and visible light (UV/VIS) through the (highly transmissive) quartz window, and allows the concentrated near infrared light (NIR) to be transmitted to alleviate thermal loading of the samples.

Figure 3 shows the layout of the IRG, cold mirror, and chamber. The chamber was designed and fabricated to allow up to four replicate samples (22 mm x 22 mm in size) each to be exposed to the same high level of accelerated solar flux at two levels each of temperature (T) and relative humidity (RH). Thus, at a given flux, sets of samples can be simultaneously exposed to four combinations of T and RH, i.e., (T_{low}, RH_{low}), (T_{low}, RH_{high}), (T_{high}, RH_{low}), and (T_{high}, RH_{high}). This allows a four-fold increase in experimental throughput at a particular exposure flux. A detailed drawing of the chamber is shown in Figure 4. During testing, the samples are mechanically attached to the top surface of the heating/cooling chamber to provide good thermal contact and consequent conductive control. The humidity chamber sets on top of the heating/cooling chamber. Humidity (and convective cooling of the samples) is supplied by introducing moist or dry air into the humidity chamber.

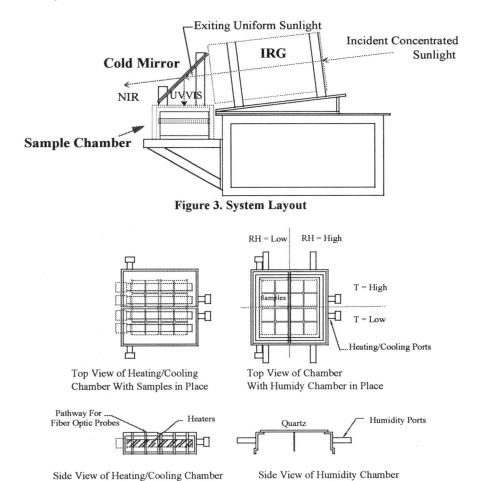

Figure 3. System Layout

Top View of Heating/Cooling
Chamber With Samples in Place

Top View of Chamber
With Humidy Chamber in Place

Side View of Heating/Cooling Chamber

Side View of Humidity Chamber

Figure 4. Detail of the Sample Exposure Chamber Design

System Characterization. To predict the spectral irradiance and to estimate the thermal loads on the samples, a model was developed for the optical performance of the system. The model uses measured spectral characteristics of each optical element and convolves those data with a measured direct-normal solar spectrum. The predicted spectral irradiance on the samples is:

$$L_{sample}(\lambda) = CF \bullet L_{d-n}(\lambda) \bullet \rho_h(\lambda) \bullet \rho_p(\lambda) \bullet \tau_a \bullet \rho_{IRG}(\lambda) \bullet \rho_{cm}(\lambda) \bullet \tau_q(\lambda) \qquad (1)$$

where:

$L_{sample}(\lambda)$	=	spectral irradiance at the sample surface, $W/m^2/\mu$,
CF	=	geometric concentration factor,
$L_{d-n}(\lambda)$	=	measured spectral direct-normal irradiance, $W/m^2/\mu$,
$\rho_h(\lambda)$	=	spectral reflectance of the heliostat,
$\rho_p(\lambda)$	=	spectral reflectance of the primary concentrator,
τ_a	=	transmittance of the attenuator,
$\rho_{IRG}(\lambda)$	=	effective spectral reflectance of the IRG,
$\rho_{cm}(\lambda)$	=	spectral reflectance of the cold mirror,
$\tau_q(\lambda)$	=	spectral transmittance of the quartz window.

The IRG represents a difficult component to characterize because some of the incident irradiance passes through directly without reflection, while much of the incident irradiance is multiply reflected. Previous results from a ray-trace analysis of IRG performance (5) gives fractions of rays with 0, 1, 2, or 3 reflections. This information, along with the known spectral reflectance of the IRG surface, can be manipulated to give the effective spectral reflectance. The total flux at the sample can be calculated as:

$$Q_{sample} = \int_{250\,nm}^{2500\,nm} L_{sample}(\lambda)\,d\lambda. \qquad (2)$$

For the measured direct-normal spectrum used, the total direct-normal flux was 993 W/m^2. With a concentration factor of 50, the resulting flux on the sample, including all of the optical losses associated with equation 1, is $Q_{sample} = 231$ kW/m^2. The nominal concentration is defined at the nominal target of the IRG, not including the impact of the cold mirror and quartz window. This is because the previous measurement of IRG performance (6) was taken at that location. The attenuator calibration was also referenced to this position. This means that the effect of $\rho_h(\lambda)$, $\rho_p(\lambda)$, and the concentration effects of both the primary and IRG are combined in that measurement. To properly use the geometric concentration factor, CF in equation 1, the nominal concentration must be divided by $\rho_h(\lambda) \bullet \rho_p(\lambda)$.

The spectral reflectance of the cold mirror was derived from measured reflectance and transmittance at normal incidence, and transmittance at the nominal 45° angle of incidence used in the system. The cold mirror exhibits a cut-on at about 300 nm and a cut-off at about 700 nm. Ideally, the cut-on wavelength should be slightly lower to assure that all UV in the solar spectrum is reflected. Although there is very little terrestrial solar irradiance near 300 nm, photons in the terrestrial UV-B bandwidth (290-320 nm) play an important role in materials degradation.

The power at the sample, obtained by convolving the spectral optical properties of the measured direct-normal spectrum and the various system elements, follows the shape of the solar spectrum closely until the cold mirror cut-off, after which very little solar irradiance reaches the sample. The integrated (total) power over a specific wavelength range is shown in Figure 5 for essentially all wavelengths up to the value on the abscissa. Flux concentration is defined as the integrated power per area on the sample compared to the integrated power per area in the solar spectrum over the wavelength range of interest. Because the cold mirror cuts off the NIR, the peak flux concentration occurs in the range of 400-700 nm.

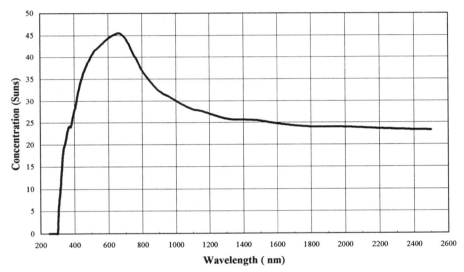

Figure 5. Calculated Concentration at Sample Surface in the Wavelength Range from 200 nm to the Wavelength Indicated Based on 50X at the IRG

Experimental Exposure Testing. Because NREL has extensive experience with durability testing of metallized polymer mirrors (7), these materials were selected for use in the present experiments. One advantage of this choice is that it allows ready comparison with historical data. In addition, because the superstrates are organic-based materials, results characteristic of other coating systems may be inferred. Thus, data meaningful to other technologies was obtained. The general construction of the test samples is:

Superstrate / Reflective Layer / Back Protective Layer / Adhesive / Substrate.

Three different material constructions (X, Y, and Z) were used in these tests. Each used approximately 100 nm silver as the reflective layer and a pressure-sensitive acrylic-based adhesive to allow lamination to ~0.9-mm thick 6061 aluminum substrates. The base resin of the superstrate film was a polymethylmethacrylate (PMMA). The three materials differed primarily in the amount of UV absorber present in the PMMA film and in the type of back protective layer, as noted in Table I.

Table I. Metallized Polymer Solar Mirror Materials Used in Experiments

Material	Weight % UV Absorber	Protective Layer
X	1	None
Y	2	None
Z	2	30 nm Cu

Three experiments were carried out. Each was intended to expose sample materials to the equivalent of one year's outdoor exposure in Colorado, at successively higher concentrations of accelerated natural sunlight. Based on the work of Marion and Wilcox (8), samples near Golden, Colorado receive an average 7227 MJ/m^2/yr of global radiation at a tilt angle equal to latitude angle (40° north). The first experiment used 50 suns concentration; at this level (50 kW/m^2, corrected for optical reflectance losses of the HFSF system), one year's equivalent exposure would be obtained after 40.2 hours. During actual exposure, it was not always possible (because of cloud transients, etc.) to maintain 50 kW/m^2. For example, in the "50 sun" experiment, the desired integrated dosage of 7227 MJ/m^2 took 56.3 hours exposure. Similarly, one year's equivalent exposure would (ideally) be accumulated after 26.8 hours at 75 suns and after 20.1 hours at 100 suns; the actual exposure times used were 32.0 and 26.4 hours, respectively.

The samples were placed in one of four areas of the chamber (Figure 4): hot and dry, hot and wet, cold and dry, or cold and wet. The average conditions for each of the exposures is given in Table II.

Table II. Average Exposure Conditions

Light Intensity Factor (suns)	T_{hot} (°C)	T_{cold} (°C)	RH_{wet} (%)	RH_{dry} (%)
50	63-71	18-23	55-75	5-10
75	60-75	17-24	55-70	5-10
100	65-75	15-25	60-75	5-10

During these experiments, the targeted nominal temperatures were 70°C on the hot side and 20°C on the cold side. The targeted nominal relative humidity on the dry side was ≤10% and ≈80% on the wet side. The intended high relative humidity could not be maintained during on-sun experiments; roughly 70% was the best that could be achieved.

The temperatures were measured using Type K thermocouples inside the cooling chamber, positioned underneath the samples. During initial testing, some thermocouples were placed above and below the samples, and a temperature gradient of 5° to 10°C was measured between the top and bottom surfaces. The maximum difference in temperature between the top and bottom ($\Delta T_{max} = T_{top}-T_{bottom}$, where T_{bottom} is held constant) of a thin polymer film (thickness, d, much less than the length and width of the film) exposed to a flux, Q_{sample}, can be expressed as (9):

$$\Delta T_{max} = 0.113 \bullet Q_{sample} \bullet \frac{d}{C_{th}} \bullet \varepsilon_{global}, \tag{3}$$

where: C_{th} =   thermal heat capacity of the polymer film,
    ε_{global} =   global emittance of the polymer film.

For PMMA, $C_{th} \approx 0.19$ Wm/m^2/°C and $\varepsilon_{global} \approx 0.8$. For a thickness of 0.09 mm, at 50 suns concentration, $\Delta T_{max} \approx 9.9$°C, in agreement with the measured data. With convective cooling associated with air/moisture flow introduced into the top chamber (to provide controlled humidity exposure), the temperature of the top surface of the film should remain somewhat cooler than $T_{bottom} + \Delta t_{max}$. From Table II, it can be seen that the hot temperature was maintained within 10°C of the desired level. The cold temperature was usually within 5°C of the set temperature.

The humidity was measured with a handheld probe, first before the samples were exposed to the sun, and then periodically during exposure. The dry humidity was maintained by a constant purge of dry air, and the humid portion of the chamber was kept within 10% of the nominal targeted level by bubbling deionized water through a dry air stream.

Figure 6 is a plot of representative, real-time stress conditions. One-minute average temperatures and flux during the 50-sun test are shown. The hot temperatures stay between 65° and 70°C most of the time, the cold temperatures are near 20°C, and the flux ranged between 0 and 50 kW/m^2. The humidity (not shown) exhibited similar control.

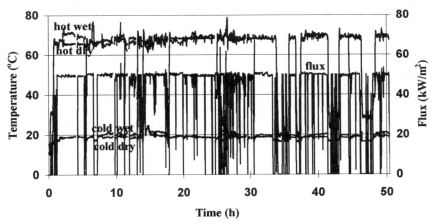

Figure 6. Sample Temperature and Flux for 50-Sun Test

In all experiments, sample locations were numbered from bottom to top starting with the right-most column and ending with the left-most column of samples (i.e., sample #1 was bottom right corner, sample #16 was top left corner, as shown in Figure 4). For the 50-sun experiment, only materials Y and Z were used; the exposed samples and temperature/relative humidity conditions are summarized in Table III.

Table III. Materials Used in Each Experiment

Sample Location	Temperature	Relative Humidity	50X Experiment	75X Experiment	100X Experiment
1	Low	High	Y	X	X
2	Low	High	Z	Z	Z
3	High	High	Y	Y	Y
4	High	High	Y	X	X
5	Low	High	Z	Y	Y
6	Low	High	Y	X	X
7	High	High	Z	X	X
8	High	High	Z	Z	Z
9	Low	Low	Z	Z	Z
10	Low	Low	Z	X	X
11	High	Low	Z	X	X
12	High	Low	Y	Y	Y
13	Low	Low	Y	X	X
14	Low	Low	Y	Y	Y
15	High	Low	Y	Z	Z
16	High	Low	Z	X	X

To allow comparison with results from previous experiments, a third material (X) was introduced into test in the 75X experiment. Two samples each of construction X were exposed in each of the four different environmental quadrants, as shown in Table III. One sample each of the two more durable materials (Y and Z) was also retested in each quadrant at this higher flux level to allow comparisons with results from the 50X experiment.

Concentrated light equivalent to 100X was used in the final experiment. The 100X experiment was identical to the 75X experiment in terms of sample selection and location, as shown in Table III.

Results

Performance Loss. Spectral hemispherical reflectance, $\rho_{2\pi}(\lambda)$, of all samples was measured before (β) and after (α) exposure using a Perkin Elmer Lambda-9 UV-VIS-NIR spectrometer. An uncertainty of ±0.5% at each wavelength is typical for these measurements. The amount of degradation experienced during exposure was computed as the difference (loss) in reflectance: $\Delta\rho = \rho_{2\pi,\beta}(\lambda) - \rho_{2\pi,\alpha}(\lambda)$. A representative plot of spectral reflectance loss is shown in Figure 7 for material X exposed at a light intensity of 100 suns. Where applicable, results for replicate samples have been averaged. Several important points can be seen in this plot. First, the reflectance loss at 400 nm, $\Delta\rho_{2\pi}(400)$, (indicated by the vertical dashed grid at $\lambda=400$ nm) is a particularly sensitive measure of degradation. As with previous work (7), $\Delta\rho_{2\pi}(400)$ was therefore chosen as the best indicator of performance loss for further analysis. The effect of the various stress factors can also be seen in Figure 7. Exposure to light at elevated temperatures is clearly the dominant stress. In general, light exposure at high temperature and high humidity results in the greatest degradation. Exposure at high temperature but low humidity produces an intermediate loss in performance. At lower temperatures, performance loss is smallest and humidity is seen to play an even less significant role. Exposure at high humidity results in only marginally increased degradation compared to low humidity when the sample temperature is kept relatively cool.

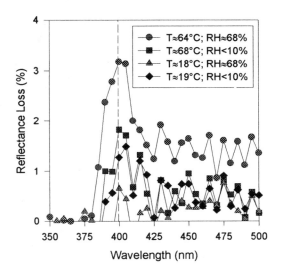

Figure 7. Reflectance Loss for Material X at 100X

In Figure 8, the spectral loss in reflectance is plotted as a function of level of sunlight exposure for material Z, tested at the most elevated temperature and humidity conditions. As before, results for replicate samples have been averaged when applicable. The change in reflectance was calculated after a targeted cumulative dose, associated with an equivalent of one year's cumulative dose of solar irradiance, had been accumulated. The intent was to investigate two important questions. First, can these types of materials be subjected to ultra-high levels of natural sunlight (50-100X) without introducing unrealistic degradation mechanisms? Second, does a reciprocity relationship exist between the level of light intensity and time of exposure? If these questions can be answered in the affirmative, the experimental procedures developed in this project can be used to allow reliable inferences to be made about the durability of these materials tested in highly abbreviated time frames. Similar relationships would need to be validated for other materials tested at highly accelerated conditions.

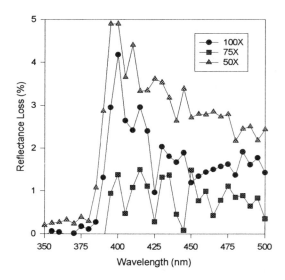

Figure 8. Reflectance Loss for Material Z at T≈70°C and RH≈65%

Figure 8 provides no evidence to suggest that the *level* of light intensity results in any systematic trend in loss of reflectance. It is the cumulative dose rather than the level of intensity (within the range 50X-100X tested) that gives rise to reflectance loss. The 50-sun exposure results in slightly greater degradation after 56.3 hours than the equivalent 100-sun exposure for material Z. The equivalent 75-sun exposure exhibits substantially less degradation than the 50X and 100X exposures. However, material Z was exposed at a temperature 10°C cooler at 75X than at 50X or 100X; when temperature effects are accounted for (as discussed in the section below on correlations between performance loss and applied stresses), the reflectance loss attributable to cumulative dose at 75X is in much better agreement with the 50X and 100X results.

FTIR Characterization of Degradation Mechanisms. The results discussed above support the claim that ultra-accelerated natural sunlight exposure can be carried out, at the light intensity levels indicated and for the materials tested, without introducing unrealistic degradation mechanisms that can be quantified in terms of loss in reflectance at the silver/PMMA interface. Although such loss is the primary performance criterion associated with these materials, other degradation may occur that would perhaps be less sensitive to this measure of performance loss. Figure 9 presents evidence that chemical/structural changes are not experienced within the polymer-film superstrate. Attenuated total reflectance (ATR) measurements of the bulk polymer were made using Fourier Transform Infrared (FTIR) spectroscopy. No appreciable changes in the spectra (especially in the sensitive 500-1700 cm^{-1} region) occur as a function of light intensity level (Figure 9), indicating that upon exposure of 50X-100X no substantial differences in the chemical structure occur in the most vulnerable top 1-2 μ of the polymer superstrate.

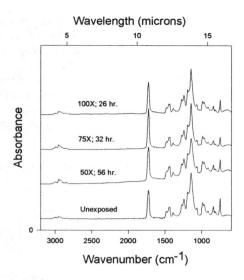

Figure 9. ATR-FTIR for Material Y as a Function of Exposure

Correlations Between Performance Loss and Applied Stresses. The ability to expose samples to the same light intensity at two (constant) levels (high and low) each of temperature and humidity allows change in performance (degradation) to be related to the environmental stress factors (damage function). As developed by Jorgensen et al. (7), such a relationship for the type of materials tested at constant stress conditions is:

$$\Delta\rho = A \bullet I_{UV-B} \bullet e^{-(E/T)} \bullet e^{C \bullet RH} \tag{4}$$

where:

$$I_{UV-B} = \int_0^t \int_{290\,nm}^{320\,nm} L_{sample}(\lambda)\,d\lambda\,dt \tag{5}$$

and: I_{UV-B} = the cumulative dose (J/m^2) in the UV-B spectral range (λ=290-320 nm),

T = Temperature (K),

RH = Relative humidity (%),

and A, C, and E are parameters to be fit from the measured data, $\Delta\rho_{2\pi}(400)$. In general, the values of these parameters will be specific to the particular material construction tested.

Table IV presents the values of I_{UV-B} corresponding to the three experiments. The data for each of the three material constructions tested were fit to equation 4; the resulting parameter estimates are given in Table V. The calculated change in performance is presented as a function of measured changes in Figure 10. Here, data for all three materials are shown, along with a composite linear regression line (which has been constrained to intersect the origin to reflect physical reality that no degradation occurs until light exposure begins). The dashed lines indicate the 95% prediction interval associated with this regression. Equation 4 is seen to provide a very good representation of the data.

Table IV. Cumulative UV-B Dose Experienced During Each Experiment

Concentration Factor	Time of Exposure (hr)	Cumulative UV-B Dose (MJ/m^2)
50	56.3	2.21
75	32.0	1.89
100	26.4	2.07

Table V. Parameter Values for Material Constructions Tested

Parameter	Material X	Material Y	Material Z
A	7.0684×10^{-4}	2.2696×10^{-5}	8.4680×10^{-4}
C	0.0139	0.0073	0.0205
E	2376	1376	2510

The parameter estimates for the material X construction are in close agreement to previously published results for a similar material construction tested by Jorgensen et al. (7). Of particular note is the excellent agreement in the thermal activation energy parameter, E, between the accelerated natural sunlight experiments (E=2376 K) and the accelerated artificial light (xenon arc) experiments (E=2339 K) in (7). This provides further confidence in the viability of both types of highly accelerated test protocols.

Comparison with Outdoor Results. A comparison of reflectance loss for material Y exposed outdoors for one year in Golden, Colorado with the same material exposed for a one-year equivalent at 75X at the HFSF is shown in Figure 11. Although both samples experienced the same cumulative UV-B dose, samples were exposed at the HFSF at nearly constant levels of temperature and relative humidity whereas these stress factors were highly variable for the samples exposed outdoors. As an approximation, samples exposed in the chamber quadrant having the closest match to the yearly average temperature and relative humidity as reported by Marion and Wilcox (8) were chosen for comparison with the outdoor samples. Equation 4 suggests that this will significantly underestimate the severity of the effect of temperature because the yearly average temperature during daylight exposure is considerably higher than when nighttime temperatures are included. Similarly, the effect of relative humidity will be overestimated. Given these complexities, it is not surprising that a difference in reflectance loss at 400 nm can be seen in Figure 11. However, at slightly higher wavelengths, good spectral agreement is evidenced between the highly accelerated and real-time test results. Ideally, a time-dependent form of equation 4 should be used for predictive purposes:

$$\Delta\rho = A \bullet \int_0^t \left[\int_{290\,nm}^{320\,nm} L_{sample}(\lambda, t)\, d\lambda \right] \bullet e^{-E/T(t)} \bullet e^{C \bullet RH(t)}\, dt,\qquad(6)$$

in which the coefficients are derived from experiments where the applied stresses are accelerated and constant. Then these coefficients can be used with time-monitored outdoor stresses to compute changes in performance from equation 6 that can be compared with measured results. This extension to variable, uncontrolled exposure conditions is planned as a future endeavor.

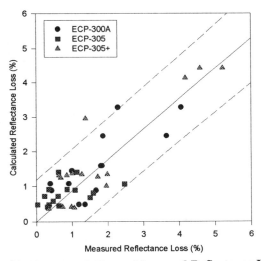

Figure 10. Calculated Versus Measured Reflectance Loss for Materials X, Y, and Z

Conclusions

This work demonstrated that the type of solar mirror materials tested can be exposed at 50-100 suns without introducing new (unrealistic) degradation mechanisms. In particular, it has been shown that the sample temperature can be controlled sufficiently well at very high levels of solar flux without adversely affecting the samples and the results. Furthermore, a reciprocity relationship was verified between time and integrated flux for the materials tested. That is, exposure at 50X for time t gives the same results (if all other stresses are the same) as exposure at 100X for time t/2. The development of this unique facility allows the universality and robustness of these findings are to be readily explored. A greater number of samples should be tested to longer accelerated exposure times to improve the statistical confidence in the results and to check reciprocity at greater levels of cumulative dose. A

184

wider variety of materials should be tested to ascertain the range of relevance of this capability, as well as to better understand its limitations. Finally, more direct comparisons must be made between ultra-accelerated test results and long-term outdoor exposure results.

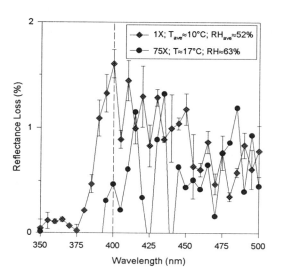

Figure 11. Comparison of the Reflectance Loss of Material Y after a One-Year Equivalent Exposure at 75X with Data Taken after One-Year of Outdoor Exposure in Golden, Colorado

Acknowledgments

This work was sponsored by NREL's FIRST Program under US Department of Energy (DOE) contract DE-AC36-83CH10093. The HFSF is a DOE National User Facility. ATR-FTIR characterization and analysis was performed by David King at NREL.

Literature Cited

1) Jorgensen, G.; Pern, J.; Kelley, S.; Czanderna, A.; Schissel, P. *Desk Reference of Functional Polymers*; Arsady, R., Ed.; American Chemical Society: Washington, DC, 1997; pp 567-588.
2) Czanderna, A. W.; Jorgensen, G. J. In *14th NREL/SNL Photovoltaics Program Review*, Witt, C. E.; Al-Jassim, M.; Gee, J. M., Eds.; AIP Conf. Proc. 394; American Institute of Physics: Woodbury, NY, 1997, pp 295-311.
3) Robbins, J. S. III; In *Accelerated and Outdoor Durability Testing of Organic Materials*, Ketola, W. D.; Grossman, D., Eds.; ASTM STP 1202; American Society for Testing and Materials: Philadelphia, PA, 1994, pp 169-182.

4) Lewandowski, A.; Bingham, C.; O'Gallagher, J.; Winston, R.; Sagie, D. *Solar Energy Materials* **1991**, Vol. 24, pp 550-563.

5) Bortz, J.; Shatz, N.; Lewandowski, A. In *Nonimaging Optics: Maximum Efficiency Light Transfer III*; SPIE Proceedings Series, 1995; Vol. 2538; pp 157-175.

6) Lewandowski, A.; Bingham, C.; Shatz, N.; Bortz, J. In *Nonimaging Optics: Maximum Efficiency Light Transfer IV*; SPIE Proceedings Series, 1997; Vol. 3139; pp 225-236.

7) Jorgensen, G. J.; Kim, H. M.; Wendelin, T. J. In *Durability Testing of Non-Metallic Materials*; Herling, R. J., Ed.; ASTM STP 1294; American Society for Testing and Materials: Philadelphia, PA, 1996; pp 121-135.

8) Marion, W.; Wilcox, S. *Solar Radiation Data Manual for Flat-Plate and Concentrating Collectors*; NREL/TP-463-5607, National Renewable Energy Laboratory: Golden, CO, 1994, p 46.

9) Kockett, D. In *Durability Testing of Non-Metallic Materials*; Herling, R. J., Ed.; ASTM STP 1294; American Society for Testing and Materials: Philadelphia, PA, 1996; pp 24-39.

Chapter 12

Reliability Engineering: The Commonality Between Airplanes, Light Bulbs, and Coated Steel

William Stephen Tait

S. C. Johnson and Son, Inc., 1525 Howe Street, Racine, WI 53403–5011

Failure comes in a variety of different forms, depending on the system and the environment to which it is exposed. However, failure has attributes that are common to all forms of failure. The attributes of failure are: 1) chemical and physical phenomena initiate small flaws in materials, 2) flaws grow under certain environmental conditions, 3) flaw growth rates can be used to estimate time of failure, 4) failure occurs when flaws grow to a critical size, or damage accumulates at a high enough level to cause failure, or the flaw can be observed by the unaided eye and considered a failure, 5) the number of observed failures in a population is not a continuous function of time, 6) the cumulative number of failures in a population is a continuous function of time, 7) the magnitude of cumulative failure in a population is influenced by the environment, and 8) cumulative failure data can be transformed into a linear form and the data extrapolated outside of its range (within reason) to estimate population failure level. These attributes can be used to estimate long-term failure times and associated levels of failure using reliability statistics. Failure levels for aerosol containers can be estimated to within an order of magnitude of actual failure levels.

Every industry strives to make products that will perform for a designated service lifetime without the level of failure exceeding a specified maximum. A specified maximum level of failure is often referred to as an acceptable quality level, or AQL.[1] An example of an AQL and its associated service lifetime is 1% failures prior to five years.

One method for estimating failure times and associated levels is a long-term exposure test. Samples are exposed to service conditions; periodically inspected for failures; and the number of failures at each inspection time are recorded.

© 1999 American Chemical Society

Long-term exposure tests can take several years to complete because they are conducted until all samples have failed. The objective of an exposure test is to determine if the desired service lifetime is achievable without exceeding acceptable failure levels.

Unfortunately global competitive pressures usually prevent conducting tests that take several years to complete, and results from a censored test must be used to estimate long-term failure levels, or verify that service lifetime will be achieved without exceeding acceptable failure levels. A censored test is one in which samples are removed from test prior to failure, or the test is terminated prior to failure of all samples.[2]

Different Types of Failure
There are many different types of failure. Figure 1 contains an example of a cracked empennage cap on a single engine airplane (a failure in progress). Mechanical force is applied to the empennage when an airplane begins to fly, and the force is removed when the airplane lands. The mechanical force applied during each flight-cycle initiates fatigue cracks like that shown in Figure 1, and causes the cracks to grow until there is enough damage to cause failure. Cyclic mechanical stress is analogous to cyclic stresses that result from daily and climatic changes in weather.

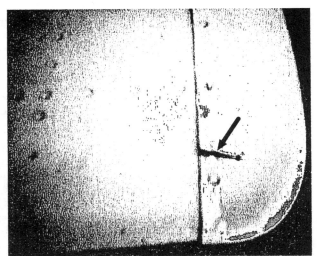

Figure 1. Cracked empennage cap on a single engine airplane

Figure 2 contains a photograph of a burned-out light bulb. A thermal stress is applied to the bulb filament when the bulb is turned-on, and the thermal stress is removed when the bulb is turned-off. Heat produced by electrical resistance also causes loss of filament material during bulb use. The cyclic thermal stress and thermal corrosion cause the bulb filament to crack, and the crack grows during each use cycle until the filament fails (breaks). Thermal cycling also occurs during weathering of coated metals.

Figure 2. Burned-out light bulb.

Figure 3 contains a photograph of a corroded automobile door panel. Water absorbs into the paint and diffuses to the paint-metal interface to cause blistering. Blistering causes separation of the paint from the door panel, and more water diffuses through the paint to accumulate inside the blister. Water accumulation in a blister causes metallic corrosion that progresses under the paint, enhancing both growth of the blister and separation of paint from the metal substrate.

Figure 3. Rusted automobile door panel.

Figure 4 contains a photograph of the bottom portion of a perforated aerosol container. The liquid in the container caused pitting corrosion in the crevice area (indicated by the arrow), and the pit grew until it perforated the container metal.

Figure 4. Perforated aerosol container

The different types of failure in Figures 1 through 4 have several attributes in common:
1. There are chemical and physical phenomena, such as electrochemical corrosion and thermal or mechanical stress, that initiate small flaws in materials
2. Flaws grow under certain environmental conditions
3. Failure occurs when flaws grow to a critical size and; a) damage accumulates at a high enough level to cause mechanical failure, or b) the flaw is considered to be a failure because it is large enough to be observed by the unaided eye
4. Flaw growth rates can be used to estimate time of failure
5. The number of observed failures in a population is not a continuous function of time
6. The cumulative number of failures in a population is a continuous function of time
7. The magnitude of cumulative failure in a population is influenced by the environment
8. Cumulative failure data can be transformed into a linear form, and the data extrapolated outside of its range (within reason) to estimate the time and level of failure for a population.

These attributes of failure can be used with reliability statistics to predict service lifetime from short-term censored tests.

Service lifetime can be mathematically expressed with the following general equation:

$$EFT_i = T_i + \frac{S*}{rate} \qquad (1)$$

Where the rate term in Equation (1) is flaw growth rate; EFT_i is the expected failure time (service life) for an individual unit; T_i is the initiation time for the flaw that causes failure; and $S*$ is the flaw size that causes or constitutes failure. Initiation time (T_i) is often a distribution of times, and defect growth rate ("rate" in Equation 1) is a function of the material and its environment. Linear growth rates have been observed for metal pitting corrosion such as that in Figure 4;[3,4,5] exponential growth rates are expected for metal cracking like the examples in Figures 1 and 2;[6] and linear[7,8,9] or exponential[10] growth rates have been reported for organic coating blistering such as that in Figure 3.

It will shown that an approximation of Equation (1) can be used to estimate failure time and cumulative failure level for aerosol container failure by pitting corrosion such as that shown in Figure 4.

The correlation between predicted and real-time results for aerosol containers will be discussed at the end of this paper, along with some key elements that are needed for any successful reliability program.

Attributes of Failure
A brief discussion of the attributes of failure will provide background for how they are used with reliability statistics to model failure levels.

1)   There are Chemical and Physical Phenomena, such as Electrochemical Corrosion and Thermal or Mechanical Stress, that Initiate Small Flaws in Materials. Phenomena like electrochemical corrosion or stress can initiate flaws such as pits, paint blisters, and cracks in materials. Flaws typically initiate in microscopic areas.[11,12,13] Consequently the onset of failure is often difficult to detect with the unaided eye, and it takes time before the flaw size can be observed with the unaided eye, or measuring equipment must be used to detect flaws in materials.

2)   Flaws Grow under Certain Environmental Conditions. It is well known that flaws like fatigue cracks, paint blisters, and corrosion pits all continue to grow as long as the environment that initiated the flaw, or conditions that enhance flaw growth, are not removed from the material.[5,14,15]

3)   Failure Occurs when Flaws Grow to a Critical Size; or Damage Accumulates at a High Enough Level to Cause Mechanical Failure; or the Flaw is Considered to be a Failure because it is Large Enough to be Observed by the Unaided Eye. Failure is defined by the type of service. For example, a steel bolt will break from tensile stresses when a pit or stress crack reduces bolt cross section to where there is no longer enough metal to support the applied load. Blisters on painted metals are considered failure when blister size is large enough to be observed with the unaided

eye. Both of these examples encompass a flaw that has reached a size at which failure occurs.

4)   Flaw Growth Rates can be used to Estimate Time of Failure. Flaws have growth rates that can be estimated or measured. Expected Failure Time (service lifetime) of an individual component is equal to the time that it takes for a defect to initiate in the material and grow to a critical size (S*) that causes failure, or is considered to be failure.

5)   The Number of Observed Failures in a Population is not a Continuous Function of Time. The number of failures observed at different times in a population are seemingly random. Figure 5 contains a histogram for the number of aerosol container failures observed at different times during a long-term test. It can be seen in Figure 5 that the number of failures do not appear to follow a continuous trend, indeed, there are several times during the test when no failures were observed.

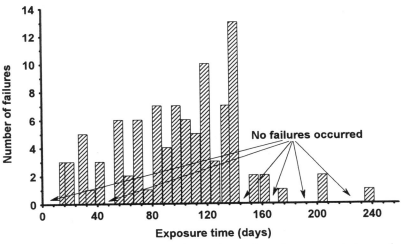

Figure 5. The number of failures observed at a given time appear to be random

6)   The Cumulative Number of Population Failures (or cumulative percent) is a Continuous Function of Time. Data like that in Figure 5 can not be extrapolated outside of test times. Fortunately the cumulative number of failures is a continuous function of time, as demonstrated in Figure 6, where data from Figure 5 are plotted in the cumulative form. Numbers of failures were normalized by converting them to percent of total test population. This conversion was done so that results from test parameters having different numbers of samples could be compared to each other.

The cumulative form of failure in Figure 6 can be extrapolated to times outside test times because failure level is a continuous function of time.

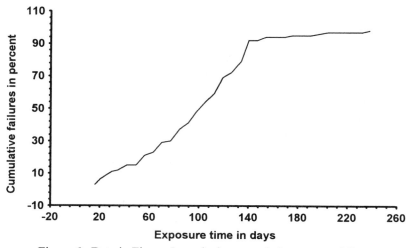

Figure 6. Data in Figure 5 graphed as cumulative percent failures

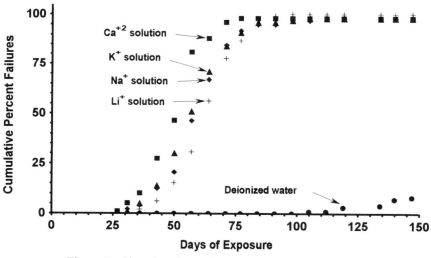

Figure 7. Changing the environment changes failure times

7)   The Magnitude of Cumulative Failure in a Population is Influenced by the Environment. The magnitude of cumulative failure is determined by the growth rate of the phenomena that causes failure, and this growth rate is determined by the material and the environment to which the material is exposed.[16] Figure 7 contains several failure curves for metal containers exposed to various salt solutions having 0.1 molar concentrations of salt in deionized water.[16] It can be seen that the failure

level for the calcium solution was higher than those for potassium, sodium or lithium solutions. Water that was essentially ion-free (deionized water) had the lowest failure levels for any given time.

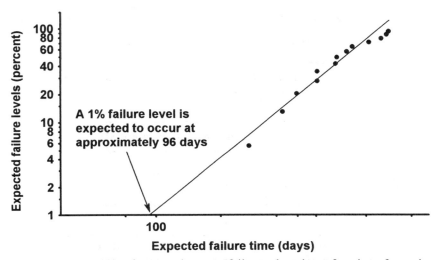

Figure 8. An example of expected percent failures plotted as a function of associated expected failure times on Weibull probability paper

8)   Cumulative Failure Data can be Transformed into a Linear Form. Data like that in Figures 6 and 7 can be transformed into a linear form with probability density functions (pdf) such as extreme value or Weibull functions. Figure 8 contains an example of failure data that have been transformed by plotting the data on Weibull probability paper. It can be seen in Figure 8 that the data are reasonably linear, and can be extrapolated within reason to test times that are outside of those for the data. For example, the data in Figure 8 are extrapolated to find the time at which 1% failures are expected to occur. There are other probability density functions that can be used to transform data into a linear form,[17] and it is advisable to use the pdf that best fits the data for an extrapolation like that illustrated in Figure 8.

Case History: Using Failure Attributes to Model Failure of Aerosol Containers
Aerosol containers are a common form of consumer packaging. Indeed, there were 2.8 billion metal aerosol containers produced in the United States in 1996.[18] Failure for aerosol containers is defined in this paper as when a pit perforates the container to cause it to leak. Figure 4 has an example of an internally coated container that failed from pitting corrosion in the bottom double seam of the container.

The algorithm used to model container failure time for pitted censored samples is contained in Figure 9. The pitting rate for a given container-formula system is estimated by dividing maximum pit depths for individual containers by the

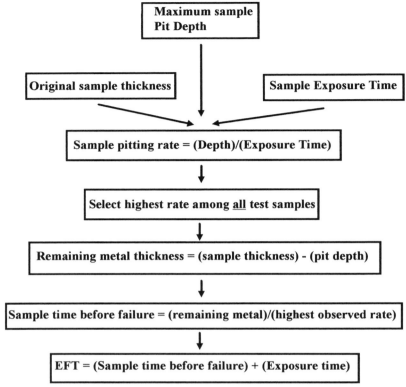

Figure 9. Algorithm used to calculate expected failure time from pit depth and
exposure time.

corresponding exposure time, and selecting the highest rate among all samples as the
best estimator for the system pitting rate. Pit depth is subtracted from metal thickness
to get the metal remaining under the deepest pit. Remaining metal thickness is
divided by the highest pitting rate to obtain the expected remaining time-before-
failure for each container. Expected failure time is calculated using a modification of
Equation (1):

$$EFT_i \ = \ ET_i \ \ + \ \ \frac{\text{remaining metal}}{\text{highest observed rate}} \qquad (2)$$

Where ET_i is the exposure time for individual samples; remaining metal is the
amount of metal left under a pit; and the highest observed rate is the highest pitting
rate observed among all samples in the test. Pitting rate is estimated by dividing
exposure time into the pit depth. Equation (2) is an approximation for Equation (1).
This approximation is necessary because it is very difficult and time consuming to

obtain the magnitude for T_i in Equation (1). It will be demonstrated that Equation (2) is a reasonable approximation for Equation (1).

The basic assumptions used to build the algorithm in Figure 9 are:[19]
1. pits nucleate at different times
2. there is an intrinsic pitting rate for a given metal in a specific environment
3. the deepest pit observed among all samples examined at a given exposure time is the oldest pit, and thus the best estimator of the intrinsic pitting rate.

Failure data were gathered from static storage tests in which aerosol containers were filled with the desired liquid, pressurized and stored in constant temperature rooms for one to several years.[20] Static storage tests were censored because containers were periodically removed from constant temperature rooms; de-pressurized; emptied; open and inspected for corrosion. Constant temperature rooms were typically maintained at 21°C and a higher temperature such as 38°C.

The algorithm used to model expected failure level for each expected failure time is contained in Figure 10. This algorithm follows the statistical procedure discussed by Kapur and Lamberson for graphical analysis of censored tests.[21]

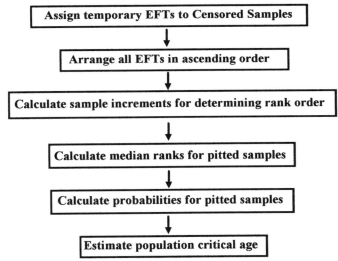

Figure 10. Algorithm for statistical treatment of censored test data.

EFT is the expected failure time that is estimated from pit depth; growth rate; and container exposure time. Critical age is the time where the specified maximum failure level for a population is observed.

Table 1 contains comparisons between expected failure levels (EFLs), and observed failure levels.[20] It can be seen that the estimated EFL is typically within an order of magnitude of the observed failure level.

The accuracy of expected failure times and their associated EFLs can be improved if the actual container pitting rate is known. Unfortunately the only way to estimate pitting rates for aerosol containers is to open containers and measure the pit depths, and assume the deepest pits were growing during the entire exposure time. The assumption about pit growth time causes inaccuracy in the estimated pitting rate, causing some of the differences between actual failure levels and EFL magnitudes observed in Table 1. The use of only the Weibull pdf to model the data also contributes to the differences between actual and EFL magnitudes. Evaluation of different probability density functions is a topic that is being pursued at this time and will be reported in a later publication.

Table 1. Comparisons between estimated and observed failure levels

EFL	Observed failure levels
5 %	11.5 %
18.5 %	16.7 %
17.7 %	33.3 %
9.2 %	22.2 %
8.1 %	8.3 %
36.8 %	30 %
55.9 %	55.6 %
64.4 %	63.6 %
5.6 %	8.3 %
12.6 %	8.3 %
13.5 %	16.7 %
8.3 %	12.5 %
6.7 %	10 %
41.3 %	41.7 %
23.8 %	25 %
23 %	14.3 %
29.8 %	33.3 %

Successfully using Reliability Statistics
There are a few key elements that should be incorporated into any reliability program:
1. Use replicate samples for each test parameter
2. flaw size that causes failure should be known or determined
3. flaw growth rate should be determined or estimated for each test parameter so that failure times can be estimated for censored samples
4. Use the pdf that best fits the data to estimate EFL

Summary
The attributes of failure can be used to obtain expected failure levels (EFLs) that are the same order of magnitude as actual failure levels.

The algorithms described in this paper can be used for a variety of failure types where flaw growth rate is known or can be estimated, and the level of damage that causes failure is known.

More accurate EFLs for aerosol container failure can be obtained if the actual pitting rate is measured without having to destroy containers to locate and measure pit depth.

There are several probability density functions (pdf) that can be used to model failure data, so that data can be extrapolated (within reason) outside of the original data set. It is highly recommended that the best fit pdf data be used for extrapolation. Carte blanche use of the same pdf for all data can lead to inaccurate estimates for EFL.

References

1. *ANSI/ASQC Z1.4-1993*; American Society for Quality Control, Milwaukee WI; **1993**; p. 2
2. Nelson, W. *Journal of Quality Technol.*; **1970**; 2(3); p. 126.
3. J. E. Strutt, J. R. Nichols, and B. Barbier, Corrosion Science, **1985**, 25(5), ppl. 305-315
4. J. W. Provan and E. S. Rodriguez III, Corrosion, **1989**, 45(3), pp. 178-192
5. D. L. Crews, *Galvanic and Pitting Corrosion - Field and Laboratory Studies*, ASTM STP 576, American Society for Testing and Materials, Philadelphia PA, **1976** pp. 217-230
6. S. T. Rolfe and J. M. Barsom, Fracture and Fatigue Control in Structures, Prentice-Hall Inc., Englewood Cliffs NJ, **1977**, p. 115
7. J. W. Martin, E. Embree, andW. Tsao, J. Coatings Tech., **1990**, 62(790), pp. 25-33
8. Leidheiser Jr., H. *J. Adhesion Sci. Tech.*; **1987**; 1(1); pp. 79-98
9. Leidheiser Jr., H. and Wang, W., J. Coatings Tech., **1981**, 53(672), pp. 77-84
10. Nguyen, T. N., J. B. Hubbard, and G. B. McFadden, J. Coatings Tech., **1991**, 63(794), pp. 43-52
11. Rolfe, S. T.; Barsom, J. M. *Fracture and Fatigue Control in Structures*; Op. Cit.; Chapter 8, p. 232
12. Szklarska-Smialowska, Z. *Pitting Corrosion of Metals*; National Association of Corrosion Engineers, Houston TX; **1986**; p. 3
13. Dickie, R. A. *Prog. in Org. Coatings*; **1994**; 25; pp. 3-22
14. Reed-Hill, R. E. *Physical Metallurgy Principles*; 2nd edition; D. Van Nostrand Company, NY; **1973**; p. 761
15. Pommersheim, J. M;. Nguyen, T.; Zhang, Z.; Hubbard, J. P. *Prog. in Org. Coatings*; **1994**; 25; pp. 23-41
16. Tait, W. S. *Corrosion*; **1994**; 50(5); pp. 373-377
17. Nelson, W. Accelerated Testing, John Wiley and Son, NY, **1990**, Chapter 3, p. 129
18. *Spray Technology & Marketing*; **June 1997**; p. 10
19. Tait, W. S. in *Corrosion Control by Coatings*; ACS symposium series, American Chemical Society, Washington, DC; **(in press)**; Chapter 18
20. Tait, W. S. *Spray Technology & Marketing*; **September 1997**
21. Kapur, K. C.; Lamberson, L. R. *Reliability in Engineering Design*; John Wiley and Sons, NY; **1977**; Chapter 11, pp. 314- 323

Chapter 13

Use of Reliability-Based Methodology for Appearance Measurements

P. Schutyser and D. Y. Perera

Coatings Research Institute (CoRI), Avenue Pierre Hologge, B-1342 Limelette, Belgium

The influence of the weathering temperature on the color change and gloss retention of a polyester/TGIC powder coating during a series of accelerated weathering experiments was studied according to the principles of the Reliability Theory. Continuous UVB-313 radiation was used throughout. The Arrhenius-Weibull model adequately described the influence of weathering temperature on color change. The gloss retention data, however, could not be modeled: increasing specimen lifetimes ranked in the order of $57 < 66 < 36 < 73°C$. The Weibull distribution may be less suitable as a lifetime distribution for the gloss data obtained in this work.

When submitting a coating for a durability test, whether outdoor or accelerated, usually only a few coating specimens are exposed. The performance of the specimens with respect to the properties of interest is measured at regular intervals. Usually the test is carried out during a preset time after which the property under study is compared with a preset threshold. If the coating performs as good as expected or better, it passes the test, otherwise the coating is considered to have failed.

Three comments can be made regarding this approach. First, it is questionable that the performance of the coating can be derived from such a small number of specimens tested. Second, it is tacitly assumed that all tested specimens behave in the same way during the test and that the coating performance is accurately described by the average performance of the specimens. Third, weathering devices are not perfect. The temperature, UV-irradiance and humidity are not always perfectly homogeneously distributed in the weathering device, nor is their stability in time perfect. This can result in a different behavior of the same product depending on the location of the specimen in the weathering device (*1*) or the time of the observations.

© 1999 American Chemical Society

Durability and reliability. The crux is how to derive a *product* characteristic (i.e. durability) by testing *individual specimens* of the product. To our knowledge, existing standards give no information on how to achieve this. In other industries [electronics (*2*), nuclear (*3*), aerospace (*4*)], facing the same problem, viz. evaluating the reliability of the product, a method, based on the Reliability Theory, is being applied successfully. Martin and McKnight (*5*) were the first to propose and use this theory in the field of organic coatings.

When durability of a coating is treated in terms of reliability, making a clear distinction between two closely related notions is necessary. The lifetime of a coating specimen is the time it complies with certain specifications (color change, gloss retention, corrosion resistance, etc.). The lifetime of specimens can be readily obtained from weathering experiments. The service life of a coating is the time a coating complies with certain specifications at a predefined probability level. This is a more useful concept as it gives information on the product itself by stating directly the percentage of the coating that is still complying with the specifications after a time equal to the service life has elapsed.

As mentioned earlier, the existing standards make no distinction between these two notions and consequently describe no procedure to obtain coating service life estimates from weathering experiments. The available literature on service life determination of coatings (*5-10*) shows what could be achieved.

Theoretical

From a sufficiently large population of "identical" coating specimens a representative sample of *n* specimens is drawn and submitted to a certain "stress" such as temperature (this work), UV-radiation, relative humidity, etc. Measuring the resulting degradation quantitatively for each individual specimen at regular intervals is essential. The curve describing the evolution of the evaluation parameter as a function of weathering time for each individual specimen is called the *sample curve* of that specimen. The *lifetime* of a specimen is then defined as the time at which its sample curve reaches a preset limit or *failure criterion*. In this way the lifetime of each specimen can be calculated for the weathering conditions used. The spread in the sample curves, due to the specific response of each specimen to the imposed stress, is also reflected in the spread of the lifetimes of these specimens. After arranging these lifetimes in ascending order, a cumulative distribution can be fitted to these values. Several types of distributions can be used in lifetime studies. In this work the cumulative Weibull distribution (*11,12*) was used:

$$F(t) = 1 - e^{-(\frac{t}{\alpha})^{\beta}} \qquad (1)$$

It is characterized by the scale parameter α and the shape parameter β, determining respectively the shift of the distribution along the time axis and the shape of the probability density function. By solving this equation for the time, t, the *maximum service life*, t_{sl}, under the weathering conditions used, can be calculated:

$$t_{sl} = \alpha \cdot [-\ln(1 - \Phi)]^{\frac{1}{\beta}} \qquad (2)$$

where Φ represents the fraction of the specimens failed after a time t_{sl}. Next, the test is repeated, each time submitting a new sample drawn from the same population to different levels of the same stress (temperature). As a result, each experiment generates its own distribution with corresponding values for the scale and shape parameter. In this way, the weathering behavior of the coating as a function of temperature is described by considering α and β as functions of the stress (temperature) applied. Equation 2 can be generalized to:

$$t_{sl} = F_1(T) \cdot [-\ln(1 - \Phi)]^{\frac{1}{F_2(T)}} \qquad (3)$$

With the help of equation 3, a *Probability of failure-Stress-Time to failure (PST)* diagram can be calculated. It describes the evolution of coating service life with stress (temperature) at a predefined reliability level (1-Φ).

From the description above it follows that the spread in the specimen lifetimes contains valuable information as long as this variation originates only in the coating itself. It is therefore very important that these data are not distorted by systematic errors due to imperfections in the weathering and measuring equipment. In the current practice of using but a few specimens, only gross systematic errors will show. However, when conducting weathering experiments according to reliability principles, systematic errors due to weathering device imperfections can become quite visible. It is the authors' view that until now, the required level of weathering device control has still not been fully appreciated by both the device users and manufacturers.

Experimental

Material. Powder coating: carboxyl-functional polyester crosslinked with triglycidyl-isocyanurate; pigment: TiO_2; PVC: 33%; $T_g \approx 78°C$ (as measured by DSC, heating rate $20°C.min^{-1}$).

Specimen Preparation. The powder coating was applied with an electrostatic spray gun on aluminum substrates treated with Alodine 1200S (The Q-Panel Co.) and cured at 200°C for 15 min. Approximately 100 identical specimens were prepared. From this population four sample sets of 18 to 32 specimens were selected using a random number generating program to avoid systematic errors due to handling or to variability in their production.

Instrumentation. The color measurements were carried out with a Minolta Chroma Meter CR-200. The color difference, expressed in ΔE units according to the CIE 1931

L*a*b* system, was calculated for each individual specimen by measuring its color before weathering as its initial (or zero exposure time) reference value. Each specimen was measured at three different spots to allow for sample heterogeneity. The gloss measurements were made with a Byk Gardner Micro Tri Gloss glossmeter at an observation angle of 20°; each specimen was measured at six different spots. Gloss retention values of the specimens during the weathering experiments were calculated with the initial gloss values of the specimens as reference values (100%). The samples were measured as such, i.e. without wiping them off before the measurements as no condensation cycles were used during weathering.

All accelerated weathering experiments were carried out in a modified "Q.U.V. Accelerated Weathering Tester" (The Q-Panel Co.) without irradiance control (9). Continuous UV irradiation with UVB-313 lamps was used at different temperatures (36, 57, 66 and 73 °C). Prior to the actual weathering experiments, the QUV was stabilized at each temperature and the temperature at each sample location was measured. The resulting temperature "maps" of the QUV allowed selection of those locations showing a standard deviation of maximum 2°C on the spatial average temperature. The long term temperature stability, expressed as the standard deviation on the average temperature during an entire weathering experiment, was in each case better than 2°C.

Accelerated Weathering Procedure. At regular intervals all the specimens were removed from the Q.U.V. for evaluation. Afterwards, the specimens were put back according to an up-and-down and left-to-right scheme based on the manufacturer's recommendations. This helps to reduce the influence of possible remaining spatial inhomogeneities of weathering conditions such as the UV-irradiance level as no means of controlling or monitoring the UV-irradiance were available.

Results and Discussion

Color Change. The 18 sample curves describing the color change at 66°C as a function of weathering time are given in Figure 1.

Nonlinear regression was used to fit an exponential model to the sample curves of the specimens to improve the accuracy on the calculation of the specimen lifetimes:

$$\Delta E = C_1 \cdot (1 - e^{-C_2 \cdot t}) \tag{4}$$

The lifetime of each specimen was determined by solving equation 4 for t for a failure criterion $\Delta E = 0.8$. Figure 2 summarizes the cumulative Weibull distributions obtained at the temperatures used. Several observations can be made. The agreement between the experimental and theoretical distributions is satisfactory, suggesting the validity of the Weibull distribution as a lifetime model. The accelerating effect of temperature on the evolution of color is obvious from the gradual shift of the distributions towards shorter lifetimes with increasing weathering temperature. An analysis of the evolution of the scale and shape parameters with temperature reveals that the shape parameter

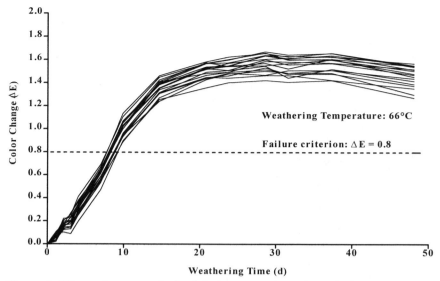

Figure 1. ΔE sample curves obtained during an accelerated weathering at 66°C (QUV; UVB-313) of a PE/TGIC powder coating.

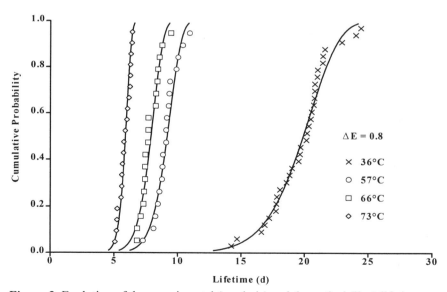

Figure 2. Evolution of the experimental (symbols) and theoretical (line) lifetime distributions for color difference with weathering temperature during an accelerated weathering (QUV; UVB-313; failure criterion ΔE = 0.8) of a PE/TGIC powder coating.

shows no significant evolution in the temperature range studied. Therefore, the mean value of the shape parameters was used in subsequent calculations. Figure 3 shows that the temperature dependence of the scale parameter can be represented by an Arrhenius-type equation in the temperature range applied. Thus, equation 3 can be modified to:

$$t_{sl} = 10^{(a + \frac{b}{T})} \cdot [-\ln(1 - \Phi)]^{\frac{1}{\beta}} \tag{5}$$

which is known as the Arrhenius-Weibull model.

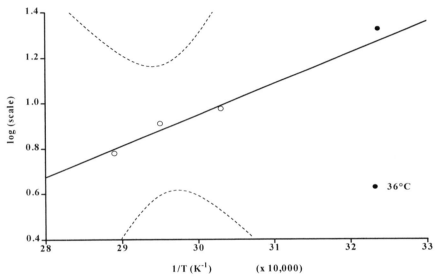

Figure 3. Arrhenius plot of Weibull scale parameter for a failure criterion $\Delta E = 0.8$ (..... 95% confidence interval).

A good agreement is found between the extrapolated and the experimentally determined service life at 36°C during a QUV weathering under the conditions used in this work: the service life at the 95% reliability level (i.e. 5% of the specimens failed, or 1.6 specimens of 32) is calculated to be 13.6 d, which is fairly close to the experimental value from Figure 2 (first specimen failure after 14.2 d, second failure after 14.9 d). These results suggest that for the powder coating studied the Arrhenius-Weibull lifetime model can be applied to the obtained ΔE data. The corresponding PST-diagram (Figure 4) represents the evolution of the service life with respect to color change as a function of temperature under the weathering conditions used. The three curves correspond to different reliability levels, viz. 99, 95 and 90%. It also shows that for the same data set, a high reliability inevitably results in a shorter service life. These diagrams hold great promise for the comparison of different types of coating, which is a prime goal of durability testing.

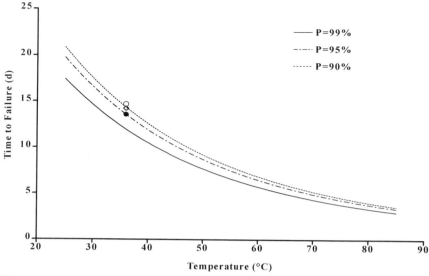

Figure 4. Probability of Failure-Stress-Time to Failure diagram for 99, 95 and 90% reliability with observed (◊: first, ○: second) and predicted (●) failure times at 36°C for a PE/TGIC powder coating during an accelerated weathering (QUV; UVB-313).

Gloss Retention. The gloss measurement data were handled as described in the previous paragraph. The applied failure criterion was 50% gloss retention; the specimen lifetimes and the Weibull lifetime distributions were calculated accordingly. Figure 5 gives an overview of the curves obtained at the different weathering temperatures used. Two important differences with respect to the corresponding ΔE data emerge. First, the distributions do not line up in the expected order: the ranking according to increasing lifetime is 57°C < 66°C < 36°C < 73°C. This reversal cannot be attributed to errors in the weathering temperature since the ΔE and gloss data were obtained from the same specimens weathered in the same experiments. Second, the agreement between the experimental and fitted lifetime distributions is not as good as found for the ΔE data. Consequently, the lifetime analysis could not be continued by the lack of a suitable model. Nevertheless, these service life experiments seem to suggest that the impact of weathering temperature on gloss retention cannot be disregarded. Specular reflection depends on the angle of reflection and on the refractive index and roughness of the reflecting surface (*13*). The question remains which of these factors are influenced by temperature and how this relates to the degradation of the coating. The comparison of gloss data obtained at different temperatures will therefore require careful examination.

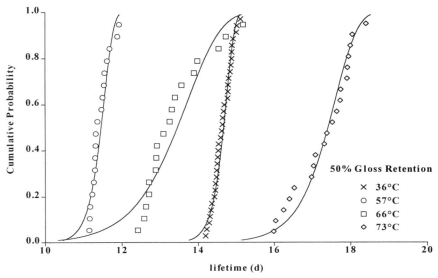

Figure 5. Evolution of the experimental (symbols) and theoretical (curve) lifetime distributions for gloss retention with weathering temperature during an accelerated weathering (QUV; UVB-313; failure criterion: 50% gloss retention) of a PE/TGIC powder coating.

Conclusions

The results reported in this work show that the evolution of the service life with temperature is described by the Weibull-Arrhenius model with respect to the color change failure mode. A good agreement was found between calculated and experimental service life values under the weathering conditions used. The gloss retention data could not be modeled due to a yet unexplained effect of the weathering temperature.

Despite the apparent simplicity of the service life experiments described in this work, it could be shown that the weathering temperature had a significant influence on color change and gloss retention.

The results presented here suggest that a thorough understanding of the underlying mechanisms of weathering be needed to ascertain that the evaluation parameters used are linked to the weathering of the coating.

The treatment of accelerated weathering data according to the principles of the Reliability Theory offers distinct advantages over the current practice of durability assessment:

- the validity of the procedure outlined in this work is not limited to ΔE vs. temperature experiments. The same method can in principle be applied to other types of weathering, as long as quantitative data are obtained;
- the method results in a quantitative, statistically meaningful statement on the service life of a coating;

- the larger number of specimens involved minimizes misinterpretations due to outlier results; instrumental and/or measurement errors will show through inconsistent data;
- in this work, the Arrhenius-Weibull model was used. However, the same data sets can be recalculated using other models, without having to repeat the weathering experiments;
- the PST diagrams hold great potential for the extrapolation of coating service life towards lower temperatures, as well as for the comparison of the service lives of different coatings, on condition, however, that the weathering conditions are chosen correctly.

Literature Cited

1. Fischer, R. M.; Ketola, W. D.; Murray, W. P., *Progr. Org. Coat.* **1991**, *19*, 165.
2. Shooman M. L. *Probabilistic Reliability : an Engineering Approach;* McGraw Hill: New York, 1968.
3. Vesely, W. E.; Goldberg, F. F.; Roberts, N. H.; Haasl, D. F. *Fault Tree Handbook;* U.S. Nuclear Regulatory Commission, NUREG-0492, 1981.
4. Whittaker, I. C.; Besumer, P. M. *A Reliability Analysis Approach to Fatigue Life Variability of Aircraft Structures;* Air Force Materials Laboratory Technical Report AFML-TR-69-65, 1969.
5. Martin, J. W.; McKnight, M. E. *J. Coatings Tech.* **1985**, *57 (No. 724)*, 39.
6. Martin, J. W. ; In *Proc. 15th International Conference in Organic Coatings Science and Technology;* Athens, 1989, 237.
7. Miller, R. N.; Schuessler, R. L. *Mod. Paint Coatings* **1990**, *80/9*, 50.
8. Schutyser, P.; Perera, D. Y.; In *Proc. XXI^t FATIPEC Congress;* Amsterdam, 1992, Vol. III, 1.
9. Schutyser, P.; Perera, D. Y. In *Proc. American Chemical Society, Division of Polymeric Materials, Science and Engineering;* American Chemical Society: Denver, CO, 1993, Vol. 68; 141.
10. Martin, J. W.; Saunders, S. C.; Floyd, F. L.;Wineburg, J. P. *Methodologies for Predicting the Service Lives of Coating Systems;* NIST Building Science Series 172, 1994.
11. Lawless, J. F. *Statistical Models and Methods for Lifetime Data;* Wiley, New York, 1982.
12. Nelson, W.; *Accelerated Testing. Statistical Models, Test Plans and Data Analyses;* Wiley, New York, 1990.
13. Braun, J. H. *J. Coatings Tech.* **1991,** *63 (No. 799)*, 43.

Chapter 14

Analysis of Coatings Appearance and Surface Defects Using Digital Image Capture-Processing-Analysis System

Fred Lee[1], Behnam Pourdeyhimi[2], and Karlis Adamsons[3]

[1]Materials Engineer, Atlas Electric Devices, 4114 North Ravenswood Road, Chicago, IL 60613
[2]Professor, Polymer Education and Research Center, School of Textile and Fiber Engineering, Georgia Tech, Atlanta, GA 30332–0295
[3]Staff Chemist, Automotive Products, Marshall R&D Lab, DuPont Company, Philadelphia, PA 19146

Durability and weathering studies of coatings involve accurate monitoring and meaningful analysis of the coating's appearance. The currently available techniques that monitor and analyze coatings are oftentimes deficient in accurate characterization of coating appearance. Furthermore, visual attributes that comprise the surface of interest are typically assessed by humans, making objective assessments laborious and difficult. Implementation of digital imaging in various scientific applications has been proven accurate and productive. The Video Image Enhanced Evaluation of Weathering(VIEEW) system has been developed at Atlas Weathering Services Group, Miami, in collaboration with Georgia Tech. In this report, applications concerning automotive top coat defects are presented using the digital imaging/image processing/image analysis instrument. The VIEEW data was also compared with conventional evaluation results. The VIEEW system demonstrates that the appearances changes and surface defects created by durability/weathering tests on automotive coatings can be visualized and quantified accurately. As regards to service life prediction(SLP) of organic coatings, meaningful visualization and precise quantitative surface measurement would undoubtedly further the effort to understand the complex kinetics of a material's chemical and physical property degradation, as well as mechanical performance changes.

Appearance property measurement is an important coatings evaluation (1,2,3,5,6,7). The word "appearance" refers to the measurement of the visual attributes of an object.

© 1999 American Chemical Society

Unlike other material properties, the appearance properties are tied to human psychology. As with other subjective visual assessments, the results are often sensitive to biases in individual's perception. Recommended protocol often requires the estimation of size, shape and distribution characteristics by eye -- perceptual judgments that are often tedious and time-consuming. The complex response of human perception can be, however, partially measured using photometric devices. While photometric devices can measure the amount of light reflected from the target area of an object, the spatial composition of the target area is still at a loss. Hence it would be desirable to develop automated instrumental methods incorporating two-dimensional imagery for evaluating coating integrity and measuring degradation.

Digital imaging and image process/analysis techniques have been available for many years in the scientific community. From biology to metallurgy, imaging has routinely been applied in surface characterization and for monitoring the kinetics of changes in surface/component morphology. Its ability to gather visual information in two (and pseudo-three) dimensional format in a transferable digital file has been fully accepted. Also, its ability to "analyze" and "archive" the imagery only furthers the applicability.

In weathering and durability studies of coatings, understanding surface deterioration and surface visualization is important. (9-20). While qualitative microscopic visualization techniques have benefited studies, quantitative macroscopic visualization techniques have not been readily available. The macro scale surface evaluation of coatings has been carried out by mostly human visual comparison techniques. Standard human visual comparison techniques used in the coatings industry include, but not limited to, visual color change, gloss change, general appearance, chalking, dirt retention, mildew growth, crazing, checking, rust, erosion, flaking, blistering, water spotting, discoloration and visual haze.

Monitoring the onset of change in a coating's appearance, and subsequent kinetics of these changes, can often provide insight into the underlying factors such as chemical or mechanical performance degradation. This information may lead to a better understanding of the adverse impact of environmental factors, material properties, material processing, and application consideration(4). Ultimately, a detailed understanding of the factors, and their respective impact, will help make meaningful service lifetime prediction a reality, as well as product design/modification more efficient.

Quantitative macroscopic visualization using digital imaging and special illumination design (Video Image Enhanced Evaluation of Weathering, VIEEW) has been studied at Atlas Weathering Services Group in collaboration with BARN Engineering in 1993 and later with the Georgia Institute of Technology. Introduced by NIST to the organic coatings industry in the early 90's, this digital imaging technology has gained interest from various materials related and, especially, the weathering and durability related field for its robust and quantitative analysis capability.

This paper describes 1) shortfalls of the conventional photometric appearance measurement techniques in weathering and durability applications, 2) the VIEEW system, and 3) the applicability of the VIEEW technology for automotive coatings durability studies, specifically in the visualization and the quantitative analysis of automotive topcoat defects such as scratch and mar defects, acid etch defects, weather induced cracking defects, and gravelometer chip defects.

Conventional Appearance Measurement Methodology I - Reflectance Based Photometric Measurements

Specular gloss is probably the most important and recognized reflectance based photometric measurement technique used in weathering studies for the analysis of the top coat of automotive coating systems. While specular gloss can quantify the reflectance characteristics of an object, it is noteworthy that gloss values do not precisely describe surface conditions of interest, in the case of automotive coatings, the surface of the topcoat. A common problem in reflectance based photometry lies in the influence of chromatic attributes (color) in the measurement value. This problem can be easily illustrated by referring to Figure 1 where idealized goniophotometric curves of the photo current as a function of viewing angle are given. The curves S and u are goniophotometric responses of a perfect diffuse reflector and a semi gloss white coating, respectfully. From the schematics, a height of gloss, h is defined as

$$h = \frac{p - u_{45}}{S_{45}} = \frac{p}{S_{45}} - \frac{A}{100}$$

$$A = \frac{u_0}{S_0}$$

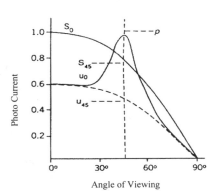

Figure 1. Schematic representation of photocurrent as a function of viewing angle. (Reproduced with permission from reference 21, Copyright 1972 Elsevier.)

Where, A is the relative brightness of the sample compared to a perfect diffuse reflector. By this definition, high gloss coatings have higher h values. From the schematics, however, one can see that the absolute value of p, the actual intensity of specular gloss, would not be the only determining factor in evaluating gloss. Here the numerator in the formula is computed by the differences between specular reflection and diffuse reflection of the sample. For example, two coatings exhibiting the same quantities of specular reflection would not necessarily have the same gloss value. A coating with a lower A value would have higher gloss value due to the diffuse dependency at the specular angle. Inspecting a black specimen and a white specimen of identical surface finish condition can easily demonstrate this phenomenon. Using a gloss meter, the black specimen would be rated at a higher gloss level than the white coating. While this result may agree with human visual perception, the results do not describe directly the materials' surface condition and may lead to inappropriate interpretation.

VIEEW Image Capture/Image Processing/Analysis System - Hardware

An imaging system is comprised of largely two separate functional components. These are 1) The optical component (Image acquisition) and 2) The computing component (Image processing/analysis). The VIEEW system includes an image acquisition software unit and separate image processing/analysis software. It has been found efficient to install the acquisition and the processing unit independently for optimum operation. The optical component is comprised of illumination sources and focusing optics.

The importance of light geometry cannot be overstated in photometric devices. For the VIEEW system, it is impossible to use conventional lighting schemes commonly used with commercial image capture and other photometric systems. Unlike the conventional photometric devices where a two dimensional optical arrangement is usually adequate, for coatings and other similar surfaces, we use a three dimensional optical arrangement where the focal plane is perpendicular to the optical axis in order to provide spatially even focusing.

The main system characteristic of VIEEW is the availability of two distinct illumination schemes enabling geometric and chromatic viewing. It is well known that it is most discriminating (as far as reflectance or gloss is concerned) when light is directed to a surface at a low incident angle. The most ideal incident angle is $0°$. This angle is particularly useful for capturing surface texture and roughness. When the light is directed to surface at $0°$ angle, it will be reflected from the surface at $0°$. When a stream of light is reflected perpendicularly ($0°$) from the surface, the direction of reflection is confined to $0°$ if the surface is optically smooth. However, when the surface is not optically smooth, the light will be reflected from the surface at angles greater than $0°$. See Figure 2.

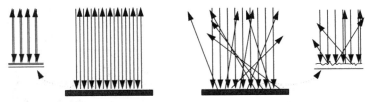

Optically Smooth Surface   Optically Rough Surface
Figure 2. The 0° incident light interaction.

This scattering effect of reflected light is most beneficial in examining surfaces in the presence of dirt, scratches, and streaks (common in weathered surfaces). These characteristics render conventional gloss measurements unreliable. When the light is reflected at large angles caused by the irregularity of the surface the scattering would not affect image formation since it is not detected by the target receptor. Figure 3 is a composite of four automotive top coat images captured using the VIEEW system showing the off-axis scattering of scratch marks appearing as dark lines.

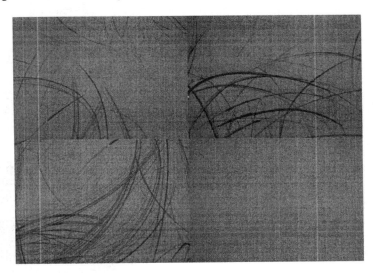

Figure 3. The scattering effect of 0° incident reflected light.

The VIEEW system uses another illumination scheme that induces chromatic reflection (diffuse reflection). When a uniform, spatially integrated radiation distribution is required, an integrating sphere technology is commonly used. An integrating sphere is composed of two perfect hemispheres that are coupled together to form a spherical cavity. Also, the inner surface of the sphere is coated with a high reflectance substance to achieve a high degree of diffuse illumination. The VIEEW system accommodates a direct viewing path through co-axially-situated entry and exit ports. For both illumination schemes (directional [0°] and diffused), the output illumination intensities are also controlled digitally. See Table 1 for the comparison of the two distinct modes of illumination conditions.

Chromatic Information (Diffuse Illumination)	Geometric Information (Directional[0°] Illumination)
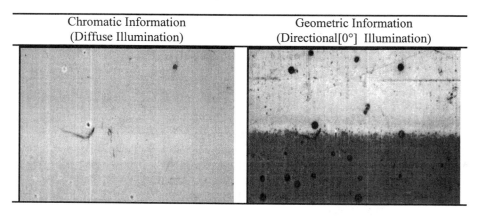	

Table 1. Chromatic and geometric viewing of a partially weathered sample. Note the difference in the ability to detect defects.

The current VIEEW system is a desktop unit as shown in Figure. 4. As part of an inspection system, the VIEEW is equipped in order to permit repeatability and reproducibility. The viewing condition of each capture session is automatically recorded for optimum repeatability. The variable intensity control of the VIEEW system enables a wide variety of surface conditions to be viewed. Similarly, the position of the sample is recorded using a stepper motor controlled X-Y stage. This allows imaging the same location of the specimen as a function of exposure time.

Figure 4. The VIEEW System.

VIEEW Image Capture/Image Processing/Analysis System - Software

The image analysis software deals with specific defect types. In dealing with each defect, the program needs to decide what attributes are relevant and how they can best be measured. Some deal with the estimation of size, shape and distribution characteristics while others derive first order as well as second order surface texture properties to estimate damage. The first category, by its nature, requires a binary image as input, thus requiring a thresholding step to convert the image to black and white. The thresholding method selected must be appropriate for the defects in question. The measurements made for each are described separately below when discussing each of the defect types.

Coatings Systems and Defect Types

1. Scratch and Mar Defects

Test:

Scratch and mar resistance testing of automotive coatings is done routinely and used as the basis for comparison of a coating's mechanical performance. Traditionally, both (lubricated) wet and dry mar testing protocols have been used (8) to inflict scratch and mar damage similar to that encountered in the field. Wet mar, for example, occurs in car washing, which is generally considered the most significant contributor to this type of damage, and in buffing operations used during repair. Dry mar comes from a broad range of materials that can contact the coating such as clothing, keys, paper, building materials, blowing sand, bushes, etc. Various imaging techniques, including optical and atomic force microscopy, have been used to determine type and extent of damage.

Marring of a commercial styrenated-acrylic/melamine automotive clearcoat was done to create six distinct levels of surface damage. Marring was accomplished using a test device that has a reciprocating motion with the arm moving. The arm is equipped with a block to which a felt pad is mounted. The surface of the panel is marred using a slurry of aluminum oxide in water and a polishing cloth pad (LECO Pan W). A polishing cloth was selected that (itself) did not damage the coating. Therefore, all of the damage was assumed to come from the grit slurry. The Daiei Rub Tester was found to be particularly useful for performing this test. Standard conditions for testing of automotive clearcoats are 500 gram weight, 10 cycles, and 320 grit aluminum oxide particles at a concentration of 0.1%. In order to generate six levels of damage we applied 10, 20, 30, 60, 75, and 100 cycles, keeping other parameters the same.

Objectives:

The main objective of the scratch and mar evaluation was to differentiate and quantify the surface condition according to the amount of damage. It was particularly interesting to see whether the image analysis would match human perception of the amount of damage visually detectable. Five images per each cycle were taken. Only 3 images per cycle are shown below. The data reported in a later section reflect all 5 readings.

Image Acquisition Condition:
1. Illumination Type: Directional Illumination
2. Illumination Intensity: Reflectance View
3. Image Dimension: 0.5 " X 0.67 "
4. Image Size: 320 X 240 pixels
6. Sample Color: Black

Revolution	Image 1	Image 2	Image 3
10			
20			
30			
60			
75			
100			

Table 2. Scratch and mar defect of automotive topcoat

Approach:

The overall effect of marring can best be measured using the overall loss in the ability to reflect light from the surface. While gloss measurements are aimed at deriving this property, the color of the underlying base coat affects the results somewhat. In addition, the uni-directional marks on the surface make gloss measurement difficult. With the VIEEW system, the $0°$ directional illumination scheme provides a better measure of the reflective properties of the surface. Furthermore, since we obtain an image of the surface, other textural attributes can easily be determined for the surface. Some of the measurement techniques are listed below.

We begin by examining the mean (x) and standard deviation (s) of image intensities. Elementary probability suggests that an image generated by randomly sampling a discrete uniform distribution, $U(G_{min}, G_{max})$ has an expected intensity value of $(G_{min} + G_{max})/2$ and a variance of $[(G_{max} - G_{min} + 1)^2 - 1]/12$. Most 'natural' images have a distribution that approaches normality, although often exhibiting a distinct skew. Since the light is kept constant for all specimens, the relative shift in the image brightness is directly a result of the surface scratching. With increasing marring, the image brightness decreases as well as a reduction in the range of the gray scales occupied in the image as measured by the breadth (variance or standard deviation) of the histogram. These provide a measure similar to gloss but are more exact. Similar to gloss, these provide a bulk measure of change.

It would also be desirable to have a measure of surface 'roughness' based on intensity variation. One such measure for a digital image is the area of surface relief. $G_{x,y}$ may be used as an altitude coordinate, i.e., the height of a column of G cubes, each having the dimensions one pixel by one pixel by one standard intensity level. We then define the gray level **area** of a digital image as equivalent to the total number of exposed faces in a landscape composed of gray level columns. Here, we are only concerned with lateral area, and may ignore the top face of each column. This quantity can be measured by comparing the intensities of pairs of edge-adjacent positions, that is, those sharing a side. The number of exposed faces on one side of a column depends on the difference, D, between the intensity of the current position, $G_{x,y}$, and the intensity of the adjacent position, $G_{x+i,y+j}$. We are only interested in non-negative differences, and stipulate that $D = G_{x,y} - G_{x+i,y+j}$ if $G_{x,y} >= G_{x+i,y+j}$, and 0, otherwise. The area of relief, A_R, is therefore the sum of the area of all sides of all columns. But A_R is also a function of the number of position pairs, P, and hence image size. To remove size effects, this quantity is normalized by computing $a_R = A_R/P$. Note that P depends on whether the image is bounded (i.e., has a perimeter) or unbounded (e.g., an inner subset of a bounded image). In an unbounded image, every column has 4 sides (P = 4XY), whereas in a rectangular bounded image the columns on the perimeter have less than 4 sides with adjacent positions, and $P = 8 + 6[(X-2) + (Y-2)] + 4[(X-2)(Y-2)]$. For the sake of simplicity, we refer to unbounded images in the following discussion.

For the sake of illustration, let us consider a_R for some idealized images. When $D = 0$ for all position pairs, i.e., the image is monochromatic or 'flat', a_R is zero. The maximum possible relief area, $a_R = 127.5$, occurs in a regular checkerboard pattern composed of alternating 0 and 255 intensities. For an image generated by randomly sampling a discrete uniform distribution, $U(0, N-1)$, the average difference, $G_{x,y} - G_{x+i,y+j} = \Sigma x_i f(x_i)$. We are only interested in cases where this difference is non-negative, which occurs about half the time for large N, $p(G_{x,y} >= G_{x+i,y+j}) = (1 + 1/N)/2$. The average normalized relief of such an image is therefore $E(a_R) = \Sigma [x_i (N - x_i)/N^2]$, where i has the range $[0, N-1]$. Another similar measure will be to examine the total volume occupied by the gray scale intensities (25-26).

Surface roughness may also be determined using the fractal dimension of the surface. Other features of the texture may be quantified using second order statistics such as co-occurrence (25-26).

2. Acid Etch Defects

Test:

Acid etch resistance testing of automotive Original Equipment Manufacturer (OEM) coatings is done regularly to determine hydrolytic stability at/near the surface. Certain types of atmospheric pollution (i.e., sulfur oxides, nitrogen oxides) in combination with water (i.e., dew, rain) can result in formation of acidic solutions/environments capable of chemical bond hydrolysis. The styrenated-acrylic/melamine clearcoat network used in this portion of the study has both ester and ether type linkages susceptible of hydrolysis under these conditions. Acid etch damage results in what appears to be random patterns of microscopic and macroscopic pitting or etching, often significantly altering the appearance. This testing provides an important tool for rating and ranking coatings according to hydrolytic stability.

Flexible clearcoats selected for study were applied to 10" X 10" thermoplastic olefin(TPO) panels that had been coated with adhesion promoter and black asecoat. These panels were fixed to horizontal tables and exposed for 14 weeks in at Jacksonville, Florida. Panels were washed and rated every two weeks with the final rating after 14 weeks of exposure, Florida. Coating A shown in Table 3 is a commercially available polyester/melamine type 1K clearcoat. Coatings B and C are experimental 1K systems that contain polyester/melamine and polyester/silane, respectively. The conventional visual by comparison with known acid etch performance provided that type C is the best performer.

Objectives:

For the analysis, it was most reasonable to examine the extent of damage by etch marks. As seen below, the amount of damage can be either interpreted from the area covered by the marks or the severity of etching effect depth wise. In addition, the differences in each type damage observed as a function of coating chemistry, exposure time and conditions, are of keen interest.

Image Acquisition Condition:
1. Illumination Type: Directional Illumination
2. Illumination Intensity : Reflectance View
3. Image Dimension : 0.5 " X 0.67 "
4. Image Size : 320 X 240 pixels
5. Sample Color : Black

Coating Type	Image 1	Image 2	Image 3	Image 4
A				
B				
C				

Table 3. Acid etching defect of automotive top coat

Approach:

This type of defect is best quantified using the size of the affected areas. Shape can also be quantified relatively easily. The algorithms for determining size and shape are well known (23-26). Since these algorithms require a binary image as input, the quality of the image and the reliability of the thresholding step required can critically affect the resulting data. The VIEEW system (shown in Figure 4), can easily help obtain high quality images required for further processing. In the resulting binary image, the etched areas will appear as black and the unaffected areas appear as white. The total area coverage provides an overall measure of the severity of the etching. More specific information on size and shape can be obtained using geometrical descriptors. These measures may be quite relevant in terms of the surface texture and the polymer system used. Note that the images shown in Table 3 contain different sized defects. For more information on the methods for determining size and shape, see references (23).

3. Weather Induced Cracking

Test:

A model styrenated-acrylic/melamine system was subjected to accelerated weathering that included UV-VIS irradiation, condensing humidity, and thermal cycling as factors in the exposure cycle. This coating did not have a UV-screener (UVA) and hindered amine light stabilizer (HALS) package in order to monitor degradation of an UV unprotected network. Panel samples were obtained after 500, 1000, 1500, 2000, 3000, 4000 and 5000 hours of exposure.

After the accelerated weathering test, the samples also underwent a series of physical tests leaving scratch type marks on the coating surface. These marks were digitally minimized using a low pass Fourier filter to recover its weathered appearance.

Objectives:

The ability to monitor the onset of surface morphology change would be the main interest in this series. In addition, the cracking needed to be quantified.

Image Acquisition Condition:

1. Illumination Type: Directional Illumination
2. Illumination Intensity : Reflectance View
3. Image Dimension : 0.5 " X 0.67 "
4. Image Size : 320 X 240 pixels
5. Sample Color : Blue Metallic

Exposure Length	Image as Captured	Processed Images	Skeletonized Image
500 hrs			
1000 hrs			

Table 4. Weathering induced cracking defect of automotive coating

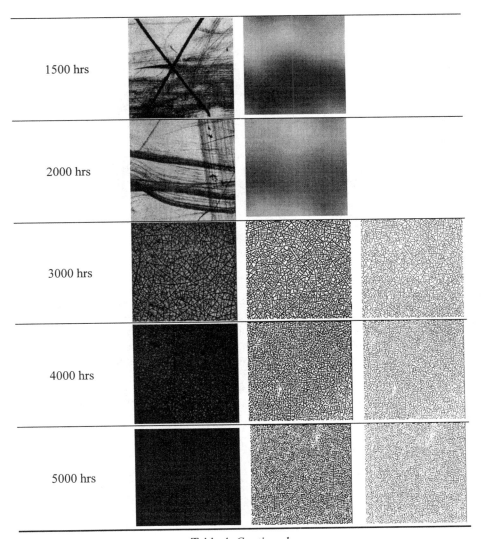

Table 4. *Continued.*

Approach:

As suggested by Martin and co-workers, the frequency and density of cracks are obviously important considerations (22). We recently proposed that crack length and orientation were also very important. Crack length is an indicator of the severity of cracking while orientation may reveal other physical or chemical attributes responsible for cracking behavior. In this context, perhaps the most significant hurdle is defining what constitutes a single crack; cracks are often not single entities and cross over other cracks (see Table 4). We demonstrated that crack length and orientation can be measured reliably (29). However, for the present discussion, we only refer to the frequency and density of cracking as measured by the total crack area coverage. Specific crack attributes such as length and orientation are meaningful when the samples' chemistry and history are known.

In dealing with cracking defects, we also need to convert the image to binary. The algorithms for carrying out the thresholding cracking images are rather more elaborate than those used in the case of defects previously discussed. Here, the features (cracks) are small entities and differ from the background only slightly in their gray level intensities. Therefore, thresholding needs to be carried out locally. This was discussed recently in full (25)

4. Gravelometer Test Chip Defects

Test:

The gravelometer test specimens (automotive type clearcoat/basecoat) were created according to the Dutch industry standard similar to the SAE J400 protocol. These specimens were solely created to demonstrate and investigate the imaging/image analysis' ability to quantify defects of this type. The chemical composition of the coating system and the gravelometer specification are, therefore, not available. One coating system was tested with varying levels of the gravelometer induced damage.

The sample numbers were assigned to each sample according to the visual assessment protocol.

Objectives:

The objective was to accurately measure the damaged area and rank according to the ascending damage amount.

Image Acquisition Condition:

1. Illumination Type: Directional Illumination
2. Illumination Intensity : Reflectance View
3. Image Dimension : 0.5 " X 0.67 "
4. Image Size : 320 X 240 pixels
5. File Format : DIB
6. Sample color : Blue

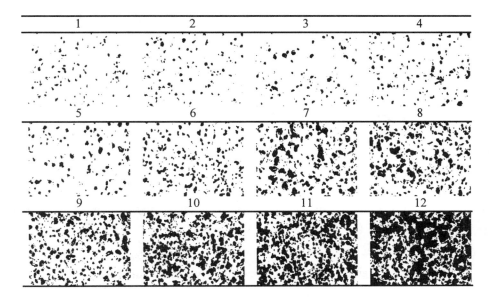

Table 5. Gravelometer test samples

Approach:

Chipping defects created by high speed impact of airborne gravel on a top coat can be imaged using the VIEEW's geometric viewing configuration. The resultant defects appear as depressions or areas of mass loss including both adhesive and cohesive failures. Note that in multi-layed automotive coatings damage often occurs at/near interfaces(i.e., adhesion failure between layers). When airborne gravel impacts the surface, chipping may occur wherein severity and formation of damage would be based on its mechanical properties. Consequently, the chipped area would loose its reflective quality enabling an accurate image capture by the VIEEW hardware. The analysis of chip resistance is similar to that of the acid etching. The image needs to be converted to binary to isolate the chipped areas from the background.

Results and Data Presentation

1. Scratch and Mar Defects

Mean intensity and variance are plotted concurrently in Figure 5 to show its significance. At the lower cycle, the images show a higher mean intensity indicating little or no damage to the top surface. Consequently, the wide range gray composition (i.e., broad gray histogram) created by marring is reflected in a high level of variance. Contrary to this, as the damage becomes more severe, the gray histogram shifts and its breadth is reduced as indicated by the low variance values. The subsequent low mean intensity also signifies a deeper and more severe marring effect.

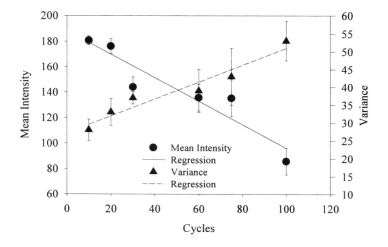

Figure 5. Mean intensity, variance vs. number of cycles.

2. Acid Etch Defects

All three type of coatings show distinct variation in etch mark formation. Type A shows a dense and the most severe etch marks among the series. Type B and Type C show a somewhat similar level of severity; however, the etch forming characteristics differ greatly in etch frequency and average etch size. It is also notable that the type A has lower reflective ability as indicated by the darker background. This characterizes its lesser smooth surface condition among the three types.

For the data presentation(Figure 6), the percent area covered by etch marks is compared with the average etch mark sizes. The severity of type A coating is clearly indicated by its high % area value. The differences in formation between type B and type C are also clearly shown by average size.

3. Weather Induced Cracking

The overall effect of weather induced surface degradation can be detected by the reflection of incident light (as seen in the scratch and mar evaluation). As the exposure increases, the geometric images became darker indicating the topcoat's diminishing light reflecting quality. After 3000 hours of exposure, presence of mud-cracking type damage greatly reduced its reflective ability, further characterizing abrupt changes on the surface. As the crack density increased, the images became darker illustrating denser crack formation.

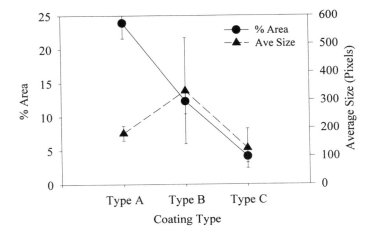

Figure 6. Percent area, average size vs. coating type.

The mean intensity clearly characterizes the coating performance in terms of the reflective property. The abrupt shift in the mean intensity also highlights the abrupt change on the surface. The abrupt change on the surface is also precisely monitored and quantified by the two measures: Mean intensity and % cracked area (Fig. 7). Also notable is the shrinking of error (mean intensity) bars during 3000, 4000 and 5000 intervals indicate uniform crack formation over the entire viewing area of the image.

4. Gravelometer Test Chip Defects

For this type of defect, a particle type analysis would be most appropriate since frequency, size and distribution are of interest. However, given the generality of particle analysis, only the ranking of the specimens regarding chipped area is presented in Figure 8.

Comparisons to Conventional Measurement Techniques

1. Scratch and Mar Defects

Conventionally, scratch and mar type defects can be evaluated via gloss measurement. Considering the high reflectance of the tested coatings system, 20° gloss readings were measured. Total of five readings per each cycle were taken and averaged. A direct comparison between VIEEW data and Gloss data is also shown in Figure 9. A correlation coefficient is presented in Figure 10.

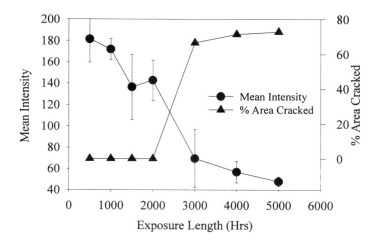

Figure 7. Mean intensity, % area cracked vs. exposure length.

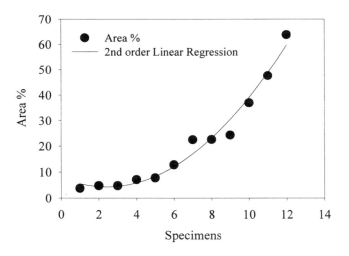

Figure 8. Percent area vs. specimen #.

Revolution	10	20	30	60	75	100
20°	82.9	78.0	64.8	60.3	58.2	49.9
Std Dev	1.4	2.3	2.1	1.8	4.4	4.0

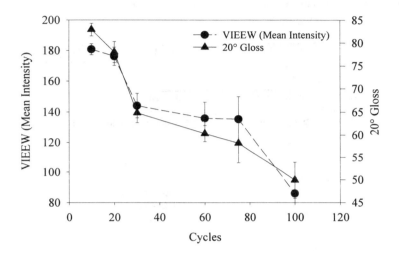

Figure 9. Direct comparison between VIEEW and gloss reading.

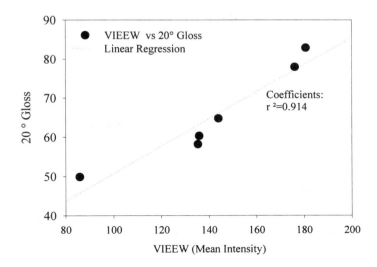

Figure 10. Correlation coefficient between VIEEW data and gloss data.

2. Acid Etch Defects

Considering the similarities of shape and formation in water spotting defects, Atlas Weathering Services Group's water spotting visual evaluation guide is used to evaluate acid etch defects. Direct comparisons between VIEEW data and the visual assessment are presented in Figure 11 and 12.

	Type A	Type B	Type C
%Area*	8	7	5
Average Size**	7	9	6

* %Area Ratings: 0, none; 2, less than 0.1% surface covered; 4, less than 1%;4, more than 10%; 8, more than 33%; 10, more than 50%.
** Average Size ratings: 0, none; 2, pin point; 4, 1 mm; 6, 2mm to 3mm; 8, 4mm to 5mm; 10, Greater than 5mm.

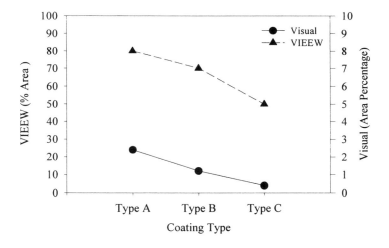

Figure 11. Direct comparison between VIEEW data and visual ratings reading for area percentage damaged.

3. Weather Induced Cracking

The visual data shown below is rated according to ASTM D661. The value 0 indicates that cracking was not detected, while 10 indicates severe crack formation.

Exposure Length (hrs)	500	1000	1500	2000	3000	4000	5000
Visual Rating	0	0	0	0	10	10	10

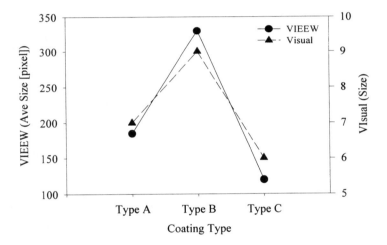

Figure 12. Direct comparison between VIEEW data and visual ratings reading for average size of etched marks.

4. Gravelometer Test Chip Defects

The visual data shown below is rated according to ASTM D610. A direction comparison between VIEEW data and visual ratings are shown in Figure 13. A correlation coefficient is calculated in Figure 14.

Specimen	1	2	3	4	5	6	7	8	9	10	11	12
Visual Ratings	6	6	6	6	5	4	4	3	3	2	2	1

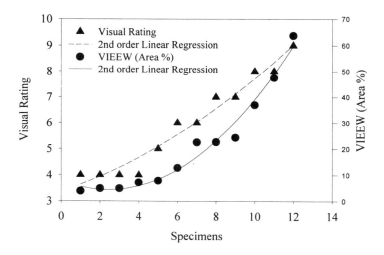

Figure 13. Comparison between VIEEW data and ASTM D610

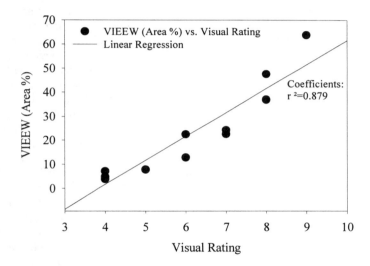

Figure 14. Correlation coefficient between VIEEW data and ASTM D610

Discussion -Micro vs. Macro Scale Quantitative Analysis

Atomic Force Microscopy provides the capability to measure nano-indentation and create nano-scratches under various controllable compressive loads and rates and allows the determination of the height and surface roughness accurately. Thus, quantitative assessment of the mechanical properties of the coating and surface damage can be reliably made. Lateral and height calibrations are checked prior to each experiment by imaging a NIST (Gaithersburg, MD) traceable standard. The measurement of micro-mechanical properties and understanding the link between viscoelastic properties of the polymer system and defect or damage tolerance are critical. These are extremely important in developing polymer systems that have optimal properties to minimize scratching without adversely affecting other properties such as acid etch and chip resistance. A variety of statistical descriptions of the surface are available using commercial software (e.g., RMS roughness, average roughness, etc.). Also, data analysis algorithms specific to a given damage type (i.e., scratch and mar, acid etch) have been developed. However, imaging by AFM is currently only practical in a laboratory setting (i.e., specialized vibration damping table/chamber, high-resolution XYZ translating stage), is often time intensive, and requires considerable training and expertise. In addition, commercial AFM units are relatively expensive. Note that the AFM data relate to the micro properties and morphology of the system and do not directly reveal the appearance characteristics of the coating systems. This is accomplished by a system such as VIEEW. The VIEEW system allows the determination of the macro properties of the surface and complements micro methods such as the AFM nano-indentation. That is, the micro-mechanical properties are quantified in terms of their effect on appearance and appearance retention. Together, these systems can reveal the mechanisms responsible for the change in appearance in coatings. The data obtained from both systems are highly correlated. Note however, that the VIEEW system has advantages over what an AFM can offer. The AFM is limited to the determination of the nano-indentation and micro scratch properties, and cannot help with other macro scale defect types described earlier. With VIEEW, we can easily quantify a large array of defect types and determine quantitatively the net result on appearance. Also, we can determine the statistical significance of a given defect.

Summary

Understanding surface deterioration is an important factor in durability and weathering studies. Quantitative understanding of surface change is fundamental to better relate various external factors resulting in coatings degradation. Meaningful visualization also aids researchers to observe the onset of surface change to further their efforts to understand kinetics of chemical, physical and mechanical degradation. In this report, several automotive coating defects were analyzed using the VIEEW (Video Image Enhanced Evaluation of Weathering) system. The imaging/image processing/analysis system provided precise and useful surface information. It was also shown that the VIEEW data was closely conformant to conventional evaluation techniques. The information obtained via the VIEEW system would also be directly beneficial to formulate a meaningful service life prediction.

Acknowledgment

The authors would like to gratefully acknowledge contributions from the following: Basil Gregorovich and Jeffrey Johnson for coated panels; Joann Paci for preparing scratch & mar damaged panels; Lesley Jacques for her contributions at the beginning of the project as R&D manager; Larry Masters for his faith and vision during the development.

References

1. Dickie, R.A., *Journal of Coatings Technology*, **1994**, 66, No. 834, 29.

2. Bauer, D.R., *Journal of Coatings Technology*, **1994**, 66, No. 835, 57.

3. Bauer, D.R., *Progress in Organic Coatings*, **1986**, 14, 193.

4. Martin, J.W.; Saunders, S.C.; Floyd, F.L.; Wineburg, J.P., *NIST Building Science Series 172, Methodologies for Predicting the Service Lives of Coating Systems*, Building & Fire Research Lab, National Institute of Standards and Technology, 1994.

5. Dickie, R.A., *Journal of Coatings Technology*, **1992**, 64, No. 809, 61.

6. Gerlock, J.L.; Smith, C.A.; Nunez, E.M.; Cooper V.A.; Liscombe, P.; Cummings, D.R., "Advances in Coating Technology: Predicting the Durability of Coatings", Proceedings of the 36-th Annual Technical Symposium, Cleveland, OH. May 18, 1993.

7. Courter, J.L., Cytec Industries Inc.; Presented at the 23-rd Annual International Waterborne, High-Solids and Powder Coatings Symposium, Feb.14-16, 1996, New Orleans, LA, USA.

8. Adamsons, K.; Blackman, G.; Gregorovich, B.; Lin L.; Matheson, R., XXIII-rd International Conference in Organic Coatings, Proceedings 23, pg. 151, Athens, Greece, July, 1997.

9. Leithner, K., *Advanced Materials & Processes*, **1993**, 11, 18.

10. Dubbeldam, G., *Progress in Organic Coatings*, **1992**, 20, 261.

11. Kody, R.; Martin, D., *Polymer Engineering and Science*, **1996**, 36, 2.

12. Quazi, R.; Bhattacharya, S.; Kosior, E.; Shanks, R., *Surface Coatings International*, **1996**, 2, 63.

13. Shen, W.; Ji, C.; Jones, F; Everson, M.; Ryntz, A., *Surface Coatings International*, **1996**, 6, 253.

14. Westburg, J., SME FC74-649, 1974.

15. Braun, J., *Journal of Coatings Technology* , **1991**, 63, 799.

16. Braun, J.; Fields, D., *Journal of Coatings Technology* , **1994**, 66, 828.

17. Braun, J.; Cobranchi, D. *Journal of Coatings Technology*, **1995**, 67, 851.

18. Wernstahl, K., *Polymer Degradation and Stabilization*, **1996**, 54, 1.

19. Coulthard, M., *Metal Finishing*, **1993**, 91, 10, 45.

20. Kahl, L.; Halpaap, R.; Wamprecht, C., *Surface Coatings International*, **1993**, 76, 10.

21. Zorll, U., *Progress in Organic Coatings*, **1972**, 1, 113.

22. Martin, J. W.; McKnight, E., *Journal of Coatings Technology*, **1985**, 57, 49.

23. Pourdeyhimi, B.; Xu, B., *International Nonwovens Journal*, **1994**, 6, 1, 26.

24. Blakey, R. R., *Progress in Organic Coatings*, **1985**, 13, 279.

25. Pourdeyhimi, B.; Sobus, J.; Xu, B., *Textile Research Journal*, **1993**, 63, 9, 523.

26. Pourdeyhimi, B.; Nayernouri, A.; Xu, B., *Textile Research Journal*, **1994**, 64, 3, 130.

27. Pourdeyhimi ,B.; Nayernouri, A., *Journal of Coatings Technology*, **1994**, 66, 834, 51.

28. Pourdeyhimi, B.; Ramanathan, R., *Polymers and Polymer Composites*, **1995**, 3, 4, 277.

29. Pourdeyhimi, B.; Lee, F. *European Coatings Journal*, **1995**, 11, 804.

Chapter 15

The Relationship Between Chemical Kinetics and Stochastic Processes in Photolytic Degradation

Sam C. Saunders[1,3] and Jonathan W. Martin[2]

[1]Department of Pure and Applied Mathematics, Washington State University, Pullman, WA 99164–3113
[2]Organic Building Materials, National Institute of Standards and Technology, Gaithersburg, MD 20899–0001

The solution of a problem in photolytic degradation, viz., determining the quantitative relationships among species concentrations as a function of irradiance and time, is presented. Stochastic methods are used and contrasted with the classical methods of chemical kinetics. The former seem conceptually simpler, technically easier and they yield solutions in certain cases which are not available otherwise.

In chemical kinetics one represents the generation of a reactant, say C, as a *zero-order* differential equation when

$$\frac{dC}{dt} = \eta, \quad \text{which implies that} \quad C(t) = \eta t \quad \text{for} \quad t > 0. \tag{1}$$

Likewise a reactant A is of *first order* if its concentration $A(t)$ at time $t \geq 0$ satisfies the differential system, for some rate constant $\eta > 0$, and initial value $a > 0$, given by

$$\frac{dA}{dt} = -\eta A \quad \text{with} \quad A(0) = a.$$

This differential equation has the solution $A(t) = ae^{-\eta t}$ for any $t > 0$. NB instead of representing the concentration of reactant A by $[A]$ we write simply $A(t)$.

A reactant B is of *second-order* when it satisfies

$$\frac{dB}{dt} = -\eta B^2 \quad \text{with} \quad B(0) = b.$$

[3]Current address: P.O. Box 458, Kirkland, WA 98083–0458

© 1999 American Chemical Society

This differential system has the solution

$$B(t) = \frac{b}{1 + b\eta t} \quad \text{or} \quad \frac{1}{B(t)} = \frac{1}{b} + bt \quad \text{for any } t > 0. \tag{2}$$

We propose simply to interprete the *duration* of a chemical reaction as a random variable, and to examine certain aspects of the photodegradation process from this non-deterministic perspective to see if quantitative prediction is more feasible and greater insight can be obtained.

The Stochastic Perspective

By a (random) reaction waiting time, hereafter abreviated rwt, is meant a non-negative random variable T; we write $T \sim \mathrm{Ep}(\eta)$ when T has an *Epstein* (memoryless or negative exponential) distribution with (hazard) rate $\eta > 0$. In this case its distribution function, label it F_1, has the form

$$F_1(t) = \Pr[T \leq t] = 1 - e^{-\eta t} \quad \text{for all } t > 0, \tag{3}$$

and we also write $F_1 \simeq \mathrm{Ep}(\eta)$ with $\bar{F}_1 = 1 - F_1$ as the complimentary distribution for any affix on F.

Thus $A(t) = a\bar{F}_1(t)$ expresses the decreasing concentration of a reactant over time when the reaction is of first-order. In the stochastic interpretation it is the *expected fraction* of material for which the reaction is completed at a given time.

In statistics a random variable $X \sim F_2$ has a *Pareto* distribution when, for some shape and scale parameters $\alpha, \beta > 0$

$$\bar{F}_2(t) = \Pr[X > t] = e^{-\alpha \ln(1 + \beta t)} \quad \text{for } t > 0. \tag{4}$$

Thus a second-order reaction has an expected concentration given by $B(t) = b\bar{F}_2(t)$, where $\alpha = 1, \beta = b\eta$.

For any two independent rwt's, say $X_1 \sim F_1, X_2 \sim F_2$, their sum $X_1 + X_2 \sim F_1 * F_2$, where the distribution $F_1 * F_2$ is called the *convolution* and is defined by the integral

$$F_1 * F_2(t) = \int_0^t F_1(t - x)\, dF_2(x) \quad \text{for any } t > 0.$$

In the stochastic view photon-flux is replaced by the random number of photon arrivals (each having one of the wave lengths specified) during the time interval $(0, t)$; call it $N(t)$. This random number is counted by the expression

$$N(t) = \sum_{n=1}^{\infty} \mathrm{I}[T_1 + \cdots + T_n \leq t] \quad \text{for } t > 0,$$

where $I[\cdot]$ is the binary indicator of whether any event is true or not. Here $T_i \sim F$ for $i = 1, 2, \cdots$ are independent and identically distributed interarrival times each with a $Ep(\eta)$ distribution. Thus, $T_1 + \cdots + T_n \sim F^{n*}$, the n-fold convolution of F, where

$$F^{n*}(t) = \int_0^{\eta t} \frac{x^{n-1} e^{-x}}{(n-1)!} \, dx.$$

From this it follows, see [2], that the expected count is

$$EN(t) = \sum_{n=1}^{\infty} F^{n*}(t) = \sum_{n=1}^{\infty} \int_0^{\eta t} \frac{x^{n-1} e^{-x}}{(n-1)!} dx = \eta t \quad \text{for all } t > 0,$$

which exactly agrees with equation 1. But here more is known, namely, the probability of exactly k photon arrivals in any time interval of length t is given by the related Poisson distribution

$$\Pr[N(t) = k] = \frac{e^{-\eta t}(\eta t)^k}{k!} \quad \text{for } k = 1, 2, \ldots,$$

(whence the term *Poisson Process*) and all of the calculus of probability applies to the consideration of ancillary events.

The function $R = \sum_{n=1}^{\infty} F^{n*}$ is called a *single-stage renewal* (function) since it involves only sums of one rwt as convolutions of its distribution F. It represents the expected number of arrivals by a given time. How does this apply to photo-degradation? Photon flux, which has long been modelled as a Poisson process, see [5], initiates, on a coating surface, complex free-radical oxidation reactions which are of such economic import they have been studied for various periods of time by different research groups. But photo-degradation is so complex it is rather like the proverbial elephant being examined by blind men; one's interpretation depends on where one stands and several correct interpretations may be possible. The one we have chosen is the Mielewski, Bauer and Gerlock [6,1993] scheme for free-radical induced photo-oxidation:

FREE-RADICAL OXIDATION REACTIONS

$$A + h\nu \xrightarrow{k_1} 2Y^\circ$$
$$Y^\circ + O_2 \xrightarrow{k_2} YOO^\circ$$
$$YOO^\circ + YH \xrightarrow{k_3} YOOH + Y^\circ$$
$$YOO^\circ + YOO^\circ \xrightarrow{k_4} \text{Terminal Products}$$
$$YOOH + h\nu \xrightarrow{k_5} YO^\circ + {}^\circ OH$$
$$YOOH + ? + h\nu \xrightarrow{k_6} \text{Terminal Products}$$

The mathematical interpretation of this set of stoichiometric equations is problematic in view of the questions about the proper quantitative chemical relationships: (1) How does one account for the cage-effect for the chromophores under photon flux? (2) What is the effect of cyclic reactions, i.e., subsequent reactions providing reactants used in preceeding steps? (3) Which, if any, of the reactants should be considered inexhaustible in relation to others? (4) Which concentrations can be determined by chemical measurement with sufficient accuracy during the process so as to provide valid data points for the determination of any unknown rate constants in any proposed dynamical model?

Furthermore the complexity of the implicit joint-solution of a differential system describing this set of reactions might necessitate only numerical investigation. If so this would further obfuscate any empirical validation of the proposed model. Moreover, the common technique used in physical chemistry of solving by assuming constant, asymptotic rates have obtained, may not be valid since the important part of the reaction may have occurred before it is achieved. For a discussion of the chemical difficulties see [6].

If all the rate equations, such as above, could be well approximated by coupled linear first-order differential equations then representing the concentrations of the n different chemical species at time $t > 0$ by $\boldsymbol{A}(t)$, an $n \times 1$ column vector having transpose $\boldsymbol{A}^\top = (A_1, \ldots, A_n)$, the chemical kinetics could be expressed by a vector equation, when $\boldsymbol{M}$ is an $n \times n$ reaction matrix,

$$\frac{d\boldsymbol{A}}{dt} = \boldsymbol{M}\boldsymbol{A}, \qquad \text{where} \quad \boldsymbol{A}(0) = \boldsymbol{a}. \tag{5}$$

The complete solution for all concentrations of reactants is expressed, theoretically, by the matrix equation $\boldsymbol{A}(t) = \exp\{\boldsymbol{M}t\} \cdot \boldsymbol{a}$ for any time $t > 0$, where $\exp\{\boldsymbol{M}t\}$ is interpreted mathematically as an infinite series of powers of the matrix $\boldsymbol{M}t$. Numerical solutions can only be obtained when enough computing power is available. The difficulty of solution increases rapidly with n and this method fails if all reactions are not first order or are coupled into cycles.

Currently PC's with sophisticated numerical software allow successive iteration whereby the calculated concentrations may be fitted to the data by adjusting the values of the unknown rate constants until the resulting curve "fits" the experimental curve. Even in simple cases the number of adjustable parameters may be sufficiently high as to allow nonsensical chemical models to provide a good fit. Just as an nth degree polynomial can pass exactly through n points, and well approximate $100n$ if they are smoothly clustered, and may provide reasonable interpolation it can provide neither valid scientific explanation nor extrapolation.

Chemical Kinetics from a Stochastic Viewpoint

Some First-Order Reactions. To fix ideas let us first examine a simple system representing the (chemical) deterioration of A to C, viz., $A \xrightarrow{k_1} B \xrightarrow{k_2} C$ by a system of coupled linear differential equations written in terms of first-order reaction rates as

$$\frac{dA}{dt} = -k_1 A, \quad \frac{dB}{dt} = k_1 A - k_2 B, \quad \frac{dC}{dt} = k_2 B, \tag{6}$$

say with initial conditions $A(0) = a, B(0) = C(0) = 0$. A stochastic interpretation is that there are two independent rwts, say $X_1 \sim F_1$ and $X_2 \sim F_2$, for which one could express equation 6 as $A \xrightarrow{X_1} B \xrightarrow{X_2} C$. Then it follows by direct interpretation that the expected fractional completions are, for any time $t > 0$,

$$
\begin{aligned}
A(t) &= a\Pr[X_1 > t] = \bar{F}_1(t) \\
B(t) &= a\Pr[X_1 < t < X_1 + X_2] = a[F_1(t) - F_1 * F_2(t)] \\
C(t) &= a\Pr[X_1 + X_2 \leq t] = aF_1 * F_2(t).
\end{aligned}
$$

The concentration $C(t)$ may be thought of as the fraction of the initial amount, a, for which both chemical reactions are complete at time t. The other expressions are equally transparent. Here any distribution of the rwt's, X_1 and X_2, can be assumed and mass is conserved since $A(t) + B(t) + C(t) = a$.

Let us assume that the reaction times are "memoryless." This means if a molecule has not reacted it is just as likely to react within the next unit of time as it was the first unit of time regardless of how how much time has passed. This assumption is equivalent with $X_i \sim \mathrm{Ep}(k_i)$ for $i = 1, 2$ since the Epstein distribution is the unique memoryless distribution, see [2]. Then from the definitions of convolution we have directly the system of solutions, for any $t > 0$,

$$
\begin{aligned}
A(t) &= ae^{-k_1 t}, \\
B(t) &= \frac{ak_1}{k_1 - k_2}\left(e^{-k_2 t} - e^{-k_1 t}\right), \\
C(t) &= a\left[1 - \frac{k_1 e^{-k_2 t}}{k_1 - k_2} + \frac{k_2 e^{-k_1 t}}{k_1 - k_2}\right].
\end{aligned}
\tag{7}
$$

One may check easily that these solutions do satisfy the differential equations given in equation 6. Clearly the concentrations equal the expected fractional completions at any time $t > 0$. NB that when $k_1 = k_2 = k$ we have, for example, that equation 7 becomes

$$C(t) = aF^{2*}(t) = a[1 - e^{-kt}(1 + kt)] \quad \text{for all } t > 0.$$

This stochastic approach generalizes to an arbitrary sequence of such successive reactions, say,

$$A_1 \xrightarrow{\eta_1} A_2 \xrightarrow{\eta_2} \cdots \xrightarrow{\eta_n} A_{n+1}, \tag{8}$$

when initially $A_j(0) = a_j$ for $j = 1, \cdots, n + 1$. As has long been recognized this scheme describes radioactive decay, see [5,1950], when the reactions are first-order. It is important and several solutions have been published, see [3, 1985] or [9, 1970]. When the reactions A_i for $1 = 1, 2 \ldots$ are consecutive, irreversible and all first-order the answer can be written down on inspection from the result in [9]:

Lemma 1 *If $X_i \sim \text{Ep}(\eta_i)$ for $i = 1, \ldots, n$ are independent rwt's then their sum $S = \sum_{i=1}^{n} X_i \sim F_n$ is given by*

$$F_n(t) = P[S \leq t] = \sum_{j=1}^{n} \begin{bmatrix} \boldsymbol{\eta} \\ j \end{bmatrix} \{1 - e^{-\eta_j t}\} \quad \text{for all } t > 0, \tag{9}$$

where the coefficients, depending upon $\boldsymbol{\eta} = (\eta_1, \ldots, \eta_n)$, are defined by

$$\begin{bmatrix} \boldsymbol{\eta} \\ j \end{bmatrix} = \prod_{i=1, i \neq j}^{n} \frac{\eta_i}{\eta_i - \eta_j}, \quad \text{with} \quad \sum_{i=1}^{n} \begin{bmatrix} \boldsymbol{\eta} \\ j \end{bmatrix} = 1. \tag{10}$$

A proof of equations 9 and 10 can be obtained by induction.

A method of solution which directly utilizes a stochastic interpretation, and equation 9, seems to be simpler for certain reactions than either the solution of differential system or the operator methods discussed in a later section and in [5].

Second-order or Compound First-order Reactions? In virtually all textbooks in physical chemistry, e.g., reference [4], is stated the caution:

> It cannot be emphasized too strongly that the order of a reaction with respect to a given substance has *no relation whatsoever* to the stoichiometric coefficient of that substance in the chemical equation . . . but reactions which take place in a single step are excepted from this statement.

The empirical method of identifying a second-order reaction, as given in [4, 1995], is to plot the reciprocal of its concentration as a function of time, to see of a linear relationship results with varying initial concentrations having different intercepts but the same slope.

However there is a mathematical result that a mixture of a few Epstein distributions (linear first-order reactions), as in equation 3 having different (hazard) rates, could approximately become a Pareto distribution (a second-order reaction) as in equation 4. The theorem is that any Gamma mixture of Epstein distributions is exactly a Pareto distribution. We now present two such cases.

CASE I: Consider a coating film in which the density of a reactant at depth $x > 0$, measured perpendicular to the surface of the film is known to be F'. For example, if the density were to fall off exponentially then for some $\eta > 0$ the reactant concentration at depth less than or equal x would be

$$F(x) = 1 - e^{-\eta x} \quad \text{for all } x > 0. \tag{11}$$

Let $C(x, t)$ be the *diffusion*, i.e., the concentration of a diffusant (penetrant catalyst) at depth x and time t. The expected number of molecules which have not reacted, the remnant, at time t would be given by

$$B(t) = \int_0^\infty \bar{F}(x)\,C(x,t)\,dx, \tag{12}$$

while the reactant concentration would be $\int_0^\infty F(x)\,C(x,t)\,dx$.

Suppose we define the propagation of the penetrant over time by the equation

$$C(x,t) = \frac{e^{-x/bt}}{t} \quad \text{for } x,t > 0. \tag{13}$$

NB the concentration of the catalyst within the coating is $\int_0^\infty C(x,t)\,dx = b$ at all times and the behaviors of $C(x,t)$ over time for fixed depths $0 < x_1 < x_2$ and for various depths x at fixed times $0 < t_1 < t_2$ are easily seen to correspond to a wave propagating from the surface.

If we substitute equations 11 and 13 into equation 12 then the remnant is of second-order, viz.,

$$B(t) = \int_0^\infty e^{-\eta x}\frac{e^{-x/bt}}{t}\,dx = \frac{b}{1 + b\eta t}. \tag{14}$$

CASE II: Consider the integral $B(t)$ in equation 14 and make the change of variable $y = x/t$; we the obtain an integrand $B(t) = \int_0^\infty e^{-t\eta y}\,e^{-y/b}dy$. This lends itself to an alternate interpretation, viz., the reaction taking place at depth y has a rwt with rate proportional to y, namely, $T_y \sim \text{Ep}(\eta y)$ while the concentration of reactant throughout the film is $e^{-y/b}$. Again $b = \int_0^\infty e^{-y/b}dy$ is the mass of reactant throughout the material.

First-order Reversible Reactions. Consider a first-order reversible reaction in terms of rwt's, say $X_i \sim F_i$ for $i = 1,2$, where

$$\text{A} \;\overset{X_1}{\underset{X_2}{\rightleftharpoons}}\; \text{B} \quad \text{with } A(0) = a > 0 \text{ and } B(0) = 0.$$

We identify the event which gives rise to B, namely,

$$B(t) = a\,\Pr[X_1 \wedge X_2 \le t, X_1 < X_2] \quad \text{where } X_1 \wedge X_2 = \min(X_1, X_2).$$

We find, frm the calculus of probability, an expression which holds for any rwt's:

$$\begin{aligned}
B(t) &= a\,\Pr[X_1 < X_2] - a\,\Pr[X_1 \wedge X_2 > t, X_1 < X_2] \tag{15}\\
&= a\int_0^\infty \bar{F}_2(x)\,dF_1(x) - a\int_t^\infty [\bar{F}_1(t) - \bar{F}_2(x)]\,dF_2(x).
\end{aligned}$$

Assuming first-order reactions, i.e. , $X_i \sim \mathrm{Ep}(\eta_i)$ for $i = 1, 2$ we obtain, since $a = A(t) + B(t)$,

$$B(t) = \frac{a\eta_1}{\eta_1 + \eta_2} \left[1 - e^{-(\eta_1 + \eta_2)t} \right], \qquad A(t) = \frac{a}{\eta_1 + \eta_2} \left[\eta_2 + \eta_1 e^{-(\eta_1 + \eta_2)t} \right] \quad \text{for } t > 0.$$

This derivation seem much simpler than those given in references [1],[3] and [4]. Surprisingly, Epstein rwt's give statistical independence in equation 15, namely,

$$\Pr[X_1 \wedge X_2 > t, X_1 < X_2] = \Pr[X_1 \wedge X_2 > t] \cdot \Pr[X_1 < X_2] \tag{16}$$

This independence, when combined with the cage effect, is the reason for the form of the quantum yield in the photo-oxidation process. This point is explicated in a later section.

First-order Two-stage Reversible Reactions. Consider a first-order two-stage reversible reaction

$$A \underset{\eta_2}{\overset{\eta_1}{\rightleftarrows}} B \underset{\eta_4}{\overset{\eta_3}{\rightleftarrows}} C$$

where the behavior is goverened by the differential system with initial values a, b, c respectively,

$$\begin{aligned}
\frac{dA}{dt} &= \eta_2 B - \eta_1 A, \\
\frac{dB}{dt} &= \eta_1 A - \eta_2 B - \eta_3 B + \eta_4 C, \tag{17} \\
\frac{dC}{dt} &= \eta_3 B - \eta_4 C, \tag{18}
\end{aligned}$$

with $A(t) + b(t) + C(t) = a + b + c$. Then the (reaction) matrix and initial conditions are, are, respectively,

$$\mathbf{M} = \begin{bmatrix} -\eta_1, & \eta_2, & 0 \\ \eta_1, & -(\eta_2 + \eta_3), & \eta_4 \\ 0, & \eta_3, & -\eta_4 \end{bmatrix}, \qquad \text{and } \mathbf{a}^{\top} = (a, b, c),$$

and by setting $\mathbf{x}^{\top} = (A, B, C)$ then the (linear) vector differential system, given in equation17, is of the type given in equation 5 with the theoretical solution, in matrix notation, $\mathbf{x}(t) = e^{\mathbf{M}t} \cdot \mathbf{a}$.

We now illustrate how an explicit solution can be obtained directly by making an appropriate identification of a rwt having distribution in equation 9. We do this by applying transforms.

By definition if $X \sim F$ then

$$F^{\ddagger}(s) = \mathrm{E}[e^{-sX}] = \int_0^{\infty} e^{-sx} \, dF(x) \tag{19}$$

is the *Laplace-Stieltjes* transform of F. In the case $f = F'$ exists then $F^\ddagger$ is the ordinary Laplace transform of f, which we denote by $f^\dagger$. The relation between the transforms of a (bounded) function and its derivative can be found through integration by parts, namely,

$$sF^\dagger(s) = F^\ddagger(s) + F(0) = f^\dagger(s) + F(0). \tag{20}$$

For example by taking the transform of equation 9 we find

$$F_n^\ddagger(s) = \prod_{j=1}^{n} \frac{\eta_j}{s + \eta_j} \quad \text{for} \quad s > 0. \tag{21}$$

Thus by taking the Laplace transforms of the differential system of equation 17 and imposing the initial conditions $a > 0, b = c = 0$, we obtain, in the usual manner regarding s as the identity function but omitting it as an argument, the system of linear equations:

$$
\begin{aligned}
sA^\dagger - a &= \eta_2 B^\dagger - \eta_1 A^\dagger \\
sB^\dagger &= \eta_1 A^\dagger - (\eta_2 + \eta_3)B^\dagger + \eta_4 C^\dagger \\
sC^\dagger &= \eta_3 B^\dagger - \eta_4 C^\dagger.
\end{aligned}
$$

Taking the transform merely effects a change of arithmetic by replacing the convolution of distributions by the multiplication of their transforms. Compare the system of equations above with equation 17. If we systematically solve for the variables (transforms) $A^\dagger, B^\dagger, C^\dagger$ we obtain, for example,

$$C^\ddagger(s) = sC^\dagger(s) = \frac{a\eta_1\eta_3}{s^2 + s\sigma_1 + \sigma_2} \quad \text{where} \quad \begin{aligned} \sigma_1 &= \sum_1^4 \eta_i \\ \sigma_2 &= \eta_1\eta_3 + \eta_2\eta_4 + \eta_1\eta_4. \end{aligned}$$

To utilize the stochastic interpretation we let $\boldsymbol{\theta} = (\theta_1, \theta_2)$, where we define $(s + \theta_1)(s + \theta_2) = s^2 + s\sigma_1 + \sigma_2$, (In this case θ's always exist as distinct positive numbers since $\sigma_1^2 > 4\sigma_2$.) so that

$$C^\ddagger(s) = \frac{a\eta_1\eta_3}{\sigma_2} \cdot \frac{\theta_1}{s + \theta_1} \cdot \frac{\theta_2}{s + \theta_2},$$

which we can invert from equation 21 by inspection; it is

$$C(t) = \frac{a\eta_1\eta_2}{\sigma_2} \left[1 - [\tfrac{\boldsymbol{\theta}}{1}]e^{-\theta_1 t} - [\tfrac{\boldsymbol{\theta}}{2}]e^{-\theta_2 t} \right]. \tag{22}$$

The term in square brackets in equation 22 is identical with the sum of two independent rwt's $X_i \sim G_i \simeq \mathrm{Ep}(\theta_i)$ for $i = 1, 2$ so one has $C \propto G_1 * G_2$ and similarly, the easily interpreted result

$$B(t) = \frac{a\eta_1\eta_4}{\sigma_2} \left[\frac{\theta_1}{\eta_4} G_2(t) + (1 - \frac{\theta_1}{\eta_4})G_1 * G_2(t) \right]$$

We see the asymptotic concentrations are

$$C(t) \longrightarrow a\eta_1\eta_3/\sigma_2, \quad B(t) \longrightarrow a\eta_1\eta_4/\sigma_2, \quad A(t) \longrightarrow a\eta_2\eta_4/\sigma_2 \quad \text{as } t \to \infty.$$

Here we see θ_1, θ_2 represent the net reaction rates from A to C. This derivation seems to provide more insight than the matrix computations.

Some Consequences

We give a simple proof of a well known result, see [7], on the decomposition of Poisson processes and apply it to stochastic behavior in the photo-oxidation process.

Lemma 2 *Let $T_i \sim \text{Ep}(\eta)$ be independent inter-occurrence times for events $i = 1, 2 \ldots$ so that $S_n = \sum_{i=1}^{n} T_i$ is the occurrence time for the nth event. Suppose each event is of type-k with probability p_k for $k = 1, 2$, where $p_1 + p_2 = 1$, and the type is independent of the occurrence times. If N_k counts the number of events, and S_{N_k} is the waiting time, until the first occurrence of an event of type-k then $S_{N_k} \sim \text{Ep}(p_k\eta)$ for $k = 1, 2$.*

PROOF: By definition N, for either affix, is a rwt such that $\Pr[N = n] = q^{n-1}p$ for $n = 1, 2 \ldots$, with $p + q = 1$, so by conditional probability we have

$$
\begin{aligned}
\Pr[S_N > t] &= \sum_{n=1}^{\infty} \Pr[N = n] \cdot \Pr[\textstyle\sum_{i=1}^{n} T_i > t] = \sum_{n=1}^{\infty} pq^{n-1} \int_{t}^{\infty} \frac{x^{n-1}\eta_n e^{-x\eta}}{(n-1)!} \, dx \\
&= p\eta \int_{t}^{\infty} e^{-x\eta} \sum_{n=1}^{\infty} \frac{(qx\eta)^{n-1}}{(n-1)!} \, dx = p\eta \int_{t}^{\infty} e^{-x\eta + qx\eta} dx = e^{-p\eta t}. \quad \blacksquare
\end{aligned}
$$

Note that the occurrence of Epstein inter-arrival times is equivalent with a Poisson process. Thus photon-arrivals, each one of which produces a scission with small probability, will generate scission events following a Poisson process with the arrival rate reduced by the probability of a scission occurrence.

The Cage Effect. We now show how this lemma explains the cage effect in photo-degradation. Let photons of fixed wavelength λ arrive with interarrival times $T(\lambda) \sim \text{Ep}[\nu(\lambda)]$. Each λ-photon cuts the polymer with probability $\phi(\lambda)$, which is called the *quantum yield*. Since a cut occurring or not is independent of photon arrival time, by Lemma 2 the interoccurrence time of scissioning λ-photons is $X(\lambda) \sim \text{Ep}[\phi(\lambda) \cdot \nu(\lambda)]$.

We now state another well known result:

Lemma 3 *If $X_i \sim \text{Ep}(\eta_i)$ are independent then $\min_{i=1}^{n} X_i \sim \text{Ep}(\sum_{1}^{n} \eta_i)$.*

From this it is easy to argue, using Riemann sums, that the interarrival time between scissioning photons of any wavelength $\lambda \in \Lambda$ is, assuming a constant irradiance $\nu(\lambda)$, over time, given by $X \sim \mathrm{Ep}(\mu)$ where $\mu = \int_{\lambda \in \Lambda} \nu(\lambda)\alpha(\lambda)\phi(\lambda)d\lambda$, with $\alpha(\lambda)$ being the absorbance of the material at wavelength λ.

Following polymer-chain scission from irradiance there are two processes, both independent of wavelength, in competition: the severed chain ends, bound by the cage effect, recombine under random molecular vibration or degradation occurs when a free-radical, Y°, combines chemically with O_2.

The rwt until recombination is assumed to be $W \sim \mathrm{Ep}(\omega)$ and the rwt until O_2 contamination is $V \sim \mathrm{Ep}(\gamma)$ both first-order independent reactions. At the time of the nth scissioning photon, $S_n = \sum_1^n X_i$ we have rwt's V_n and W_n which are competing and memoryless.

The total number of YOO° radicals formed is given by

$$N_{YOO^\circ}(t) = \sum_{n=1}^{\infty} \mathrm{I}(V_n + S_n \leq t)\mathrm{I}(V_n < W_n),$$

from which can be found the expected number of photon-initiated radicals, viz.,

$$\mathrm{E}N_{YOO^\circ}(t) = \frac{\gamma\nu}{\omega + \gamma}\left(t - \frac{1 - e^{-(\omega+\gamma)t}}{\omega + \gamma}\right).$$

Again note that here $\Pr[V_n < W_n] = \gamma/(\omega + \gamma)$ and the fact that two Y° radicals are generated at each scission is accounted for by doubling this expectation.

We now show an important result for multiple competing contingencies, when the rwt between each contingency is Epstein. The rwt until the first occurrence of a contingency of any type will also be Epstein and the outcome will be each contingency with a separate, independent probability.

Lemma 4 *If $T_i \sim \mathrm{Ep}(\eta_i)$ are independent then for any index i we have*

$$\Pr[t < T_i < \min_{j \neq i} T_j] = \Pr[T_i < \min_{j \neq i} T_j] \cdot \Pr[t < \min_{j=1}^{k} T_j].$$

PROOF: Fix i and set $X_2 = T_i \sim \mathrm{Ep}(\eta_i)$ and $X_1 = \min_{j \neq i} T_j \sim \mathrm{Ep}[\sum_{j \neq i} \eta_i]$ and then substitute into equation (14) and the result is immediate. ∎

Under the assumptions of Lemma 4, setting $T_{(1)} = \min_{i=1}^{k} T_i$, then $\Pr[T_{(1)} = T_i, T_i \leq t] = \Pr[T_{(1)} = T_i] \cdot \Pr[T_{(1)} \leq t]$.

Cycles of Chemical Reactions.

We now apply these ideas and results to the photo-oxidation wheel. Consider a cycle of chemical reactions, which consists of k stages as in equation 6 but where reactant A_1 is the same as reactant A_{n+1}. Furthermore at the jth stage there are $n_j \geq 1$ possible reaction paths, all of which are competing and each reaction has a different rate. Let the reaction times be $X_{j:i} \sim \mathrm{Ep}(\eta_{j:i})$ for $i = 1, \ldots, n_j$ and $j = 1, \ldots, k$. Here $\eta_{j:i}$ corresponds to the rate of occurrence of the ith alternative reaction at the jth stage. Let $p_j = \eta_{j:1}/\sum_{i=1}^{n_j} \eta_{j:i}$ for $j = 1, \ldots, k$ be the probability that the reaction having rate $\eta_{j:1}$ occurs at the jth step in the cycle, this being necessary for the degradation cycle to continue. Thus $p = \prod_{j=1}^{k} p_j$ is the probability of completing all the necessary steps (reactions) in the k-cycle, or it may also be interpreted as the fraction of the initial species A_1 which will be regenerated each time the cycle repeats.

WE now consider the rwt for a cycle completion: $T = \sum_{j=1}^{k} Z_j$ where $Z_j = \min_{i=1}^{n_j} X_{j:i}$ for $j = 1, \ldots, k$. The rwt for the ith cycle completion is $T_i \sim F$, independent for all $i = 1, 2 \ldots$ with the same distribution. So $S_n = \sum_{1}^{n} T_i$ is the total rwt until the cycle has repeated n times. We now seek an expression for the expected concentration of A_1, assuming unit concentration initially, at any later time. It is given by the renewal function

$$R(t) = E N_{A_1}(t) = E \sum_{n=1}^{\infty} I[S_n \leq t] \cdot p^n = \sum p^n F^{n*}(t).$$

Since the transform changes convolution into multiplication we have

$$R^{\ddagger}(s) = \sum_{n=1}^{\infty} [p F^{\ddagger}(s)]^n = \frac{p F^{\ddagger}(s)}{1 - p F^{\ddagger}(s)},$$

where, by writing $Z_i \sim G_i$, we obtain as before

$$F(t) = \Pr[\sum_{1}^{k} Z_j \leq t] = G_1 * G_2 * \cdots * G_k(t). \tag{23}$$

We recall that if $G_j \simeq \mathrm{Ep}(\eta_j)$ then $G_j^{\ddagger}(s) = \eta_j/(\eta_j + s)$ and since $\eta_j = \sum_{i=1}^{n_j} \eta_{j:i}$, it follows from equation 23 that

$$F^{\ddagger}(s) = \prod_{j=1}^{k} \frac{\eta_j}{\eta_j + s}.$$

After some algebraic simplification we obtain the transform of the renewal

$$R^{\ddagger}(s) = \frac{p \prod_{j=1}^{k} \eta_j}{\prod_{j=1}^{k} (\eta_j + s) - p \prod_{j=1}^{k} \eta_j} = \frac{p \sigma_k}{\sum_{i=0}^{k} \sigma_i s^{k-i} - p \sigma_k} = \frac{p \sigma_k}{\sum_{i=0}^{k-1} \sigma_i s^{k-i} + q \sigma_k},$$

where, for short we set $p = [\prod_{j=1}^{k} \eta_{j:1}]/[\prod_{j=1}^{k} \eta_j]$ with $q = 1 - p$ and the σ_i are the symmetric polynomials in the variables $\eta_1, \ldots, \eta_k$. An exact inversion has been given earlier for $k = 2$; the greater the number of stages the more difficult the inversion since in the denominator one must obtain the roots of polynomials high order. But higher orders give solutions in which pulsing in R can occur as well as monotone behavior. Whether such solutions are observed will be discussed in a subsequent paper.

Conclusion and Supposition

The difficulty with deriving a dynamical model for polymer degradation through photo-oxidation is the exact mathematical accounting for: (i) the effect of photon-flux in the chemistry, and (ii) the consequence of cyclic reactions which provide reagents for subsequent regeneration. Here a new perspective is provided by following the mathematical consequences of the Poisson irradiance compounded with combinations of first-order reactions which cycle chemically. This was done by applying concepts and results from stochastic processes to the physical model. The "stochastic" answers agree in all simple cases with the differential equations used in physical chemistry. With empirical verification of the fundamental mechanisms and definitive data a comprehensive stochastic model could provide an accurate method of service life prediction.

References

[1] Atkins, P.W., *Physical Chemistry*, Third Edition, W.H. Freeman and Company, New York, NY, 1986, Chapter 28, pp 687-712.

[2] Barlow, R.E.; Proschan, F., *Statistical Theory of Reliability and Life Testing*, To Begin With, Silver Spring, MD, 1981

[3] Bogdanoff, J.L.; Kozin, F.,*Probabilistic Models of Cumulative Damage,* John Wiley & Sons, New York, NY, 1985,p 84.

[4] Castellan, G.W., *Physical Chemistry*, Second Edition, Addison-Wesley Publishing Company, Reading, MA, 1995, Chapter Thirty-One, Chemical Kinetics.

[5] Capellos, C.; Bielski, B.H.J, *Kinetic Systems*, Wiley-Interscience,New York, NY, 1972

[6] Fano, U., *Radiation Biology*, Chapter 1, Principles of Radiological Physics, 1950.

[7] Mielewski, D.F.; Bauer, D.R,; Gerlock, J.L., *Polymer Degradation and Stability*,**1993**; 41, pp323-331.

[8] Ross, Sheldon M.,*Stochastic Processes*, Second Edition, John Wiley& Sons, Inc., New York, NY, 1996

[9] Saunders, S.C.,*On Confidence Limits for the Performance of a System when few Failures are Encountered,* Proceedings of the Fifteenth Conference on the Design of Experiments in Army Research, Development and Testing, 1970, ARO-D Report 70-2, pp 797-833.

Chapter 16

Prediction of Coating Lifetime Based on FTIR Microspectrophotometric Analysis of Chemical Evolutions

Jacques Lemaire and Narcisse Siampiringue

Centre National d'Evaluation de Photoprotection, Ensemble Scientifique des Cézeaux, 63177 Aubière Cedex, France

Most of the world activity on research, development and control of polymer durability is still based on empirical techniques developed in the early ages of polymer uses. Those techniques should be critically analysed considering the advances of the fundamental understanding of these complex phenomena. In the field of coating ageing, empirism is prevalent in the mode of application of environmental stresses in laboratory conditions and in the degradation criteria used. A more rational approach is described which is based on the recognition of the chemical evolution mechanisms. Applications of the physico-chemical stresses and definition of the degradation criteria should be consistent with the identified mechanisms. Aliphatic and cycloaliphatic hydroxylated polyesters crosslinked with substituted melamins or condensed isocyanates are presented as examples of the "mechanistic approach" ; the identification of the two chemical routes which account for the mechanical detriment is based on FTIR and micro-FTIR spectrophotometric analysis, the later technique being used to observe specifically the chemical evolution of the elementary layers of the clear-coat and base-coat.

In the field of polymeric materials ageing, coatings appear as very special systems for many reasons. (i) The large values of the ratio surface to volume of these systems favour the detrimental effect of the main environmental stress *i.e.* the UV light. (ii) The geometrical characteristics of coatings have important consequences on the efficiency of photostabilisers. (iii) Coatings are often exposed in the form of multilayers systems, each elementary thin layer presenting a different photochemical evolution. (iv) As organic layers located on inorganic or metallic substrates, these systems are not easily analysed *in situ*, especially when the chemical composition varies heterogeneously along the light penetration. (v) Since coatings should protect the substrates, their own durability is therefore an essential property.

© 1999 American Chemical Society

In the past 40 years, those special features have promoted a large use of empirical techniques at two levels.

1. the application of the physico-chemical stresses on the coatings, in laboratory conditions, has been essentially carried out as an exact simulation of environmental stresses to insure the relevancy of the artificial ageing ;

2. the degradation criteria have been based on the variations of macroscopic physical properties (gloss loss, microfissurations, aspect change, colorimetric variations, chalking ...).

Although the empiricism is present at a large extent in the field of polymer ageing, the part of empiricism has been so important in the prediction of coating lifetimes that the state of the art in this particular field has not progressed like for other polymeric systems.

A critical analysis of the ageing units based on stresses simulation

In the early 1950s, when failures of polymeric materials in use were observed, particularly in outdoor conditions, laboratory testing, that could reproduce the phenomena causing the degradation of the polymer, was urgently developed. In those years, the relevancy of the laboratory experiment was based on an exact qualitative and quantitative simulation of the physico-chemical stresses which exist in environmental conditions, applied during a time scale which was only somewhat shorter than the real lifetimes of the materials in use conditions. Indeed more or less exact simulation of the stresses implies that any acceleration of the degradation was unacceptable since it was anticipated that an accelerated degradation would be irrelevant. In the ageing units of the 50s, daylight was simulated using a carbon source, then a Xenon source emitting a continuous spectrum, rain was simulated by periodic water aspersion, day and night by dark and light periods ... Those principles are still respected in more than 90% ageing units functioning in 1997 (*i.e.* Weather-O-Meter and Xenotest), they however prompted the following critical comments :

i) A high pressure Xenon arc emits a continuum from 240 nm to the IR, that continuum is normally filtered to avoid any radiation from 240 to 300 nm. Using Bisphenol A Polycarbonate as a solid actinometer sensitive to short wavelengths (300-330 nm) and to long wavelengths ($\lambda \geq 300$ nm) with two different chemical consequences, it is very easy to demonstrate that the filtered Xenon source contains an excess of short wavelengths relatively to daylight [1]. A medium pressure Mercury arc filtered by a borosilicate envelop and emitting discrete lines appears to be more relevant (this experimental result is not unexpected when vibrational relaxation in the condensed phase is considered). Xenon sources present a second inconvenience due to their short lifetimes (... and cost).

ii) In the fifties, it was not understood that the complex chemical evolution of polymeric systems exposed to UV, heat, O_2 (and H_2O) involved some non-photochemical processes and presented fairly large apparent activation energy. The control of the temperature of the exposed surface is therefore a strict prerequisite of any laboratory experiment. In simulators, an external "black body" more or less insulated indicates only a temperature which is not related to the actual temperature of the exposed surfaces of the samples, especially of clear samples.

iii) Experimental results have often shown that the water aspersion in most standardised cycles simulates very heavy and frequent rains. In moderate weathering conditions, the oxidised layers are built up from March to September. Hydrolysis or mechanical abrasion of these layers occurs rather from September to March when the oxidation progressed only very slowly. Moreover water aspersion simulates mostly the mechanical role of rain and far less the physical and chemical influences of water (hydrolysis of some oxidation photoproducts, extraction of hydrosoluble stabilisers, matrices swelling etc. ....).

iv) In most cycles, dark period has been introduced to favour the migration to the surface of stabilisers whose depopulation in the superficial layers could be too fast during UV exposure. Dark periods have been fitted with additives which were generally migrating fast. Presently, the most migrating additives are not migrating that fast (compare for example monomeric HALS and BHT). During the fairly short dark periods, the depopulation at the surface could not be any more compensated. It should be emphasised too that photo-oxidation during the light period and migration during the dark period are two dynamic processes which could not be generally accelerated within the same factor.

v) Since every environmental stress is applied in the same experiment, analysis of the degradation origin remains difficult. Only correlations (which means relationships no one could account for) can be observed between weathering data and laboratory data.

From those comments, it could be concluded that any experimental procedure based on simulate units could be only pragmatically accepted, considering the large number of the corresponding units in function. An objective analysis of that testing of organic coatings is made even more difficult by the empiricism prevalent too in the choice of degradation criteria. To evaluate the degradation of coating, the following criteria are extensively used :

– loss of gloss related to the extent of superficial microfissuration ;
– aspect changes evaluated through visual inspection, colorimetry or spectrocolorimetry. Aspect changes result indeed from superficial microfissurations, formation and bleaching of yellowing photoproducts, formation and photo-oxidation of fluorescent photoproducts and bleaching of dyes and pigments. Any aspect change could be a very complex information since each phenomenon should obey a different kinetic low .
– fissuration of the coating film (mosaic formation) ;
– chalking.

As illustrated in the next section, a better understanding of the complex ageing phenomena could be gained considering the chemical evolution of polymeric systems through ageing. The macroscopic physical changes listed above could be indeed accounted for by the different chemical evaluations identified. Up to the recent years, however, it appears that the analytical difficulties which were inherent to coatings, have inhibited such an effort. The physical changes have not been often related to the chemical changes.

Towards a rational approach based on the analysis of chemical evolutions

When a polymeric material is submitted to the environmental stresses, which should be ranked as rather moderate, the degradation of most physical properties is due to chemical ageing. Therefore, through weathering, the polymer should be considered as a "photochemical reactor" in the presence of light and as a "thermal reactor" in its absence [2-5]. The exact nature of the chemical events responsible for the physical detriment should be understood (in the sense which will be explained later on). The analysis and the follow-up of the chemical evolutions, occurring as well in artificial conditions as in natural weathering or in real use conditions, have two advantages :
i)   it affords a very strict relevancy control based on the comparison of the on-going chemistry ;
ii) since the chemistry extent could be related to the exposure duration either in average conditions (weathering) or in well-defined conditions (artificial ageing), the lifetime in laboratory conditions could be converted into lifetime in use conditions.

Analysis of the chemical evolution of a polymeric material submitted to light, heat, O_2, H_2O, etc. ... , appears to be complex for the following reasons :
-   only evolutions in the solid state are relevant and analysis should be carried out in that solid state (specially to examine the stability of intermediate products) ;
-   chemical evolution at very small extent should only be considered. When the extent of the chemistry exceeds 0.5 to 1%, the loss of physical properties is complete and it becomes useless to study the further reactions occurring in the fragments (except when the "ultimate" fate of polymeric materials is examined for the safe of environmental protection) ;
-   chemical evolution includes indeed many mechanisms of various importance and it is required to identify the transformations which account for the physical detriment. Usually, the most important route involves a photo-oxidation or a thermo-oxidation mechanism whose products are formed in concentrations high enough to be observed in vibrational spectrophotometries, even in local zones (FTIR, micro-FTIR, FT-Raman, micro-FT Raman, FTIR with photoacoustic detector, ATR, ATR-H).

The extent of the chemical evolution is determined from the accumulation in the matrix of a "critical product" which :
-   should measure the main detrimental route of the matrix and therefore, should be formed from a chain scission process ;
-   should be chemically and photochemically inert in the matrix ;
-   should not diffuse out ;
-   should accumulate linearly with the exposure time until the complete loss of physical properties insuring the function ; since the complete physical detriment is observed at low extent, the system is "at initial time".

From the correlation which appears in accelerated conditions between the variations of physical properties and the variations of the critical product concentration, the lifetime of the polymeric materials is determined in artificial conditions and afterwards, converted into the lifetime in weathering conditions using a predetermined acceleration factor of the corresponding chemistry.

Acceleration of ageing cannot be fundamentally justified by practical reasons although users of polymeric materials are prompted to demand and accept quick testing. Indeed acceleration should be recommended for a very different reason associated with the impossibility to extrapolate the data collected in non-accelerated conditions. The usual techniques of homogeneous kinetics cannot be applied to handle the chemical transformation of a polymer matrix through ageing. It is indeed very difficult to obtain a reliable expression of the evolution rate of a solid polymer because of the complexities of the occuring reactions (short chain reactions, solid state processes, conjugation between chain reactions and step reactions, copropagaters, etc ...) and more important because of heterogeneity. Kinetic treatments were more used to rationalise the data collected in laboratory conditions than to predict the lifetime in use conditions [6-7]. As consequence of the limitation of any kinetic treatment, it is strongly recommended to observe the chemical and the physical evolution of the polymer until its life end. This life end can only be reached after a reasonable period of time in artificial accelerated conditions.

After many years of experiments on polymer photo-ageing in accelerated and in environmental conditions and examinations of the corresponding mechanisms, the following principles could be proposed :
a) it is possible to provoke accelerated chemical evolutions in solid polymers which obey the same mechanisms as the non-accelerated evolutions.
b) acceleration should be limited to the level where chemical distortions are observed (like enhancement of cross-linking due to biradical recombination).
c) photo-ageing acceleration should be due only to high light intensity (and not to shorter wavelengths) and to high temperature ; the temperature increase should however be limited by the fact that photothermal transformation should exceed largely any pure thermal conversion.
d) acceleration is only allowed when one unique dynamic process controls the ageing in natural and in artificial conditions. In many outdoor uses of polymeric systems, photo-oxidation is the main detrimental mechanism. It is therefore required to accelerate photo-oxidation without observing, in artificial conditions, some irrelevant control due to an oxygen starvation effect or to a stabiliser migration.

The example of cross-linked saturated polyesters

Presently, aliphatic or cycloaliphatic saturated polyesters with lateral or terminal hydroxyl groups are often used as clear-coats or base-coats in the automotive industry. They are processed at high temperature through cross-linking with melamine derivatives or with condensed isocyanates. The replacement of the aromatic units by saturated units in the polyester chain favours a photostability increase, either in the form of transparent matrix (clear-coat), or in the form of pigmented matrix (base-coat). This photostability can indeed be improved through the optimisation of the macromolecular structure or through additivation. However, durability problems have appeared in the automotive industry and this failure has prompted some fundamental research on the detriment mechanisms [8-11] and some developments on photostabilisation [12-14].

Non-hydroxylated saturated aliphatic polyesters were observed to have a very long lifetime, exposures during several thousand hours in the severe conditions of a SEPAP 12.24 unit have not induced any significant oxidation. Industrial hydroxylated saturated polyesters, used as coatings, contain indeed either some aromatic units as contaminants, or some aliphatic double bonds added on purpose. These reactive sites are able to induce some fast oxidation of the polyester chain. Cross-linked polyesters are even more reactive since the new bonds formed in the cross-linking reactions are highly photo-oxidable, as shown in the results presented in the next sections.

The more usual cross-linkers are substituted melamines, like methoxy-methylated melamines :

or condensed isocyanates with low volatility like trimers (isocyanurates) or oligomers. The isocyanate groups could be blocked and the blocking agents BH are eliminated at high temperature during the cross-linking process :

R is aliphatic or cycloaliphat

Based on micro-FTIR spectrophotometry, micro-Raman spectrophotometry, or FTIR spectrophotometry with photoacoustic detection (FTIR-PAS), the control of the condensation reaction of the hydroxyl groups either through trans-etherification* (melamines) or through reactions with the isocyanate groups, could be carried out in the different elementary layers of the clear-coat and the base-coat. Using FTIR-PAS technique, a non -destructive analysis can be used on the solid substrate (polymeric or metallic).

Photo-oxidation mechanism of melamine-cross-linked polyesters. The photo-oxidation mechanism of aliphatic and cycloaliphatic polyesters cross-linked with methoxy-methylated melamine has been recently examined in Laboratoire de Photochimie** [15]. A simplified structure of the network is presented on scheme 1, the maxima of the main IR absorption bands are indicated on :

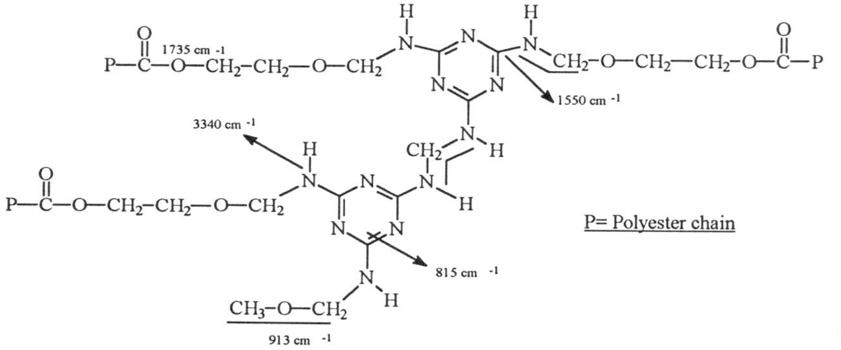

P = Polyester chain

1735 cm^{-1}, 3340 cm^{-1}, 1550 cm^{-1}, 815 cm^{-1}, 913 cm^{-1}

Scheme 1

In the experimental conditions of the artificial accelerated photo-ageing of the SEPAP 12.24 unit, the polymeric network is involved in a photo-induced oxidation, the chemical chromophoric defects absorbing the UV light. The most reactive site, in the presence of peroxy radicals formed photochemically, was observed to be the ethylene group -CH$_2$- located in the α-position of the oxygen and the nitrogen atoms :

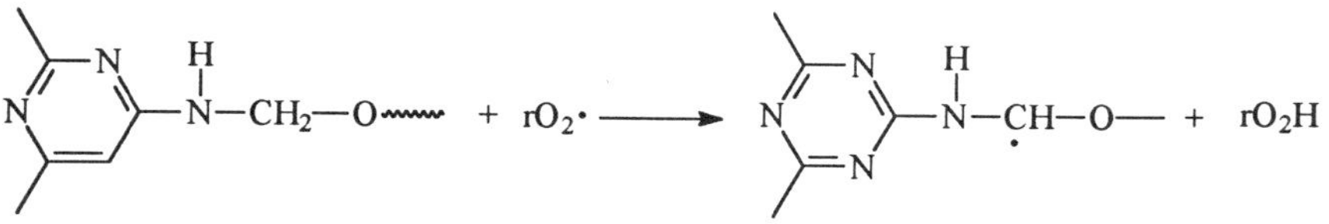

$$-N-CH_2-O\sim + rO_2^\bullet \longrightarrow -N-CH-O- + rO_2H$$

The radical-initiated oxidation mechanism is presented on scheme 2, the various intermediate photoproducts being determined by FTIR spectrophotometry (with or

---

* Meanwhile the trans-etherification proceeds in the cross-linking operation, self condensation of melamine occurs, the solid network is indeed very complex like an interpenetrated network (IPN)
** P. Delorme, PhD Doctorate, Université Blaise Pascal (Clermont-Ferrand, France), Dec. 1995

without derivatization reactions). This mechanism implies the oxidation scission of the cross-links -NH-CH$_2$-O- and a radicalar attack on the polyesters chains with the formation of acidic and anhydride groups.

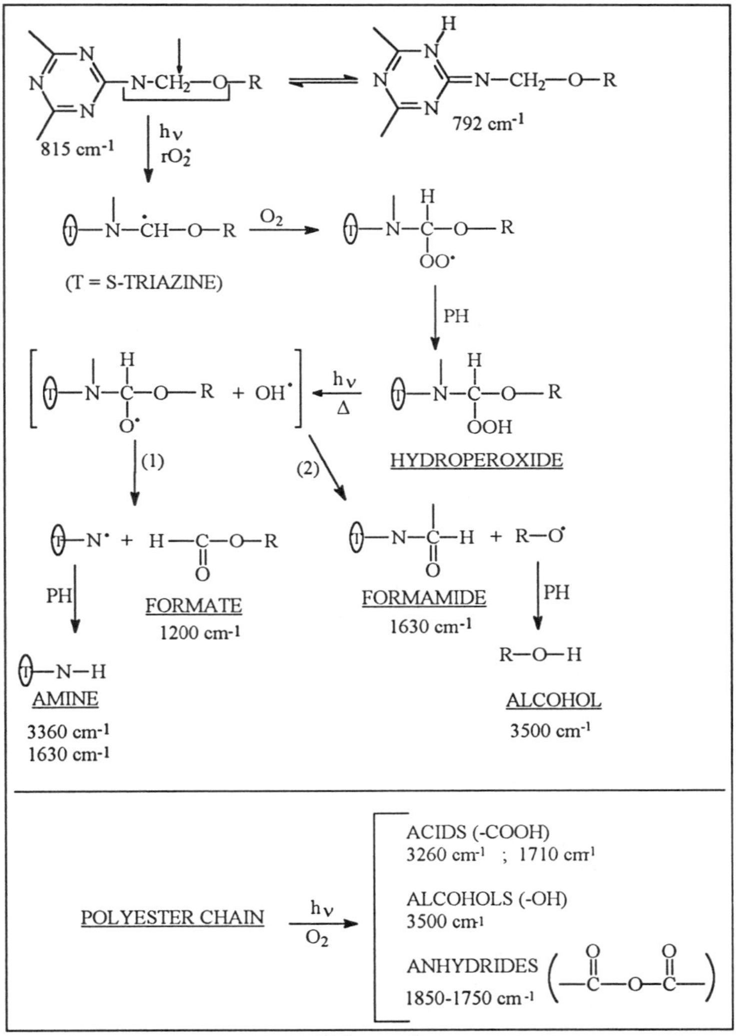

Scheme 2

As shown by the photo-oxidation mechanism, the evaluation of the network detriment could be assessed from the scission percent of the cross-links and from the accumulation of the acidic and anhydride groups. To account for some properties which are less detrimented than expected, an oxidative cross-linking has been proposed as a restoration process. Such a process, which would be more favoured in the most severe conditions, where a high accumulation of oxidized radicals provoked some recombinations, should not be considered as important in environmental conditions.

Photo-oxidation mechanism of isocyanate-cross-linked polyesters. A simplified structure of the cross-linked network could be represented on scheme 3 :

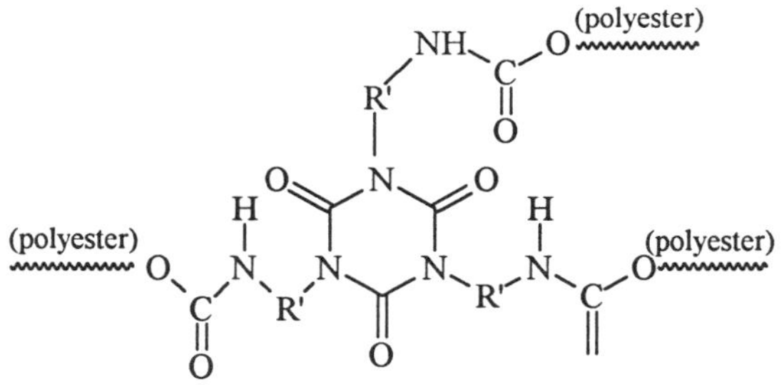

Scheme 3

where R' is an aliphatic or cycloaliphatic unit presenting a CH_2 or CH group in the α position of the urethane groups.

The mechanism of photo-oxidation again in the experimental conditions of the SEPAP 12.24 unit is presented on scheme 4 and involves both the oxidation of the urethane groups and the oxidation of the polyester chains.

As indicated in this scheme 4, acylurethanes are formed in the oxidation of urethane groups. These groups are easily hydrolysed, at room temperature, water has therefore a complementary detrimental effect since it destructs a cross-link bond [16].

It should be however emphasized that the primary degradation occurs as oxidation of the urethane groups, water is just revealing a potential detriment. In a testing protocol, when degradation is evaluated from the consumption of urethane groups, experimentation in "dry" conditions is relevant. However, if degradation is evaluated from the variation of any physical property (mechanical, or aspect ...), application of water as a chemical stress is a pre-requisite for relevancy. It should be added, that the influence of water is not conjugated with the influence of UV, heat or O_2, since water is hydrolysing the pre-formed oxidation products. A simple immersion in water following a "dry" photo-ageing test should reveal the detrimental effect of water. It is therefore shown that the understanding of the ongoing chemical phenomena makes easier any testing in laboratory conditions.

The photo-oxidation mechanism observed in accelerated artificial ageing has been observed too throughout weathering in climatic station, or in normal use conditions.

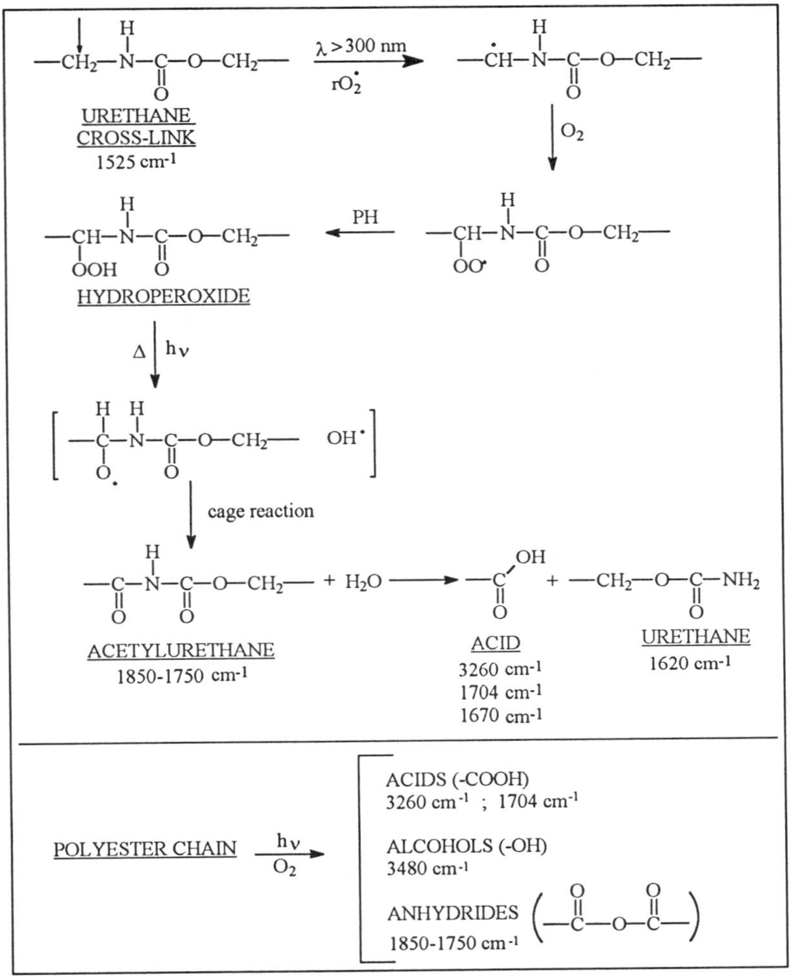

URETHANE CROSS-LINK
1525 cm⁻¹

λ > 300 nm
rO₂•

O₂

PH

HYDROPEROXIDE

Δ | hν

OH•

cage reaction

ACETYLURETHANE
1850-1750 cm⁻¹

+ H₂O →

ACID
3260 cm⁻¹
1704 cm⁻¹
1670 cm⁻¹

URETHANE
1620 cm⁻¹

POLYESTER CHAIN →(hν/O₂)

ACIDS (-COOH)
3260 cm⁻¹ ; 1704 cm⁻¹

ALCOHOLS (-OH)
3480 cm⁻¹

ANHYDRIDES
1850-1750 cm⁻¹

Scheme 4

Conclusions

Photoageing experiments carried out in laboratory conditions on coating would afford relevant data and fairly correct lifetime prediction (within the errors range of environmental lifetimes), when 3 conditions are obeyed :

- the chemical evolution mechanisms, and especially the chemical route, which is controlling the detriment of the functional physical property, should be identified ;
- the physico-chemical stresses which provoke the identified chemical evolution should be applied at levels which are not modifying the observed mechanism ;
- the detriment should be evaluated from the extent of the controlling reactions

through the accumulation of a selected critical photoproduct. In most coatings exposed in outdoor conditions, photo-oxidation is the most important degradation process. Therefore, the extent of photo-oxidation should be evaluated using vibrational spectrophotometries (IR or Raman) and measuring the formation rate of the critical products specifically in the clear-coat or in the base-coat.

It is fairly important not to consider the variations of the aspect of the samples exposed in artificial conditions more or less accelerated. As detailed in the previous sections, aspect variations result from different types of photo-reactions which are not accelerated within the same factor. Variations of aspect are extremely important in actual use conditions, variations of aspect could be very misleading in artificial testing.

References

[1]   RIVATON A., SALLET D. and LEMAIRE J., Polym. Degrad. Stab. 1986, 14, 1
[2]   LEMAIRE J., Pure and Appl. Chem. 1982, 15, 1432
[3]   LEMAIRE J., ARNAUD R., GARDETTE J.L., LACOSTE J. and SEINERA H.,
       Kunststoffe 1986, 76, 149
[4]   LEMAIRE J., ARNAUD R. and LACOSTE J., Acta Polym. 1988, 39, 27
[5]   LEMAIRE J., Chemtech 1996, 10, 42
[6]   SOMERSALL A.C. and GUILLET J.E., Polym. Stab. and Degrad., ACS Symposium
       Series, P. Klemchuk Ed. 1985, 211
[7]   GEUSKENS G., DEBIE F., KABANKA M.S. and NEDELKOS G., Polym.
       Photochem. 1984, 5, 313
[8]   MIELEWSKI D.F., BAUER D.R. and GERLOCK J.L., Polym. Degrad. Stab. 1991,
       33, 93
[9]   GERLOCK J.L. and BAUER D.R., J. Polym. Sci. Polym. Phys. 1984, 22, 447
[10]   GERLOCK J.L., MIELEWSKI D.F. and BAUER D.R., Polym. Degrad. Stab. 1988,
       20, 123
[11]   BAUER D.R. and MIELEWSKI D.F., Polym. Degrad. Stab. 1993, 40, 349
[12]   BAUER D.R., GERLOCK J.L., MIELEWSKI D.F., PAPUTA PECK M.C. and
       CARTER III R.O., Polym. Degrad. Stab. 1990, 27, 271
[13]   BAUER D.R., J. Coat. Technol. 1994, 66(57), 835
[14]   MIELEWSKI D.F., BAUER D.R. and GERLOCK J.L., Polym. Degrad. Stab. 1993,
       41, 323
[15]   DELORME P., PhD Doctorate, Université Blaise Pascal, Clermont-Ferrand
       (France), December 12. 1995
[16]   WILHELM C., PhD Doctorate, Université Blaise Pascal, Clermont-Ferrand
       (France), January 7. 1994

Chapter 17

Depth Profiling of Automotive Coating Systems on the Micrometer Scale

Karlis Adamsons[1], Lance Litty[2], Kathryn Lloyd[2], Katherine Stika[2],
Dennis Swartzfager[2], Dennis Walls[2], and Barbara Wood[2]

[1]Marshall R&D Laboratory, DuPont Company, Philadelphia, PA 19146
[2]Central Research & Development, Corporate Center for Analytical
Sciences, Experimental Station, Wilmington, DE 19898

Techniques for obtaining chemical composition and component distribution depth profiles for automotive coating systems have been identified, developed and applied. Primary reasons for this work were to determine general system composition, as well as component or additive distribution, as a function of locus (i.e., surfaces, interface/interphase regions, depth profiles) and exposure history (i.e., outdoor or accelerated exposures).

Depth profiling studies have been applied to both freshly prepared systems, as well as samples subjected to outdoor and/or accelerated weathering. *In-plane (or slab) microtomy* is demonstrated as a powerful tool for preparing samples for depth resolved analysis using techniques which otherwise would not have direct depth specificity. (1) IR-microscopy using transmission mode analysis provides general chemical composition. (2) Solvent extraction followed by HPLC chromatography gives information on the extractable materials, such as UVA's, surfactants, and degradation products. *Cross-section microtomy* is used as another technique for preparing samples for depth profiling. These cross-sections are used in a variety of techniques, including Raman-microscopy, Time-of-Flight Secondary Ion Mass Spectrometry (ToF-SIMS), optical-microscopy, and transmission electron microscopy (TEM) analyses. (1) Raman microscopy provides chemical information with high spatial resolution. Raman imaging approaches can be used to map chemical heterogeneity at surfaces or cross sections. (2) ToF-SIMS analysis provides ion-imaging capability with isotopic sensitivity to establish chemical distribution and spacially resolved durability profiles. (3) Optical microscopy gives information on component distribution and coating layer thickness. (4) TEM analysis of very thin ($< 1 \mu$) multi-coat cross-sections, together with appropriate functional group or sub-

© 1999 American Chemical Society

structure staining techniques, provides information on component distribution and coating layer thickness.

INTRODUCTION

Tools for effective surface/near-surface and interface/interphase characterization, as well as depth profiling, are critical for effective development of current and future automotive coating systems, including those targeted for original equipment manufacturer (OEM) and refinish markets [1, 2]. The detailed study of chemical composition, physical properties, mechanical performance, and appearance characteristics of freshly prepared and weathered (either outdoor or accelerated) coating systems has been our primary focus [3]. A better understanding of the overall chemistry, including network heterogeneity, component stratification or segregation, additive migration, and/or formation of degradation products, is believed to be essential for the creation and service lifetime prediction of high performance products [4, 5, 6].

Current automotive coating systems are complex in design, being both multi-component and multi-layered [7]. For example, a typical automotive coating on a metal substrate usually has four layers - Electrocoat (E-coat), Primer (PR), Basecoat (BC), and Clearcoat (CC) - each designed with specific function. Each of these layers is a complex mixture of different components, designed to impart a specific (i.e., pigment, latex, microgel, mica and/or metallic flake) distribution, and binder chemistry. As a system, they are carefully designed to provide customers with a desired set of properties and acceptable durability under a range of weather exposure conditions [3, 8, 9]. Note, however, that the ultimate customer - the vehicle owner - typically has only one major concern with regard to the coating on their vehicle: *the appearance*. The responsibility for providing the customer with what they want belongs to the supplier(s) of the different coatings and to the manufacturer applying them according to specifications. The challenge of living up to our responsibility is quite substantial; we must understand the individual components, as well as the integration of components in the architecture of the overall coating system. The enormity of the challenge has started to bring together the combined measurement technology resources of raw material suppliers, coating system developers, manufacturers, universities, and government/private laboratories [5, 10]. Work that is reported herein derives, in large part, from efforts that evolved from many of these relationships.

There are many potential contributing factors in the failure of a coating system [11, 12, 13]. The common root and basic faults associated with the failure of coating systems have been previously detailed [5]. The study of the effects of weathering exposure on the properties of coating systems is a very challenging problem, due to the number of contributing variables [14, 15, 16]. The ultimate goal of a weathering study is to identify the changes that occur in a coating system that eventually lead to systemic failure and the environmental factors that contribute [17, 18, 19, 20]. The environmental factors can be studied in a variety of ways, including natural weathering in a given climate or with the use of accelerated testing equipment which usually allows for precise control of a subset of desired exposure factors (i.e., temperature, relative humidity, and irradiation conditions). From these tests, we hope to develop protocols for predicting service lifetime [21]. However, the task of reliably predicting service lifetimes is an enormous challenge that is not yet well understood. This approach assumes that the variability of material properties, material processing, design, and application consideration is minimal, relative to the impact of environmental factors [5].

Understanding the chemical and morphological changes that accompany weathering could play an important role in understanding the relationship between weathering exposure and predicting, and ultimately improving, service lifetime. We currently are examining this approach to understand the complex interrelationships inherent to these complex problems. Some of these studies highlight the utility of multi-technique approaches in monitoring changes in binder chemistry as a function of outdoor or accelerated exposure conditions. Other studies demonstrate the utility to obtain information at a given locus in a coating system (i.e., surface/near-surface, bulk, interface/interphase, or as a depth profile) [7].

Sampling for depth profiling includes in-plane [1,2,4] and cross-sectional [7,22] microtomy techniques. In-plane microtomy is typically done with a large-scale, or slab type, microtome. The sampling in the reported work was done co-planar to the surface of the coatings. Cross-sectional microtomy is readily done by large- or small-scale microtoming techniques. TEM cross-sectional sampling is unique in that cryo-microtomy is used to generate very thin sections.

Table 1.   **Methods & Locus of Automotive Coating System Analysis.**

Analytical Depth Profiling Technique Applied (Chem/Phys)	Cross-Sectional Morphology (Physical)	Surface Analysis (Chem/Phys)	Interface Analysis (Chem/Phys)
IR-Micro [2,23,24] Chem θ	---	Chem θ	Chem Δ
Raman-Micro [7,25-28,36] Chem θ	---	Chem θ	Chem θ
ToF-SIMS [7,29-31] Chem Δ	---	Chem Σ	Chem Δ
Opt-Micro [32,33] Phys Δ	Phys Σ	Phys Σ	Phys Δ
TEM [7,34,35] Phys Δ	Phys Δ	Phys Δ	Phys Δ
Extr/HPLC Chem Δ	---	Chem Δ	Chem Δ

cutting, coating systems were potted in an acrylic or epoxy based polymer to provide mechanical support during handling. For practical considerations, in order to avoid, or minimize, contamination from the potting polymer a composition unique to the coating system layer(s) under study is chosen (i.e., epoxy for acrylic-based CC/BC bilayers). In addition, microtoming is done along the direction of the interface, thus minimizing cross-contamination. . Note that potting was not used in the more surface specific ToF-SIMS experiments due to the risk of cross-contamination. Such contamination can result in misleading data where a significant amount of the spectral information is due the potting compound. Instead, specialized cryo-ultra-microtomy techniques were employed to generate the ultrathin ($< 1 \mu$) sections required.

IR-Microscopy Analysis:

Transmission mode analyses were done on a Nicolet Magna-IR 760 FT-IR Spectrophotometer equipped with a Nicolet Nic-Plan microscope. Attenuated total reflectance (ATR) mode analyses were done on a Nicolet 20-SXC FT-IR Spectrophotometer equipped with a Spectra Tech IR-Plan microscope and an ATR-objective using a ZnSe internal reflection element.

Solvent Extraction & Chromatography Analysis:

Methylene chloride (CH_2Cl_2) was used as solvent to extract benzotriazole type ultra-violet screeners (UVAs) from in-plane (slab) microtomed sections. At typical loading levels for these additives (~1-3 % by weight), it was observed that a 1"x 1.5"x 10μ sized section extracted by 5 mL solvent resulted in UVA concentrations sufficient for easy detection and identification by HPLC methods. Solvent extraction was done for a period of at least 24 hours using a lab benchtop rocker/mixer device.

Raman Microscopy (Microprobe) Analysis:

A Renishaw Raman imaging microscope was used in depth profiling of specific coating system components (i.e., organic pigments used in BC's). This instrument consists of a conventional optical microscope integrated to a Raman spectrometer with a laser for excitation. The coating systems reported herein were studied using an argon-ion-laser-pumped Ti-sapphire laser (Coherent Inc.) for excitation. The wavelength was tuned to 780 nm, corresponding to the center frequency of the notch rejection filters available in our lab. This wavelength is presently the best compromise available to maintain a high degree of fluorescence rejection while still maintaining an adequate spectral range and sensitivity. Most of this work on automotive coatings would not be possible without the availability of near-infrared excitation sources, due to characteristic sample fluorescence. Power levels of ~1 mW were typical in these experiments. High spatial resolution analyses were performed using a high numerical aperture 100X objective, which provides a diffraction limited laser spot of approximately 1μ in diameter. These spot sizes are much smaller than can be provided by infrared microscopy, which, at best, provides a diffraction limited spot size on the order of ~10μ (with a more practical minimum spot size of ~30μ with conventional laboratory equipment). . A motorized X-Y translating stage (Prior) was used for area point mapping measurements. Raman maps were generated by taking spectra point-wise across an area of a paint film cross-section at 1 μm intervals with the use of the mapping stage which is under software control. Sample cross-sections were prepared by potting the coating systems in an epoxy matrix and microtoming, polishing, or by simply cracking the coating after immersion in liquid nitrogen. Background intensity due to fluorescence is subtracted out using spline baseline fitting routines (GRAMS/386 [TM], Galactic Industries). Peak area integration was performed on the successive Raman spectra to generate pigment concentration Raman images.

Time-of-Flight Secondary Ion Mass Spectroscopy (ToF-SIMS) Analysis:

The Time-of-Flight Secondary Ion Mass Spectrometry (ToF-SIMS) analyses were performed on a TRIFT II Spectrometer using a 15 keV pulsed (~5 nsec) and bunched (~1 nsec) Ga ion gun. The beam size is about 1 μm and the rastered area is 200x200 μm. Typically the mass resolution after Oxygen-16 (negative ion mode) is ~2000. During the 30 min. acquisition (dose ~5×10^{12} ions/cm^2) a mass spectrum and 16 mass selected ion images (256x256 pixels) can be obtained.

The experiments described herein illustrate the sensitivity and high spatial resolution of the technique which can be used to determine the in depth distributions of elements and molecular species in automotive finishes.

Optical Microscopy Analysis:

The images were obtained with a Leica DM RXA Microscope equipped with an MTI 3CCO color video camera and Kodak Digital Science 8650 PS color printer. A 200X magnification of the sample was used.

Transmission Electron Microscopy (TEM) Analysis:

The images reported in these studies were obtained using a JEOL 2000FX Transmission Electron Microscope (TEM) operated at various accelerating voltages (in the 80-200 KV range) and recorded on large format sheet film. Automotive coating systems were found to be sufficiently stable in the TEM electron beam permitting imaging of ultra-thin cryo-microtomed (<1μ thick) sections at 50-400X magnifications. TEM was successfully applied to cured films of paint containing binder components as well as organic and/or inorganic pigments, latex, metal flake and other particles.

Techniques such as TEM effectively employ ultra-thin cross-sections of both single-layer and multi-layered coatings. These cross-sections are typically prepared at a nominal thickness of 100 nm, and cut normal to the surface, using cryo-microtomy sectioning. The sectioning is done on free films that are supported (embedded) in an epoxy matrix not found in the coating system.

RESULTS & DISCUSSION

Characterization of Automotive Coating Systems:

The multi-component and multi-layered design of most of today's automotive coating systems presents many challenges in their characterization. Designers of new high performance coating systems are now beginning to appreciate the need to understand the various correlations between chemical composition, physical property, mechanical performance, and appearance. Another level of complexity is observed in component distributions in a given coating system layer, as well as component mixing at interface/interphase regions. The techniques used and results reported herein are selected to demonstrate the utility of a more holistic, multi-technique approach in study of these systems.

Component Morphology via Imaging:

Optical Microscopy Imaging:

Imaging of surfaces, revealed surfaces, or cross-sections has become a technique to routinely identify coating defects, degradation, morphology, or thickness. Lighting conditions (i.e., optics configuration, masking, polarization, and/or direction) have been adapted to optimize imaging of defects, component distribution, and interfaces. For purposes of practical thin coating or coating system handling, samples for cross-section analysis are usually supported upright by clamps and potted in an epoxy matrix. Since the majority of samples studied are acrylic-, polyester-, urethane- and/or silane-based, a standard quick drying epoxy is used for the potting. The epoxy chemistry is readily differentiated from most coatings or coating system layers. A polishing technique is then used to prepare a flat, clean surface. The potted sample is anchored on the X-Y translating stage for convenient positioning. Samples for surface or revealed surface analysis are often anchored in-plane (flat) to the stage with clips. Magnifications of 50-400X are commonly used in imaging.

Figure [1-A/F] shows an epoxy potted cross-section of an automotive coating system under dark-field illumination and back-lighting conditions. Note the easily distinguished coating layers, as well as the distribution of particles (i.e., pigment, latex) in BC and PR layers. Figure [2] shows an entire exposure series potted in chronological sequence. This allows one to monitor changes in surface morphology (i.e., roughness, cracking) and thickness changes (i.e., mass loss) in a given layer as a function of exposure time and conditions. Potting a series of sample cross-sections together allows for greater convenience, and even some degree of automation in stage positioning, during analysis.

The importance of optical microscopy is sometimes undervalued. It is often an integral part of the other experimental approaches (i.e., Raman-microscopy) and is frequently needed to identify regions for analysis and to provide sample registry.

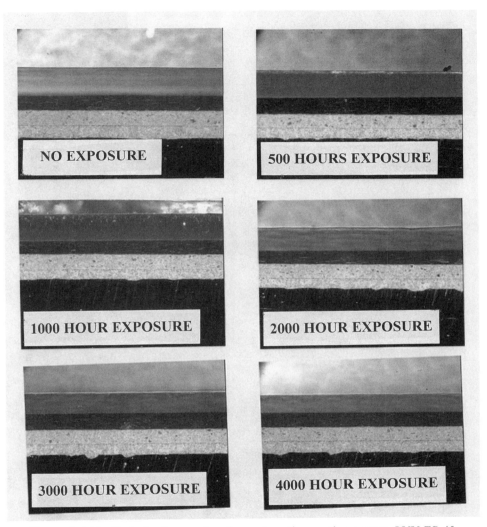

Figure 1-A/F. Optical micrographs of an automotive coating system QUV FS-40 exposure series. The images show CC (topmost layer), BC, PR, and E-coat. The upper left image is coating system without QUV exposure.

264

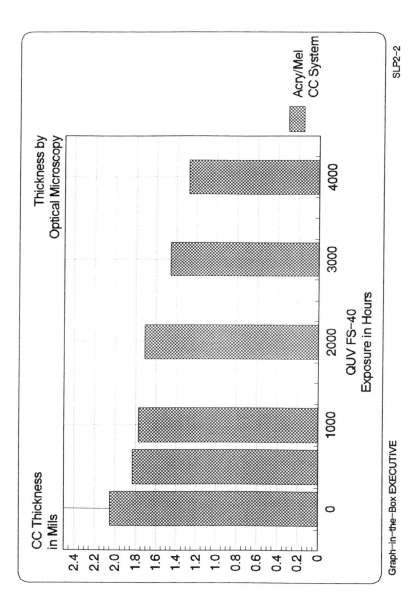

SLP2–2

Figure 2. CC thickness monitored *via* optical microscopy of automotive coating cross-sections for a QUV FS-40 exposure series. The thickness was determined over a 4000-hour period. The coating is an acrylic/melamine type system.

TEM Imaging:

Figure [3-A/B] shows a cross-section of a BC layer containing latex and several types of organic pigment, and exploded view of the CC/BC interface, respectively. The pigment particles are easily distinguished from the polymer matrix and the latex particles. Both pigment particle shape and size distribution can be determined in certain cases. Image analysis techniques can be used to obtain individual pigment (type) particle size distributions, and their distribution across the entire BC. In this example the pigment particle distributions appear to be generally uniform with minor aggregation, suggesting good mixing during application and during the cure stage.

Figure [4] shows a CC/BC/PR tri-layer. The CC/BC and BC/PR interfaces can be described as sharp and well organized. This sample has been subjected to several years of outdoor Florida exposure conditions. No degradation (i.e., local debonding) of the interface/interphase regions was apparent from the image. Several small voids were created as a result of the microtome sectioning. Each of these voids appears to be lined with mica flake aggregates, which tend to be mechanically weak regions, particularly in the context of the forces applied during microtoming. In fact, this system was later characterized as very durable in the Florida environment.

Figure [5] demonstrates imaging of the core/shell structure of latex particles used in a BC. The BC is shown after exposure to RuO_4 vapor. Selective staining of the micron-sized latex-rich regions results, where individual acrylic latex particles are well defined white circles ~70 nm in diameter. Here a cross-section of a lightly pigmented BC that has melamine removed from the binder formulation is shown. The strong staining of the latex particle shells is apparent, and thus the pigment particles are easily identified (and distinguished). The latex shell staining is due to presence of melamine there. The latex core does not stain since melamine is absent. The observed matrix staining is minimal.

Figure [6] shows a CC/BC bi-layer cross-section that has been potted in an epoxy matrix. The microtoming axis is along the direction of the interface, but relatively close to the interface. This example is used to illustrate a type of de-bonding where the fracture occurs on the BC side of the CC/BC interface. This is referred to as a cohesive (i.e. in a given coating layer), rather than adhesive (between coating layers), type of failure. An example of where an IR surface analysis technique (i.e., ATR microscopy) is used to identify the locus of this failure is detailed in an earlier publication [7]. The reason for de-bonding very close to the interface is not understood. Figure [7] shows an exploded view of this kind of cohesive BC failure. Primarily latex particles appear along the BC interface boundary, highlighted by RuO4 staining. Note that this type of failure tends to occur within 500 μ of the interface.

TEM images compliment the chemical composition, physical property, or mechanical performance information obtained by techniques such as IR-

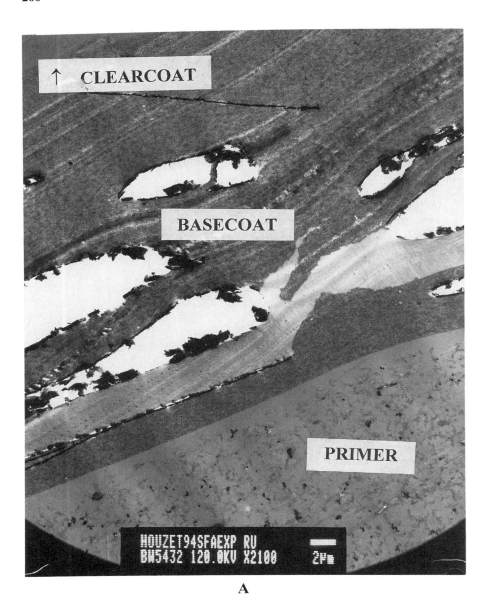

Figure 3-A/B.   TEM cross-section of an automotive coating system BC layer containing latex and several types of organic pigment, and exploded view of the CC/BC interface, respectively.

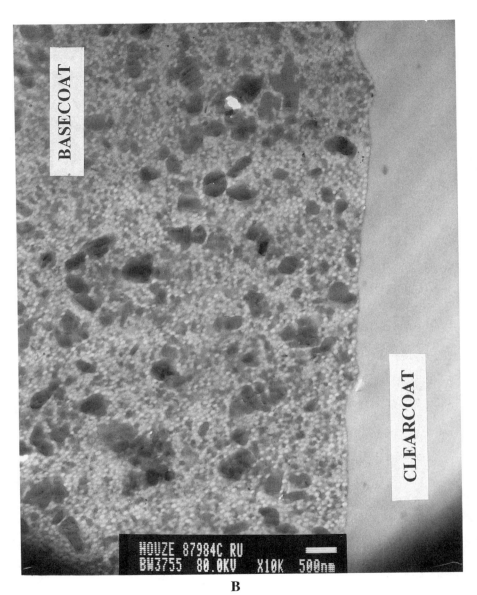

BASECOAT

CLEARCOAT

HOUZE 87984C RU
BW3755 80.0KV X10K 500nm

B

Figure 3- A/B. Continued

Figure 4.   TEM cross-section of an automotive coating system CC/BC/PR tri-layer.

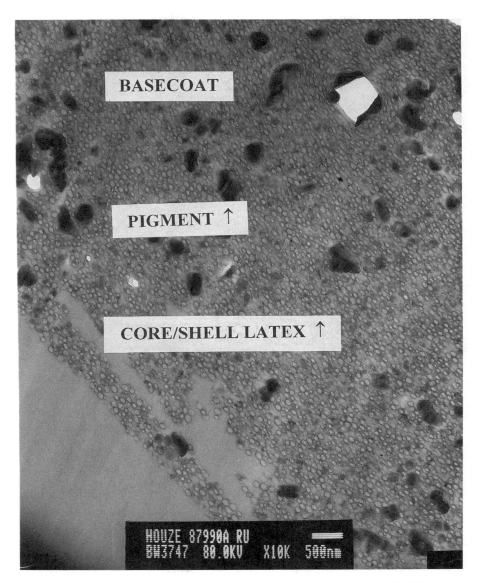

Figure 5. TEM cross-section of an automotive coating system BC; highlight of core/shell structure of latex particles.

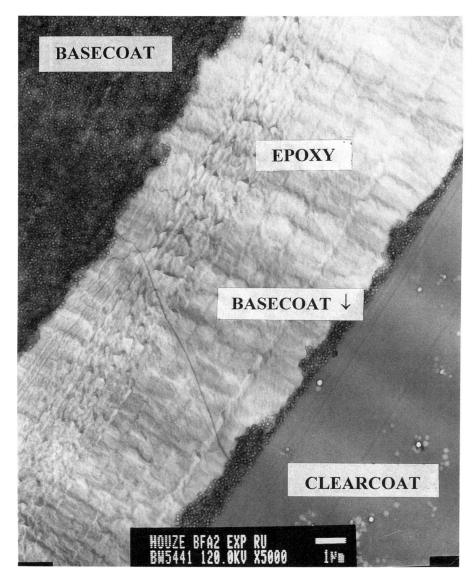

BASECOAT

EPOXY

BASECOAT ↓

CLEARCOAT

HOUZE BFA2 EXP RU
BN5441 120.0KV X5000   1µm

Figure 6.   TEM cross-section of an automotive coating system CC/BC bi-layer; highlight of type of de-bonding wherein a fracture occurs on the BC side of the CC/BC interface.

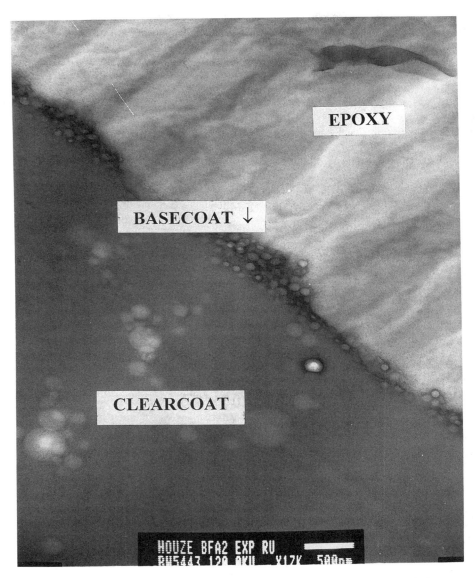

Figure 7. TEM cross-section of an automotive coating system de-bonded CC layer; highlight of latex and pigment particles along CC/BC interface showing cohesive failure in the BC layer.

microscopy, Raman-microprobe, ToF-SIMS, laser desorption mass spectrometry (LD/FTMS), ESCA, DSC, micro-hardness, et. al.. The utility of a number of these techniques will be discussed elsewhere in this manuscript. As additional multi-technique studies are completed on coating systems, inter-property correlations will become more apparent. In the future it will be possible to chemically and morphologically define a coating system to give an end-user (i.e., customer) a desired appearance and durability.

The distribution and concentration of particles as revealed by TEM images provides insight into expected physical properties, mechanical performance, and appearance of a given coating system. Factors known to influence appearance, which are of primary importance to the automotive vehicle owner, include the pigment particle loading, shape, size, color, and orientation in the vicinity of the CC/BC interface. During coating system development, comparisons of TEM images can effectively reveal the morphological consequences of changing the solvent mixture, matrix polymer blend, or in using additives such as pigment dispersants, adhesion promoters, leveling agents, UV absorbers (UVA's), and hindered amine light stabilizers (HALS).

Note that although optical microscopy techniques, such as reflected light imaging of polished layers viewed in cross-section, are useful for defect screening and identification/ characterization of these coating systems, TEM techniques offer better spatial resolution. Sub-micron particles, domains, and the detailed structure of interface/interphase regions can be easily imaged without approaching the spatial resolution limit, which is below 0.2 nm for most modern TEM instruments.

Monitoring particle distribution in a polymer matrix is rapidly becoming a routine analysis in our laboratory. Pigment, latex, non-aqueous dispersion (NAD), microgel, silica, mica and metal flake are commonly observed particles. This type of imaging is often done on cross-sections without need for chemical staining (or tagging). On occasion, however, the need exists for differentially imaging particles possessing unique chemistry (i.e., pigment vs latex), core/shell structure (i.e., some types of latex and microgel), surface treatments, or polymer domains, stratification, segregation, and component mixing. This is done with appropriate selection of heavy metal stains (i.e., RuO_4, OsO_4) which facilitate differentiation based on chemical functionality. Monitoring interface/interphase regions between adjacent coating layers is also possible. This permits a high-resolution inspection of the boundary morphology. The CC/BC bilayers, and associated interface/interphase regions, are studied regularly for purposes of determining the sharpness of the particle rich boundary, excessive concentration of particles near/at the boundary, and mixing of matrix components between adjacent CC and BC layers.

Component Distribution via Spectroscopic & Spectrometric Techniques:

IR Analyses:

Slab microtomy was used to prepare in-plane sections of the CC layer from an automotive coating system. IR transmission mode analysis was used on each of these slab sections to obtain a chemical composition depth profile. Figure [8] highlights a Hialeah, Florida (USA), exposure study. Here an unexposed control of a CC (over a black BC) is compared to 2, 3, and 4 years outdoor exposure. The ratio of cross-linker to backbone polymer (i.e., melamine to acrylic carbonyl) is shown versus exposure time and as a function of depth. Figure [9] highlights an analogous Australia exposure study. Here the same unexposed control is compared to 1, 2, 3, and 4 years outdoor exposure. The same ratio is plotted. This study permits several comparisons for the coating system exposed to different climates. (1) The ratio was effectively constant, regardless of the depth, for a given climate and exposure time.   (2) The Australian climate accelerated the decrease in the ratio versus the Florida climate.   (3) The kinetics of the change in this ratio was readily determined.

Raman Analyses:

Raman spectroscopy has become an increasingly important tool for polymer and coating analyses in the past 10 years, due to rapid development of a number of supporting technologies that has led to the development of a new generation of highly efficient and sensitive sampling techniques. These new instruments are easily configured with optical microscopes to produce Raman microscopes with unprecedented sensitivity. Raman microscopy is a technique that combines the high spatial resolution inherent to optical microscopy with the inherent chemical specificity of Raman spectroscopy. Chemical specificity is derived from the fact that Raman spectroscopy is a vibrational spectroscopic technique and is thus sensitive to the same type of molecular level information about composition, morphology, and micro-structural organization that is more commonly associated with infrared spectroscopy. Spatial resolution can be very high, since the laser can easily be focused on spots (sampling areas) as small as 1 μ in diameter.

Raman microscopy can be used in a number of ways in the characterization of coating systems before and after weathering. Microprobe measurements, where specific locations on or within a coating layer are identified with optical microscopy and subsequently analyzed by overlapping that location with the excitation laser focus, are the most basic measurements available. Defects or sites that have obviously degraded as a result of weathering that can be located visually can be easily analyzed with this approach. Using confocal imaging approaches, even sub-surface defects and interfaces can be analyzed with minimal contribution from the matrix or over-layers, since the confocal arrangement adds depth selectivity to the measurement. Confocal imaging approaches provide one strategy for depth profiling changes in composition as a function of depth resulting from weathering exposure in clear (i.e., particle free or optically transparent) coatings, such as the CC used in virtually all modern automotive finish systems.

Raman microscopy also has tremendous potential as a tool for mapping and imaging chemical heterogeneity in material systems such as automotive coating systems. In many systems, it is not enough to obtain chemical information from

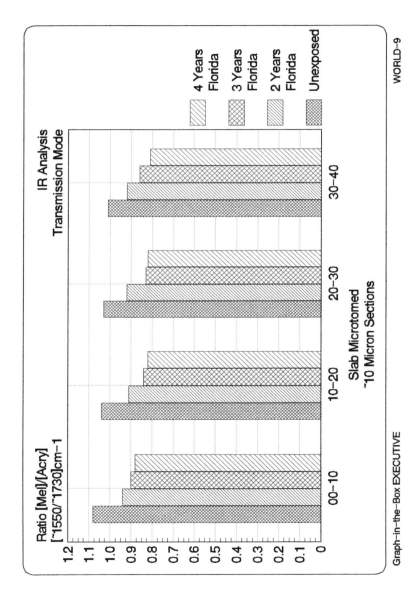

Figure 8. IR transmission-mode analysis of an acrylic/melamine automotive coating system exposed in Hialeah, Florida, as function of time and CC depth; Control (un-exposed) and 2, 3, and 4 year outdoor exposures.

275

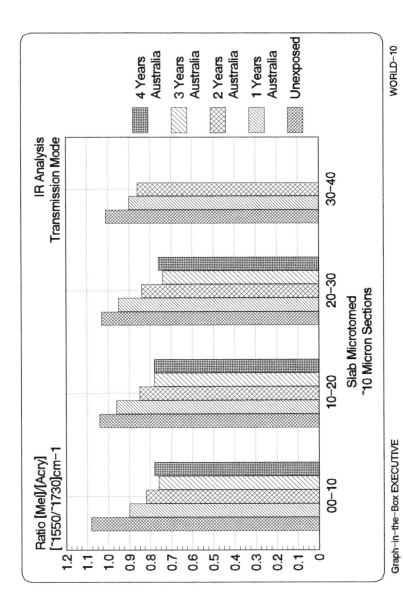

Figure 9. IR transmission-mode analysis of acrylic/melamine automotive coating system (same as studied in Fig. 8) exposed in Australia (outside Melbourne), as function of time and CC depth; Control (un-exposed) and 1, 2, 3, and 4 year outdoor exposures.

discrete spots on a sample to learn more about the relationship of that measurement to the overall material properties; the overall architecture and the specific assembly of the components that comprise the sample must be known to fully understand the system. Imaging approaches capable of providing direct chemical-specific image contrast are needed to fill this gap. These types of measurements can be used as either a technique for line profiling composition as a function of distance in one dimension or as a tool for imaging chemical heterogeneity on a surface in two dimensions (and even three dimensions with confocal imaging approaches). The most common applications rely on line or area mapping approaches, were spectra are collected as a function of position in one or two dimensions using a computer controlled XY (translation) mapping stage to position different locations of the sample in the laser focus. The step size and area mapped, as well as the spectral range collected from each scan area, are determined to fit the needs of the measurement. Spatial resolution in the micron range is easily attainable. Full spectral information is available at each pixel of the image for subsequent image processing. With all of this information, it is possible to extract images of the sample which reveal image contrast based on differences in chemistry. This is a highly desired attribute of these measurements and can be exploited in many ways, including mapping of surface chemistry and depth profiling by mapping material heterogeneity in cross section.

A specific application of Raman mapping analysis to issues of interest to weathering and durability issues in automotive coatings is in their utility for depth profiling material heterogeneity. Cross sections of the coating of interest are prepared by microtomy after delaminating the coating from the substrate. Sometimes, a cross section of sufficient quality for mapping can be obtained by cracking the delaminated coating after cooling with liquid nitrogen. One specific issue of keen interest is the distribution of pigments in automotive BC's as a function of depth in the coating. In formulating and applying a BC to a substrate, it is hoped that the pigments are evenly dispersed throughout the coating thickness. Spatial inhomogeneities in pigment concentration could lead to potential mechanical and durability performance issues. We have found that Raman spectroscopy can be an effective tool for mapping pigment heterogeneity in BC systems.

Figure [10-A] serves to demonstrate how a Raman image can be constructed and shows an example of the type of cross-sectional chemical heterogeneity that can be observed in these systems. Figure [10-B] shows a typical spectrum obtained from a standard red BC layer. The strong Raman bands observed are associated with one of the pigment components in this coating. We can use this spectral information to track the relative concentration of this pigment component in a cross-sectional slice of such a coating. We will focus on the integrated intensity of the band near 1345 cm-1 to follow the relative amount of this pigment present at each point in the Raman image. The Raman image shown on the top of this figure was created mapping a cross-sectional sample across the entire thickness of the BC layer over a length of 36 microns. In this image, the CC layer would be on the left side and the PR layer would be on the right. The CC/BC and the BC/PR interfaces

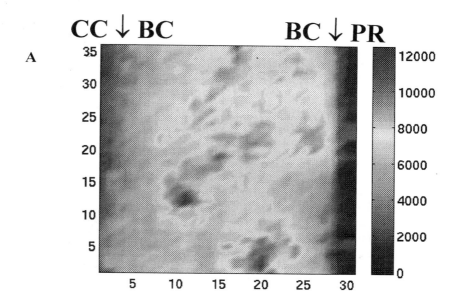

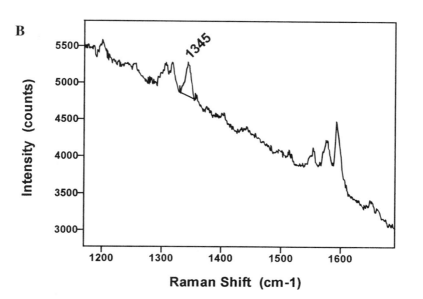

Figure 10-A/B.   [**A**]. Raman image (or map) of an automotive coating system CC/BC/PR tri-layer; Example of cross-sectional chemical heterogeneity based on mapping one of several red pigments used as colorant mixture to provide BC colour. [**B**]. Typical spectrum obtained from a standard red BC layer; the peak at 1345 cm[-1] is associated with one of the red pigments.

would be at approximately 4 and 28 microns on the X-axis of this figure. The intensity at each pixel in the image is associated with the baseline-corrected integrated intensity over the spectral range encompassing the full width of the pigment peak at 1345 cm-1. The intensity at each point in the image can be read from the colorbar displayed to the left of the image. Note that only in the central part of the image (in the vacinity of the CC) is appreciable intensity observed; very low integrated intensities are obtained in both the CC and PR films because there is no comparable band at this same location in their spectra.

The Raman image in Figure [10-A] clearly shows that this particular pigment is not uniformly distributed throughout the BC. There is evidence for local regions of abnormally high pigment level that may suggest pigment particle aggregation (i.e., poor dispersion in pigment). Pigment rich regions such as these could correspond to regions lacking the appropriate levels of binder needed to provide mechanical integrity. Such areas are of concern because they could indicate regions of this coating that are prone to premature failure.

The information obtained in the Raman image can be used in a number of different ways. By averaging the information along the plane of the substrate, the information in this image can be compressed to provide an average depth profile of the composition in this region. Figure [11] shows an example of using an image to create an average depth profile of the pigment mapped in Figure [10-A]. This figure reveals additional details concerning the pigment heterogeneity of these systems. Going from the CC layer (left side) of the profile, the level of pigment rises at a steady, but not sharp rate. It reaches a uniform level at approximately 10 microns on this figure, which corresponds to about 6 microns into this layer. A fairly uniform concentration is observed for the next 15 microns into the coating. Note, though, as one approaches the PR layer, the pigment concentration sharply drops. The differences in fall-off in pigment concentration at the two interfaces clearly indicate that the BC layer interacts with the PR and CC layers differently. Work is continuing to try and understand this phenomena. One likely hypothesis is that the CC/BC interface is smeared due to significant interpenetration of components from these two layers when the CC is applied to the BC. The degree of interpenetration is an extremely important factor that almost certainly has important implications on the interlayer adhesion performance in these systems.

Potentially much more information is available from these Raman imaging approaches. This example dealt with the mapping of one single pigment in a BC layer. However, all commercial BC layers typically incorporate colorant systems incorporating mixtures of different pigments. The chemical specificity of Raman spectroscopy offers the potential of mapping the distribution of each component in a colorant package selectively. Thus, for example, we can use this approach to look for evidence of demixing of the different components of the colorant package. Also, these approaches are general and can be extended to the examination of chemical heterogeneity in other regions of these coating systems. One area of current interest is in the use of Raman mapping techniques to study the effects of accelerated weathering on the localized structure of the CC layers in these systems.

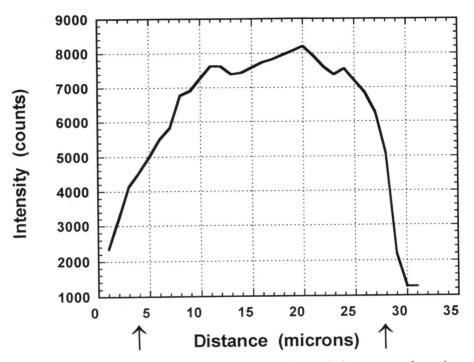

Figure 11. Raman average depth profile derived from red pigment map shown in Fig. 10. The arrow pointing up at ~4 μ is the approximate CC/BC boundary. The arrow pointing up at ~28 μ is the approximate BC/PR boundary.

ToF-SIMS Analyses:

Shown in Figure [12] are four of the mass selected ion images obtained from a cross-section of an automotive coating system after exposure to a nitrogen atmosphere saturated with ^{18}O labeled water (H_2O^{18}) at 60°C for 1200 hours. The CC is at the bottom of the multi-layered cross-section in each case. This is followed by the BC, which is clearly delineated by the $^{35}Cl^-$ ion image (bottom-left). There are two chlorinated organic pigments in the BC, accounting for this signal. The bright band in the $^{16}O^-$ ion image (top-left) delineates the PR layer. The ion image at mass 80 (not shown) is due to SO_3^-, a fragment ion from dodecylbenzenesulfonic acid (DDBSA). The ion image at mass 18 (top-right) clearly shows substantial incorporation of ^{18}O in the CC, as well as the PR and E-coat, but much less is observed in the BC. In the case of the CC, the ^{18}O uptake is believed to be the result of a reaction with unreacted silanol cross-linker. For the PR and E-coat the ^{18}O uptake is primarily the result of hydrolytic exchange with hydroxyl groups on the surfaces of the inorganic fillers.

In order to quantify the depth distribution of the various species the 2-dimensional images are converted to line scans by summing the intensities of all the pixels in every line parallel to the film interfaces. The line scans produced in this fashion are shown in Figure [13]: the distribution labeled "P" is the (M-1)$^-$ species for one of the organic pigment at mass 355; and the curve labeled DDBSA is the (M-1)$^-$ species for DDBSA at mass 325.

Shown in Figure [14] are line scans obtained in similar fashion from the same paint film after 1500 hours exposure to a filtered Xenon light source (simulated sunlight) in an $^{18}O_2$ atmosphere. This particular paint film was formulated with ultraviolet light absorbers (UVA's) and hindered amine light absorbers (HALS). Here the ^{18}O uptake is primarily the result of photo-oxidation, but some hydrolytic exchange with the H_2O^{18} produced during the photolysis may also be possible.

Precise quantitative results can generally be obtained during isotopic enrichment experiments (i.e., ^{18}O uptake) by proper manipulation of the pertinent line scans (e.g., referencing the ^{18}O line scan to the ^{16}O line scan) if there are no matrix effects (or yield differences). Such effects are expected in the primer layer where a significant fraction of the oxygen signal is due to the inorganic fillers. Caution should be exercised when interpreting line scan data, which cannot be unambiguously referenced.

Line scans derived from ToF-SIMS mass selected images obtained from cross-sectioned films (using a 90° cut) can provide depth profiles with a resolution of ~1 μ while maintaining good mass resolution. It is possible to slightly improve the resolution with oblique angle cuts.

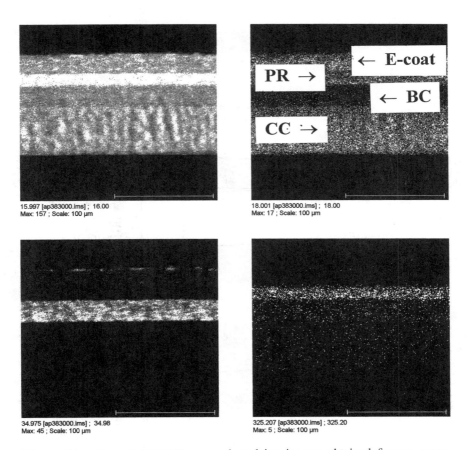

15.997 [ap383000.ims] ; 16.00
Max: 157 ; Scale: 100 μm

18.001 [ap383000.ims] ; 18.00
Max: 17 ; Scale: 100 μm

34.975 [ap383000.ims] ; 34.98
Max: 45 ; Scale: 100 μm

325.207 [ap383000.ims] ; 325.20
Max: 5 ; Scale: 100 μm

Figure 12. Four ToF-SIMS mass selected ion images obtained from a cross-section of an automotive coating system after exposure to nitrogen atmosphere saturated with H2O^{18} at 60°C for 1200 hours; Mass 16 corresponds to ^{16}O, mass 18 corresponds to ^{18}O, mass ~35 corresponds to ^{35}Cl, and mass 325 corresponds to DDBSA.

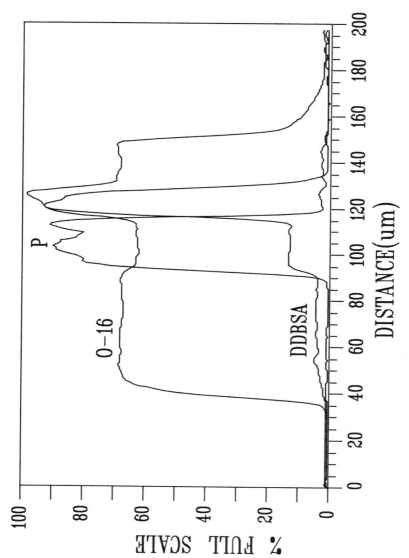

Figure 13. ToF-SIMS line scans determined from mass selected ion images obtained from a cross-section of an automotive coating system; line scans of ^{16}O, Red pigment, and DDBSA are plotted.

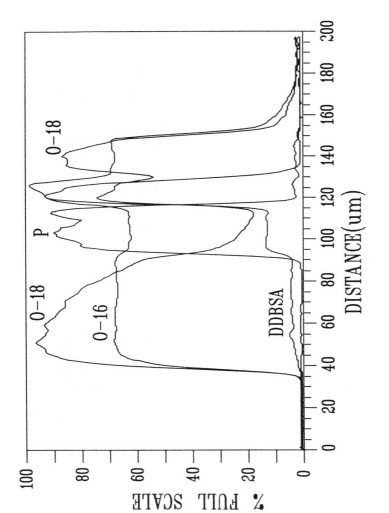

Figure 14. ToF-SIMS line scans determined from mass selected ion images obtained from a cross-section of an automotive coating system (same as used in Fig. 13) after 1500 hours exposure to a boro-silicate filtered Xenon light source (simulated sunlight) in an $^{18}O_2$ a atmosphere; line scans of ^{16}O, ^{18}O, Red pigment, and DDBSA are plotted.

Component Distribution via Slab Extraction & Chromatographic Techniques:

CH₂Cl₂ Solvent Extraction / HPLC Analyses:

Figure [15] shows the benzotriazole UVA depth profile (histogram) for an acrylic/melamine type CC that was obtained with the solvent extraction technique. The CC had experienced several years of outdoor exposure in the Denver, Colorado, area. The plot compares the observed UVA loss from two exposure conditions, "under bra" (i.e., covered and not subject to direct sunlight) and "no bra" (i.e., horizontal hood area exposed to direct sunlight). The primary difference between these two environments is the UV/VIS radiation exposure; it is assumed that the humidity, direct wetting, temperature, and atmospheric pollution factors associated with the exposures are very similar. The weight percent of this UVA in CC is ~1% when newly formulated. The depth profiling done "under bra" indicates a nearly uniform UVA loss throughout the CC, with a slight increase after ~30 μ depth. A loss of ~17-19% is observed, relatively uniform across the CC. The depth profiling done under "no bra" conditions results in a very different UVA distribution. A loss of ~60% UVA is observed in the first 10 μ section, gradually going to ~40% loss after 20 μ depth. The gradient observed suggests that sunlight have a significant impact on UVA depletion in these coatings. The potential modes of UVA loss, either via migration to the surface (and thus subject to removal by rain or dew) or into underlying layers (i.e., BC, PR), is currently under investigation. Published studies [1, 2, 4] suggest that the UVA's can migrate into underlying pigmented layers, wherein they are adsorbed onto pigment surfaces.

SUMMARY

Various combinations of sampling and analysis techniques have proven to be very useful in elucidating chemical composition and morphological detail as a function of depth in automotive coating systems. Surface and interface/interphase analysis, as well as depth profiling, can now be done routinely on both simple monolayer coatings and complex multi-component, multi-layered coating systems. Of particular significance is the ability to study coating systems subjected to outdoor or accelerated exposures to assist in identifying degradation pathways, degradation product concentrations, component distribution changes, and associated kinetics. This is expected to be invaluable in service lifetime predictions. Also, it is our belief that in future coating product design this level of knowledge will be critical to effectively incorporate the desired chemical composition, physical properties, mechanical performance, and appearance characteristics.

ACKNOWLEDGEMENTS

The authors thank the following scientists for assisting in sample preparation and experiments: Jim Halpin and Lenny Abbott in IR-microscopy, Scott Peacock in IR-microscopy and Raman-microscopy, Wendy Justison in ToF-SIMS, John McLaughlin in optical microscopy, and William Wright for TEM studies. Eric Houze, Ray Polovich and Bob Barsotti are acknowledged for providing some of the automotive coating systems studied.

285

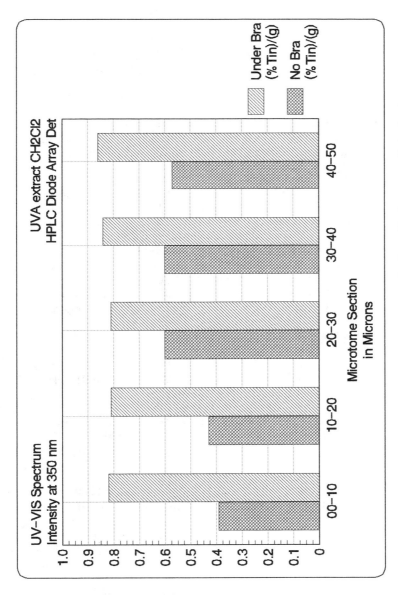

Figure 15. Benzotriazole UVA-screener depth profile for an acrylic/melamine type CC that was obtained with the solvent (CH₂Cl₂) extraction technique; CC exposed for several years in Denver, Colorado, area; Comparison of "Under Bra" and "No Bra" exposure conditions.

REFERENCES

1. Haacke, G, Book of Abstracts, 215th ACS National Meeting, Dallas, March 29-April 2 (1998), PMSE-195 Publisher: American Chemical Society, Washington, D.C.
2. Haacke, G; Brinen, J.S.; Larkin, P.J., J. Coat. Technol. (1995), **67** (843), 29-34.
3. Adamsons, K.; Blackman, G.; Gregorovich, B.; Lin, L.; Matheson, R.R.; XXIII-rd International Conference in Organic Coatings, Proceedings 23, pg. 151, Athens, Greece (1997).
4. Haacke, G; Andrawes, F.F.; Campbell, B.H.; J. Coat. Technol. (1996), **68** (855), 57.
5. Martin, J.W.; Saunders, S.C.; Floyd, F.L.; Wineburg, J.P.; NIST Building Science Series **172**, Methodologies for Predicting the Service Lives of Coating Systems, Building & Fire Research Lab, National Institute of Standards and Technology (1994).
6. Dickie, R.A.; J. Coat. Technol., **66**, No. 834, pg. 29 (1994).
7. Adamsons, K.; Lloyd, K.; Stika, K.; Swartzfager, D., Walls, D.; Wood, B.; Characterization of Multi-layered Automotive Paint Systems Including Depth Profiling and Interface Analysis, *Interfacial Aspects of Multicomponent Polymer Materials*, Ed. Lohse et. Al., Plenum Press, New York (1997).
8. Bauer, D.R.; Poly. Degrad. and Stab., **48**:2, pg. 259 (1995).
9. Courter, J.L.; 23-rd Annual International Waterborne, High-Solids and Powder Coatings Symposium, Feb 14-16, New Orleans, LA (1996).
10. Gerlock, J.L.; Smith, C.A.; Nunez, E.M., Cooper, V.A.; Liscombe, P.; Cummings, D.R.; Proceedings of the **36**[th] Annual Technical Symposium, Cleveland, OH, May 18 (1993).
11. Pospisil, J.; Adv. Chem. Ser., **249**, pg. 271 (1996).
12. Crouch, B.A.; Am. Soc. Mech. Eng., **46**, pg. 339 (1993).
13. Magaino, S.; Fukazawa, Y.; Uchida, H.; Kawaguchi, A.; Yamazaki, R.; Ind. Res. Inst. Kanagawa Prefect., Yokohama, 236, Japan, Kenkyu Hokoku – Kanagawa-ken Kogyo Shikensho, 65, pg. **47** (1994).
14. Tanabe, H.; Kagaku to Kogyo (Osaka), **68** (11), Pg. 558 (1994).
15. Perera, D.Y.; Oosterbroek, M.; J. Coat. Technol., **66** (833), pg. 83 (1994).
16. Schutyser, P.; Perera, D.Y.; Ind. Lackierbetr, **60** (11), pg. 382 (1992).
17. Bauer, D.R.; Gerlock, J.L.; Mielewski, D.F.; Peck, M.C.; Carter, R.O.; Ind. Eng. Chem. Res., **30** (11), pg. 2482 (1991).
18. Bauer, D.R.; Polym. Degrad. Stab., **48** (2), pg. 259 (1995).
19. Bauer, D.R.; Mielewski, D.F.; Polym. Degrad. Stab., **40** (3), pg. 349 (1993).
20. Bauer, D.R.; Mielewski, D.F.; Gerlock, J.L.; Polym. Degrad. Stab. **38** (1), pg. 57 (1992).
21. Braun, J.H.; Cobranchi, D.P.; J. Coat. Technol. **67** (851), pg. 55 (1995).
22. Adamsons, K.; Gregorovich, B.; Lin, L.; McGonigal, P.; Blackman, G.; Lloyd, K.; Stika, K.; Swartzfager, D.; Walls, D.; Proceedings of Automotive Finishing '96, SME, May 1-2, Cobo Conference Hall, Detroit, MI (1996).
23. Nishioka, T.; Teramae, N.; Bunseki Kagaku, **40** (11), pg. 723 (1991).
24. Nishioka, T.; Nishikawa, T.; Teramae, N.; Kobunshi Ronbunshu, **46** (12), pg. 801 (1989).

25. Katon, J.E., Review, Vibr. Spectros., **7**, pg. 201 (1994).

26. Tabaksblat, R.; Meier, R.J.; Kip, B.J.; Appl. Spectrosc., **46**, 6 (1992).

27. Hajatdoost, S.; Yarwood, J.; Appl. Spectrosc., **50**, 5 (1996).

28. Williams, K.P.J.; Pitt, G.D.; Smith, B.J.E.; Whitley, A.; Batchelder, D.N.; Hayward, I.P.; Raman Spectrosc., 25, pg. 131 (1994).

29. Briggs, D.; Brown, A.; Vickerman, J.C.; *Handbook of Static Secondary Ion Mass Spectrometry*, John Wiley & Sons, Chichester (1989).

30. Benninghoven, A.; Rudenauer, F.G.; Werner, H.W.; *Secondary Ion Mass Spectrometry*, John Wiley & Sons, New York (1987).

31. Gerlock, J.L.; Prater, T.J.; Kaberline, S.L.; deVries, J.E.; Poly. Degrad. and Stab., **47**, pg. 405 (1995).

32. Hoenigman, J.R.; NIST Spec. Publ. **801**, pg. 350 (1990).

33. Bach, H.; Proc. SPIE-Int. Soc. Opt. Eng., **381**, pg. 113 (1983).

34. Lancin, M.; Le Strat, E.; Miloche, M.; Int. J. Mass Spectrom. Ion Processes, **163** (1,2), pg. 69 (1997).

35. Rickerby, D.G.; Friesen, T.; Proc. Int. Metallogr. Conf., pg. 379 (1996).

36. Meier, R.J.; Kip, B.J.; Microbeam Analysis, **3**, pg. 61 (1994).

Chapter 18

Interfacial Composition of Metallized Polymer Materials after Accelerated Weathering

D. E. King and G. J. Jorgensen

National Renewable Energy Laboratory, 1617 Cole Boulevard, Golden, CO 80401

Correlations between changes in the macroscopic performance and in the microscopic properties experienced by weathered materials can be used to identify degradation mechanisms and suggest new formulations that exhibit improved durability. Of particular interest is the relationship between optical reflectance and interfacial composition of metallized polymer solar mirrors. A sample test matrix of silvered polymethyl methacrylate (PMMA) mirrors was prepared to systematically study changes in interfacial composition with accelerated exposure and how the chemical changes influence the reflectance of the mirrors. The samples were subjected to accelerated exposure testing in a XENOTEST 1200 LM environmental chamber. Spectral hemispherical reflectance was measured as a function of exposure time, and selected samples were removed at various exposure times to allow surface analysis of the silver/PMMA interface.

There is a large potential market for lightweight, highly reflective mirrors for use in solar energy concentrating systems. Metallized polymer constructions can provide low-cost, highly reliable reflectors. To be cost effective for commercial power production, the solar-weighted specular reflectance of the mirrors must exceed 90%. The reflective layer of choice for this application is silver (1). Ideal polymer reflectors must remain stable for more than 20 years in an outdoor service environment, be easily cleaned, and require little maintenance. Commercially available silvered polymethyl methacrylate (PMMA) films (in which, for example, PMMA is used as the superstrate) combine excellent outdoor weatherability with high transparency and UV stability (2). Understanding the correlation between accelerated weathering and real time in-service environmental exposure will allow accurate prediction of service lifetime for these materials. The determination of failure mechanisms as a function of accelerated weathering

© 1999 American Chemical Society

is a starting point in understanding real time field failures and can help direct future reflector research efforts. Research to improve the performance of this class of reflector materials has been a continuing priority within the solar thermal electric industry and at NREL (*3*). This study was initiated to elucidate the mechanism of loss in reflectance at the silver/polymer interface of two sample constructions. An extensive review of possible reactions and failure mechanisms at the silver/polymer interface is provided in (*4*).

A variety of mirror materials currently being tested include two different PMMA constructions: construction I (PMMA/Ag/Adhesive), and construction II (PMMA/Ag/Cu/Adhesive), shown in Figures 1 and 2, respectively.

Construction I

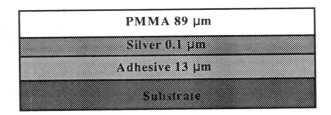

Figure 1. Mirror construction I is produced by evaporating 0.1 mm of silver onto PMMA sheet. The metallized film is then coated with an acrylic adhesive. By removing a release liner the film can be laminated to any surface producing a highly specular mirror.

The transparent PMMA typically has up to 2 wt% UV absorbers that effectively screen out most high energy photons. However, because silver is transparent to UV light around 320 nm, the addition of a metallic copper layer behind the silver further reduces the UV flux incident on the unstabilized acrylic adhesive, thus providing further protection against possible light-induced reactions in the adhesive layer that could have deleterious effects on the reflective layer.

These different constructions can fail during real-world deployment in two different ways: the first type of failure is corrosion of the metallic silver, which typically results in a steady loss of reflectance as a function of exposure time; and the second failure is a catastrophic delamination that occurs at the silver/polymer interface upon exposure to excessive moisture. Construction II is much more resistant to delamination failure than construction I. During accelerated weathering, both constructions fail because of loss in reflectance. It is possible that for the copper backed mirror (construction II), copper metal diffuses to the polymer interface, resulting in improved metal/polymer adhesion at the reflective interface.

Construction II

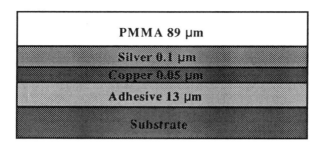

Figure 2. Mirror construction II is similar to construction I except that 0.05 mm of copper is evaporated onto the silver before the acrylic adhesive is applied to the metallized film.

This behavior is important as it significantly reduces delamination failure of the material in the field. Construction II also has an extended effective lifetime relative to construction I, as shown in Figure 3, where the solar weighted hemispherical reflectance remains above 90% for nine weeks longer in the accelerated exposure test than construction I. The two constructions have been found to exhibit different corrosion failure modes. The copper backed material tends to be highly delamination resistant and displays increased lifetime, but after a time the mirror does fail. To produce an improved reflector, it is important to understand why construction II exhibits an increased lifetime and what the ultimate failure mechanisms are for both constructions.

A "corrosion" failure mode of the copper backed material may involve the diffusion of copper metal through the silver reflective layer and subsequent buildup at the silver/polymer interface as a function of accelerated exposure. This copper buildup may then contribute to the loss in reflectance of the mirror. In addition, the copper metal may then react with the oxygen functionality of the PMMA, or perhaps simply oxidize in the presence of water and oxygen that diffuse through the PMMA. The copper oxides formed would then further contribute to the drop in performance of the mirror.

An increase in the sulfur concentration was found, as a function of weathering, at the silver/PMMA interface in construction I (NREL internal reports, 1989-1997). It is well known that silver metal does not oxidize in air, but sulfur compounds will react with and corrode silver surfaces (4-6). The origin of the sulfur and the precise nature of the chemical interaction at the reflective interface is not well understood. It is suspected that sulfur species migrate to the silver/PMMA interface from the PMMA, the bulk adhesive, or both. The metallic copper layer in construction II may serve as a diffusion barrier or getter and slow the transport of corrosive sulfur species to the reflective silver interface. This may explain the increased lifetime of construction II mirrors. To correlate the observed optical degradation and interfacial composition with accelerated weathering, X-ray photoelectron spectroscopy (XPS) was used to determine the chemical composition, including copper and sulfur concentrations, at the reflective interface as a function of exposure time.

Experimental

Two sample sets of 50 samples (44.5 mm x 66.7 mm coupons) each consisting of the two different reflector constructions were prepared, optically characterized, and introduced into an accelerated weathering chamber. The XENO chamber uses a filtered xenon arc light source to closely reproduce a terrestrial air-mass 1.5 solar spectrum at an intensity of about two suns. Samples were exposed at 60°C and 80% relative humidity (RH). Some samples were positioned in the weathering chamber with half of the reflector surface shielded from the light source to provide control samples exposed to the same temperature and RH, but without any incident light. The sample set consists of construction I and construction II both laminated to bare aluminum substrates, as shown in Figures 1 and 2, respectively. For both constructions, the thickness of the acrylic adhesive layer is 13 μm and the silver reflective layer is 0.1 μm, and for construction II the copper layer is 0.05 μm thick. Both constructions were produced by vacuum evaporating the metals onto biaxially stretched 89 μm thick PMMA film.

All samples were removed from test and the reflectance was measured every two weeks. A random subset of samples was optically characterized during the intervening week. Reflectance was measured with a Perkin-Elmer UV/VIS/NIR Lambda 9 spectrophotometer. In addition, selected samples were also periodically removed from test to allow XPS analysis of the reflective (silver/polymer) interface. This interface was accessed by cutting small (10 mm x 14 mm) coupons from the reflector samples and soaking them in deionized (DI) water for a few minutes to several hours. The water intrudes between the evaporated silver layer and the polymer substrate resulting in a significant reduction in the silver-polymer adhesion. After the water soak, the clear PMMA film can be removed intact, exposing the reflective surface. Once this interface is exposed the samples were immediately placed into the vacuum chamber for analysis to prevent atmospheric contamination of the silver metal surface. Although the water soak introduces some uncertainty as to the identity of the actual interface being analyzed, it is the only practical way of exposing the silver/polymer interfacial region in these mirrors.

XPS data were acquired with an LHS-10 surface analysis system using a 250-W non-monochromatic Mg Kα X-ray source, previously described (7). The base pressure of the vacuum chamber typically was in the 10^{-10} Torr range, but during the analysis of these polymer samples the system pressure increased to about 3×10^{-8} Torr from the outgassing of the polymers. The peak heights from the survey XPS spectra were used to determine the elemental composition of the surface of the reflector and thus to follow changes in the interfacial composition of the mirrors as a function of accelerated exposure testing.

Results and Discussion

The two reflector constructions perform equally well until about the seventh week of accelerated testing when construction I begins to degrade at a faster rate than construction II. This trend continues until about week 20 when the two constructions again degrade at equivalent rates. Construction I falls below the critical reflectance value of 90% by week 14, while construction II remains above 90% for 22 weeks. The copper backed construction clearly has an increased stability during accelerated exposure, as shown in Figure 3.

Reflectance as a Function of XENO Exposure

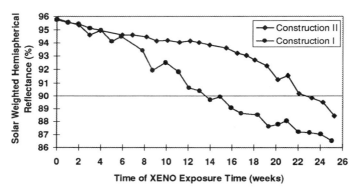

Figure 3. The two mirror constructions perform equally well until about the seventh week of accelerated weathering, when I begins to degrade at a greater rate than II. After about 20 weeks, the rates of degradation again become roughly equivalent. Construction II has a reflectance above 90% for about 9 weeks longer than construction I.

XPS survey spectra of the silver interface of both constructions are shown in Figures 4 and 5 after 20 weeks of accelerated weathering. In general, there is an increase in the surface carbon content and a concomitant decrease in silver. The increase in carbon appears to result from the loss of the sharp silver/PMMA (reflector) interface. Decomposition of the PMMA or migration of decomposed adhesive components through pinholes in the silver layer may be responsible for the increase in carbon at the reflector interface. The parts of the samples that were shielded from the light showed very little change in reflectance or surface composition as a function of time in the accelerated test.

XPS Survey Spectrum of the Reflector Surface of Construction I After 20 Weeks Exposure

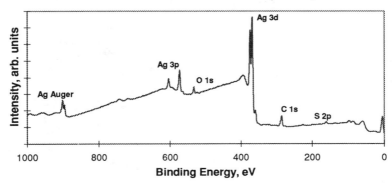

Figure 4. After a brief soak in deionized water, the transparent PMMA is easily removed from the reflective layer allowing XPS analysis of the silver surface. Atomic concentrations from the survey spectra indicate that after 20 weeks of weathering there is a 20% decrease in surface silver, a 5% loss in oxygen, a 5% increase in sulfur, and a 20% increase in carbon.

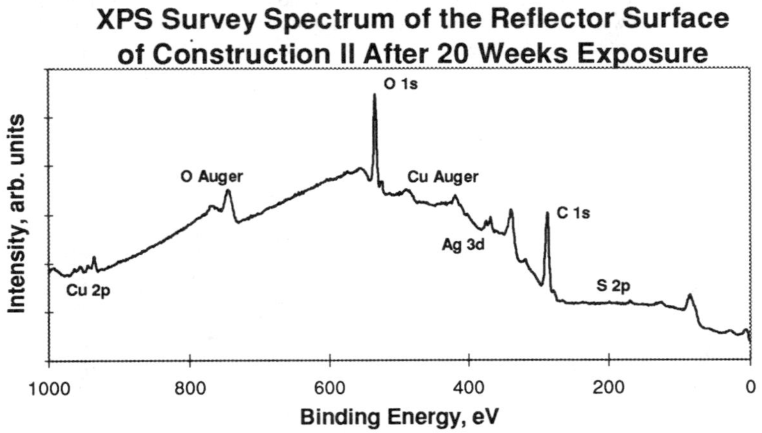

XPS Survey Spectrum of the Reflector Surface of Construction II After 20 Weeks Exposure

Figure 5. The reflective interface of weathered construction II has a different surface atomic composition than construction I. After 20 weeks, the surface carbon has increased by only 6%, sulfur has decreased 3%, and silver has decreased 35%, whereas the surface copper remains constant. The minimal loss of reflectance (5%) after 20 weeks and the low surface silver content may be the result of degradation of the silver/polymer interface.

There is a direct correlation between reflectance and the surface concentration of silver in construction I for the first 11 weeks of accelerated weathering. After this time other mechanisms of degradation appear to dominate the loss in reflectance. This is evident from Figure 6, where the surface silver concentration tends to stabilize after 11 weeks of weathering but the reflectance continues to decrease.

Construction I

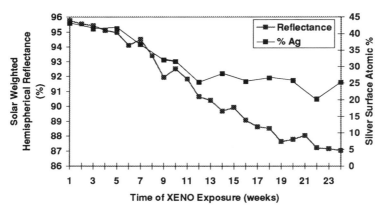

Figure 6. For construction I, the decrease in the surface silver atomic % correlates with the loss in reflectance until about week 13 when the silver concentration tends to plateau at about 25% whereas the reflectance continues to degrade.

Construction II

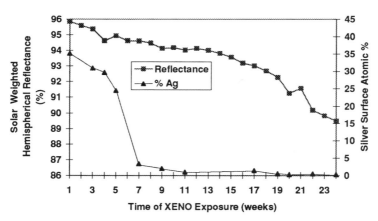

Figure 7. Construction II shows a correlation between reflectance and surface silver atomic % for only the first 4 weeks. The silver atomic concentration rapidly decreases to below 10% by week 7, while the reflectance decreases at a much slower rate. This suggests that the silver/polymer interface is broadened and that the interface exposed for XPS analysis is not the reflector interface.

For construction II no correlation exists between the surface silver concentration and reflectance after 4 weeks of accelerated weathering, suggesting that degradation mechanisms other than the simple loss of silver play a major role in the loss in performance of construction II (Figure 7).

The reflectance of construction II remains above 90% for 15 more weeks of accelerated weathering even though there is less than 5 atomic % silver detected at the "polymer/metal" interface. Thus, the actual reflective interface may be different from the interface exposed by the DI water soak and analyzed with XPS.

Figures 8 and 9 show plots of the atomic concentrations of C, O, Ag, S, and Cu as a function of exposure time in the XENO chamber. There is clearly more sulfur at the reflective interface of construction I relative to construction II. This sulfur concentration may be responsible for the corrosion of the silver and thus the loss in reflectance in construction I. Other than the loss of silver and an increase in surface carbon there are no other clear chemical contaminants or mechanisms evident in the weathering/analysis profile that may be responsible for the change in performance of construction I. Some sulfur is always detected on the surface of the PMMA and at the PMMA/silver interface in both constructions even before accelerated weathering. The source of the sulfur is likely the PMMA layer, which has a bulk sulfur content of from 500-1000 ppm or from the adhesive which contains from 50 to 100 ppm sulfur. (The polymer materials were analyzed for bulk sulfur content by Galbraith Laboratories, Inc., Knoxville, TN).

XPS Analysis of the Silver Interface of XENO Exposed Construction I

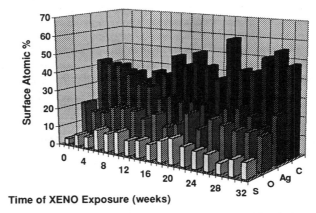

Figure 8. XPS-determined surface composition clearly shows compositional changes of the silver reflective surface as a function of accelerated weathering. Sulfur increases to about 10% by week 4 then remains constant; oxygen remains essentially constant, and silver decreases as carbon increases.

XPS Analysis of Silver Interface of XENO Exposed Construction II

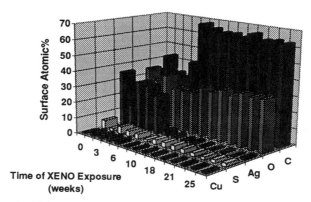

Figure 9. Construction II shows a similar carbon/silver relationship as construction I, but there is a significant decrease in sulfur on the silver surface. Copper concentration initially increases slightly then decreases and remains constant from week 8 through 27. Other than less surface sulfur, there are no clear compositional changes evident from the XPS data to explain the improved performance of construction II.

For the copper containing construction, copper diffuses to the silver surface early in the weathering process, with some copper detected on unweathered, as deposited samples (Figure 9). The Cu concentration increases slightly during the first 6 weeks of exposure and then becomes constant from week 8 through 27 weeks of exposure. The surface sulfur concentration appears to stabilize early in the weathering profile and remains almost constant through week 26. The lower sulfur concentration found on the interface of construction II relative to construction I may result from the formation of a copper passivation layer between the silver and the PMMA. A surface copper-sulfur species may then be responsible for preserving the reflectance of the silver layer. However, after a significant initial loss of surface silver and a subsequent increase in carbon, there is little change in the composition at the Ag/PMMA interface of construction II that can be used to explain the continued loss of optical performance. These data further suggest that the composition of the interface accessed by a DI water soak may not represent the true reflective interface in either construction after accelerated weathering.

Conclusions

From the XPS data taken from the accelerated-weathered samples of constructions I and II, there is no clear compositional information evident to explain the different reflectance losses of the two constructions. Copper may be successful in gettering sulfur species present in the PMMA or adhesive, thus changing the rate in loss of reflectance of construction II. Copper/sulfur species may also serve to passivate the reflective surface at the PMMA/Ag interface. From the XPS data, the main loss in reflectance in both constructions appears to involve the loss in a distinct metallic silver layer and the accumulation of carbon species at the reflective interface. This may be the result of polymer/metal reactions of the PMMA and/or the acrylic adhesive that result in low molecular weight species covering what remains of the metallic silver, thus degrading the once sharp polymer/metal interface. The XPS data may also have been taken at a different interface from the silver reflective surface because of the water soak method used to delaminate the PMMA from the mirror. Water intrusion may mask some of the compositional changes at the interface by dissolving some species, or the actual interface exposed by the water soak may change as the materials degrade as a function of accelerated exposure. The lack of a consistent interface for analysis is indicated by construction II where the reflectance does not correlate well with the amount of silver found on the interface that is exposed by the water soak.

Future work will include further interfacial chemical analysis studies with more controlled samples using XPS and Fourier transform infrared spectroscopy to further elucidate the interfacial species present.

Acknowledgments

This work was supported under Contract DE-AC36-83CH10093 to the U.S. Department of Energy.

Literature Cited

1) U.S. Department of Energy, *National Solar Thermal Technology Program, Five-Year Research and Development Plan 1986-1990*, DOE/CE-0160, 1986.
2) Schissel, P; Jorgensen, G.; Kennedy, C.; Goggin, R., *J. Sol. Energy Matls. and Solar Cells* **1994**, Vol. 33, pp. 183-197.
3) Kennedy, C. E.; Smilgys, R. V.; Kirkpatrice, D. A.; Ross, J. S., *Thin Solid Films* **1997**, Vol. 304, pp. 303-309.
4) Schissel, P.; Czanderna, A. W., *J. Sol. Energy Matls.* **1980**, Vol. 3, pp. 225-245.
5) Cotton, F.A. and Wilkinson, G. In *Advanced Inorganic Chemistry*, 4th Ed., 1980, pp. 968-975.
6) Graedel, T.E., *J. Electrochem. Soc.* **1992**, Vol. 139, No.7, pp. 1963-1970.
7) Pitts, J.R., Thomas, T.M., and Czanderna, A.W., *Appl. Surf. Sci.* **1986,** Vol. 26, pp. 107-113.

Chapter 19

Low Molecular Weight and Polymer Bound UV Stabilizers

Otto Vogl[1], Lutz Stoeber, Andres Sustic, and John Crawford

Polytechnic University, Brooklyn, NY 11201

ABSTRACT

   Most polymers must be stabilized against the impact of the environment. Antioxidants and ultraviolet (UV) stabilizers are of vital importance to protect polymeric materials from deterioration. To be effective UV absorbers, these compounds must also have the capability to dissipate the energy without causing damage. Long range and efficient protection of organic polymeric materials from the environment requires carefully designed and stabilizing systems including polymer-bound UV stabilizers.

INTRODUCTION

   It is widely recognized that sunlight is an essential factor in the deteriorative aging and weathering processes which occur in polymers. Following light absorption, bond dissociation occurs spontaneously to produce free radicals which can start the radical photooxidation process via hydrogen atom abstraction from the carbon-hydrogen bond [1]. The absorption of energy and its transfer to the bond to be broken may are the photophysical aspect of the photodegradation process.

   The first chemical step in photodegradation is usually a homolytic bond scission whereby free radicals are formed. If oxygen is present, these free radicals will rapidly react with molecular oxygen to form peroxy radicals which are very reactive and capable of abstracting a neighboring hydrogen atom generating a new free radical and a hydroperoxide. In this way, visible, and especially UV radiation, are effective initiators of photooxidation. Such degradation is noticeable in a plastic as a change in color, such as yellowing and/or the deterioration of its physical properties.

   It is evident that for most polymers, some form of photostabilization is necessary if proper protection against the damaging effects of solar radiation is to be

[1]Present address: Department of Polymer Science and Engineering, University of Massachusetts, Amherst MA 01003

© 1999 American Chemical Society

achieved. The photostabilization of light sensitive polymers involves the inhibition, retardation or elimination of various photophysical and photochemical processes that take place during photodegradation.

LOW MOLECULAR WEIGHT ULTRAVIOLET ABSORBERS

Decades ago it was found that various materials, even foodstuffs , undergo deterioration under the influence of the environment [2]. Of these deteriorations, autooxidation, the slow oxidative process of organic materials, occurs slowly at room temperature and its progress depends on the chemical structure of these materials. The need for protection and the prevention or at least the slowing down of the oxidative deterioration was probably first recognized for natural and synthetic rubbers where the butadiene units in the polymers are especially sensitive to autooxidation.

The autooxidation processes are accelerated under the influence of light and certain light absorbers were found to be capable of absorbing the light, reaching an exited state and dissipating the light energy by a non-radiative way. Stabilization of polymers against the detrimental effects of the exposure to the environment has been of great importance for the use of polymeric materials outdoor or exposure or to higher energy sources indoors. Consequently, polymers must be stabilized against photodegradation.

Practically speaking, this is done by adding long lasting UV screens and absorbers, or hindered amine light stabilizers which interfere with the photo-decomposition of hydroperoxides. Early stabilizers were aminobenzoate derivatives found in sunscreen formulations. Other stabilizing additives included salicylate esters, particularly phenyl esters, 2-hydroxybenzophenones, hindered amine light stabilizers (HALS) and substituted oxanilides [3].

It was later found that substitute phenols provided a better protection against photooxidation of the polymeric materials because such molecules have a mechanism of reversible energy dissipation which was found to be effective. Salicylates, where R=COOR with R preferentially is a phenyl group provided a better group of compounds and 2-hydroxybenzophenones, where R is a COC_6H_5 group. In actual fact salicylates can be converted under the influence of light by photo-Fries rearrangement into 2-hydroxybenzophenones. This is an important but not quantitative reaction, because the cost of salicylates is only ca. 20 % of that of 2-hydroxybenzophenones, but the photo-Fries reaction is not quantitative and it also forms yellow by products; colored by-products are undesirable for most applications, even though they, themselves, might be effective UV stabilizers (Scheme 1).

R=COOR'
R=COPh
R=2H-Benzotriazole

Scheme 1

These 2-substituted phenols very often have additional substityuents in the 4- and/or 6 position. To have a substituent in the 4-position is essential, because electrophilic reactions on phenols, often needed for the synthesis of these compounds, are *para*-directing. Tertiary-butyl groups, on the other hand, in the 6 position are often used (because tertiary butylation utilizes isobutylene as alkylation agent which is a facile reaction) because the compounds produced by tertiary butylation are more soluble and make the phenyl ring more electron rich which promotes the desired hydrogen bonding the UV stabilizer. For many years the choice of the most effective stabilizers had been by trial and error and was very often based on what chemical compounds of compositions were commercially available. In fact, even today, we depend on decisions of producers of UV stabilizers for what they are interested in commercializing. Finally, the successful application of a stabilizing system often depends on the skill and experience of the formulator.

Over the years, much understanding of the action of UV stabilizers was accumulated by studying the mechanism of the photochemistry and the photophysics of these molecules It was recognized that not only is the structure of the stabilizer important for its effectiveness, but also the distribution of the stabilizer in the polymer matrix, the location of the stabilizer molecules in the polymer matrix and the possible aggregation of the stabilizer molecules [4,].

For reasons of simplicity and convenience most commercial applications of UV stabilized compositions are based on low molecular weight compounds. Compositions with low molecular weight stabilizers have the advantage of potentially lower cost and ease of incorporation on the stabilizer by melt blending. They have the disadvantage of volatility of the stabilizer (important during high temperature melt fabrication) limited compatibility of the stabilizer within the polymer matrix which could lead to diffusion, migration out of the polymer article and of aggregation of the stabilizer molecules which might change the local distribution of the stabilizer.

All these types of photostabilizers have limited solubilities in the polymers, especially in semicrystalline material where solubility is limited to the amorphous phase. In hydrocarbon polymers, the solubility of polyaromatic compounds are limited, they form clusters and are not evenly distributed in the matrix and, furthermore, because they are clusters and not individual molecules and the individual molecules lose their full effectiveness in the photophysical process.

Such changes might deprive other parts of the polymer matrix of the stabilizers. Some of these disadvantages that low molecular weight stabilizers might have, have been overcome by increasing the molecular weight of the stabilizers molecules , especially with the objective of adding compatibilizing groups as a means of increasing the molecular weight of the stabilizer molecule.

Because they are excellent UV absorbers, 2(2-hydroxyphenyl)2H-benzotriazole derivatives are, at present, the most important and widely used and effective photostabilizers for polymeric materials and constitute about over one third of all the photostabilizers sold and used. The costs are, however also by a factor of about 5 higher than the 2-hydrobenzophenones.. The stabilizing mechanism of these types of compounds is based on the formation of intramolecular hydrogen bonds between the *o*-hydroxyl group of the phenyl ring and the nitrogen atom of the benzotriazole moiety. In other words, for the 2(2-hydroxyphenyl)2H-benzotriazoles, the hydrogen bonded form provides a facile route for harmlessly deactivating the excited state induced by light absorption. For example, the absorption maximum of 2(2-hydroxy-5-methylphenyl)2H-benzotriazole is found at about 340 nm with a molar absorptivity coefficient of about 25,000 L/mol x cm. The capability of intramolecular hydrogen bond formation was found to be the most important structural element that provides photostability.The mechanism of the 2(2-hydroxyphenyl)2H-benzotriazole energy dissipation is shown in Scheme 2.

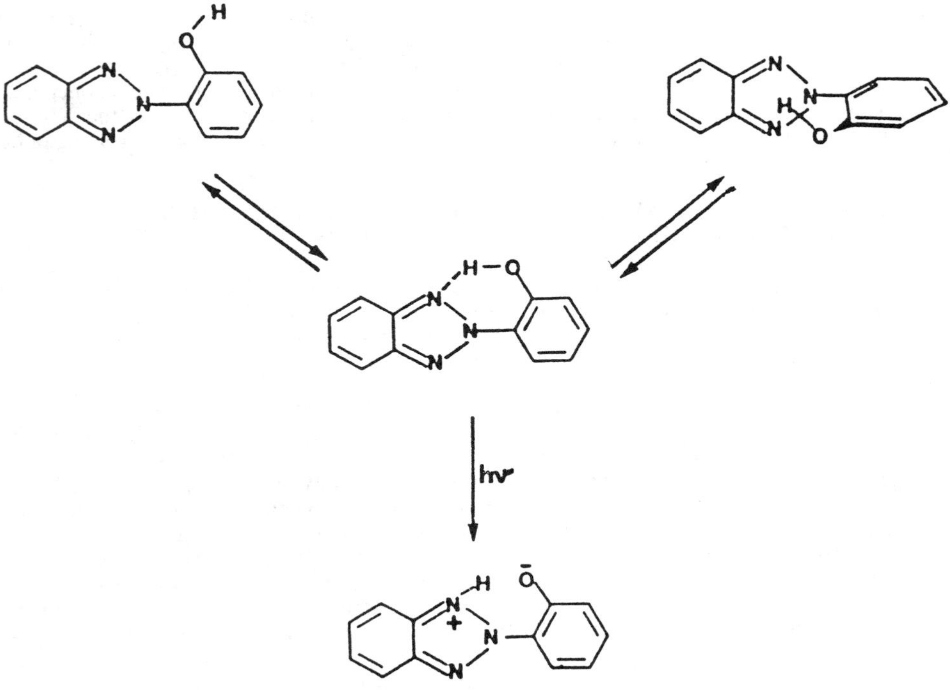

Scheme 2

Throughout the years, there has been an intense activity in the synthesis of 2(2-hydroxyphenyl)2H-benzotriazoles, starting with the first description of the synthesis of the first 2(2-hydroxyphenyl)2H-benzotriazole by K. Elbs [5,6]. But it was not until the mid-fifties that it was realized that compounds such as 2(2-hydroxyphenyl)2H-benzotria-zole, were not only good UV absorbers but were very stable compounds in photophysical processes and therefore, were excellent candidates as UV. stabilizers for polymers [7]. Early synthetic work was carried out in the 1960's [8,9].

With the increasing demands on the performance of the final molded plastic product, it is important to incorporate into polymeric compositions the most efficient UV. absorbers. Several approaches have been undertaken towards that goal. One of them is to use molecules that incorporate more than one benzotriazole moiety, specifically, multichromophoric 2(2-hydroxyphenyl)2H-benzotriazoles.

With this in mind, several of these type of compounds has been synthesized [10,11]: 2,4[di(2H-benzotriazol-2-yl)]1,3-dihydroxybenzene, 2,4[di(2H-benzotriazol-2-yl)]1,3,5-trihydroxybenzene, and 2,4,6[tri(2H-benzotriazol-2-yl)]1,3,5-trihydroxybenzene. These three compounds have high molecular weights and melting points while showing unusually high absorption at wavelength maxima of 325, 340 and 337 nm, respectively.

All of the above multibenzotriazoles are products of the benzotriazolization of polyhydroxylated compounds such as resorcinol and phloroglucinol. In a similar approach [12,13], benzotriazolization of 2,4-dihydroxyacetophenone and 2,4-dihydroxybenzophenone gave the dibenzotriazolized compounds 3,5[di(2H- benzotriazol-2-yl)]2,4-di-hydroxyacetophenone, and 3,5[di(2H-benzotriazol-2-yl)]-2,4-dihydroxybenzophenone. These compounds are effective and useful UV stabilizers as they both have the 2(2-hydroxyphenyl)2H-benzotriazole moiety and the 2-hydroxybenzophenone (or acetophenone) moiety in the molecule.

ULTRAVIOLET STABILIZERS OF MEDIUM MOLECULAR WEIGHT WITH COMPATIBILIZING CHARACTERISTICS.

Requirements for lower volatility and better compatibility of UV stabilizers and the proposal to incorporate such stabilizers directly into the polymer led to the comercialization of UV stabilizers with lower volatility combined with functionalities that allow compatibilization.

Such stabilizers could still be added similarly to low molecular weight stabilizers, by melt blending. The commercial stabilizer (Tinuvin 1130, Ciba-Geigy Co) of choice became the poly(ethylene oxide) ester of 2[2-hydroxy-{3-carboxypropionato}]2H-benzotriazole with a tertiary-butyl group in 6 position [14]. It is important to realize that the key compound, 4-hydroxyphenyl-3-propionic acid is made from phenol and acrylic acid. Tertiary butylation with isobutylene gives 3-tertiary butoxy-4-hydroxyphenyl propionic acid. and "benzotriazolization" gives the very important UV stabilizer intermediate, For the commercial product the acid was esterified with poly(ethylene glycol). We have actually used the methyl ester and the poly(ethylene oxide) ester for the synthesis of our own intermediate. Reduction of the commercial poly(ethylene glycol)ester with lithium aluminum hydride gave the corresponding alcohol, 2[2-hydroxy-{3-hydroxypropyl}]2H-benzotriazole [15](Scheme 3).

2[2-Hydroxy-3 tertiary butyl-5-(3-carboxyethyl}phenyl]2H-benzotriazole, 2[2-hydroxy-{2-hydroxyethyl}]2H-benzotriazole and 2[2-hydroxy-(3 tertiary butyl(3-hydroxypropyl}]2H-benzotriazole were excellent intermediates for further studies .

2[2-Hydroxy-3 tertiary butyl-5-(3-carboxypropyl)- phenyl]2H-benzotriazole was used for the UV stabilization of polymer blends. This is the first time a study of photostabilization was undertaken. Blends of polystyrene with poly(methyl vinyl ether) were subjected to stabilization with 2[2-hydroxy-3 tertiary butyl-5-(3-carboxyethyl}phenyl]2H-benzotriazole. It was found that when 2[2-hydroxy-3 tertiary butyl-5-(3-carboxyethyl}phenyl]2H-benzotriazole was added to the polymer blend, 2[2-hydroxy-3 tertiary butyl-5-(3-carboxypropyl}phenyl]2H-benzotriazole preferentially was associated with poly(methyl vinyl ether) and assisted in the stabilization of the blend, in spite of the fact that polystyrene/poly(methyl vinyl ether) blends were miscible as far as their DSC behavior was concerned [15].

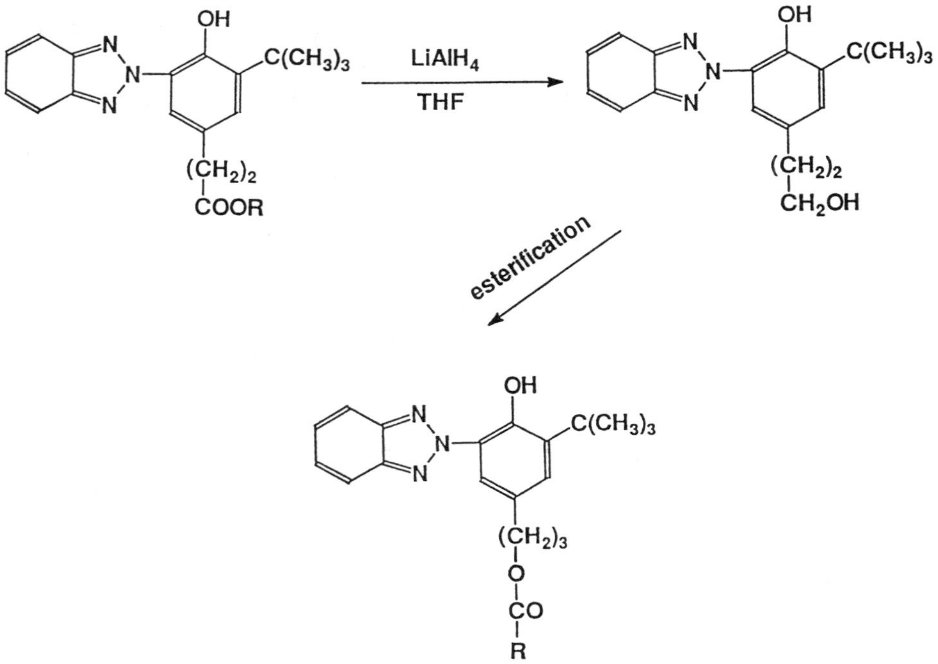

Scheme 3

2[2-hydroxy-(3-hydroxypropyl))]phenyl]2H-benzotriazole was a useful intermediate that allowed the smooth esterification with carboxylic acids, which was the basis for the preparation of UV stabilizers with long aliphatic chain. Esterification with stearic and palmitic acid, and with unsaturated long chain aliphatic hydrocarbon acids, such as oleic, linoleic and linolenic acids, gave UV stabilizers with potentially polymerizable groups.

UV stabilizers with polysiloxane groups or with long fluorocarbon groups were synthesized from 2[2-hydroxy-(2-hydroxyethyl)phenyl]2H-benzotriazole and the corresponding hydroxy terminated silicones or fluorocarbon alcohols. Reaction with (meth)acrylamidomethanol gave a stabilizer with a polymerizable group in 3 position. Usually a chain length in excess of 12 fluorocarbons or siloxane units are necessary for their effectiveness [16].

POLYMER BOUND ULTRAVIOLET STABILIZERS

All these considerations led to the idea of having UV stabilizers incorporated in the polymer chain, which then would be less readily lost during fabrication and exposure to the environment. The myth that stabilizers must have relatively high mobility to be effective has, however, persisted, and only recently have indications become compelling that high mobility of low molecular compounds is not essential for the effectiveness of stabilizers in polymeric materials. In fact, incorporation of polymerizable stabilizers into polymer chains has been found to make them effective, especially in long term exposure. Optimally speaking, a flexible spacer group between the polymer backbone and the active stabilizer moiety seems to be desirable and has become the most recent target of investigations [17].

When UV absorbing groups are introduced into polymers, this can be accomplished in principle the following ways: a.) Synthesis of a polymerizable UV stabilizer monomer followed by polymerization or copolymerization; b.) By polymer attaching the stabilizing group to the polymer by polymer reaction and c.) Using an initiator whose initiating fragment contains the stabilizing group [using, for example, an azo initiator which introduces a (2-hdroxybenzophenone) stabilizer at the end of the polymer chain].

About 25 years ago we started to undertake a different approach to the problem - the direct incorporation of the UV stabilizing moiety into the polymer chain. Initially we incorporated p-aminobenzoate units, either randomly into the side group of the polymer or as end groups of the polymer chain [18]. Later we prepared polymerizable stabilizers of the class of vinyl-salicylates [19], vinyl-2-hydroxybenzophenones [20] and vinyl a-cyano-b-phenyl-cinnamates [21]. These styrene type monomers could very readily be polymerized and copolymerized with the right initiating systems. It was found that the hydrogen bonded 2-hydroxyl groups of most of these monomers did not noticeably interfere with the radical polymerization of the polymerizable stabilizers.

For the last 20 years our work on polymerizable UV stabilizers focused on the synthesis, characterization and study of the efficiency of polymerizable 2(2-hydroxyphenyl)2H-benzotriazole stabilizers. [22,23].

The design of polymerizable 2(2-hydroxyphenyl)2H-benzotriazole UV stabilizers has led to a number of 2(2-hydroxyphenyl)2H-benzotriazole derivatives with various functional and polymerizable groups in the molecules. These 2(2-hydroxyphenyl)2H-benzotriazoles with vinyl- [24,25], (meth)acryloxy- [26], hydroxy-, acetoxy- and carbomethoxy-groups have been synthesized [27] (Scheme 4). The polymerizable group allowed them to be incorporated into acrylic, styrenic addition polymers, into unsaturated polyesters [28] and ABS type [29] polymers and also into condensation polymers such as polyesters, polycarbonates and polyamides [30-32].

Recently, a thorough study of the reactivity ratios of 2(2-hydroxyphenyl)2H-benzotriazoles (vinyl and isopropenyl) respectively in their copolymerization with styrene, methyl methacrylate and n-butylacrylate has been published [33].

The incorporation of UV absorbers into polymers by free radical grafting is another attractive alternative because it can be carried out on commercially available polymers [34,35]. Once the stabilizer monomer has been grafted onto the polymer backbone, the grafted copolymer can be either molded to the end use product or it can be added as a master batch to the unstabilized polymer so that the molar concentration of stabilizer is sufficient to protect the host polymer.

Grafting of vinyl monomers onto polymers with a hydrocarbon backbone chain of a polymer was first accomplished many years ago. More recently polymerizable 2(2-hydroxy-4-vinyl-phenyl)2H-benzotriazole has been successfully grafted onto saturated aliphatic C-H groups of polymers [36]

More recently the surface photo grafting of polymerizable 2(2-hydroxyphenyl)2H-benzotriazoles such as 2(2-hydroxy-4-methacryloxyphenyl)2H-benzotriazole and 2(2-hydroxy-4-acryloxyphenyl)2H-benzotriazole 2(2-hydroxy-4-

Polymerizable Groups	Position
- CR=CH$_2$	5, 5'
- OCO-CR=CH$_2$	4
- OCH$_2$CHOHCH$_2$OCOCR=CH$_2$	4
- CH$_2$NHCOCH=CH$_2$	2

R = H, CH$_3$

Substituents	Position
- OH	4, 5'
- OCOCH$_3$	4, 5'
- F	5'
- Cl	5', 6'
- Br	5'
- Morpholino	5'
- N(CH$_2$CH$_3$)$_2$	5'
- (CH$_2$)$_3$OCOR	5
- (CH$_2$)$_2$COOR	5

R = H, CH$_3$

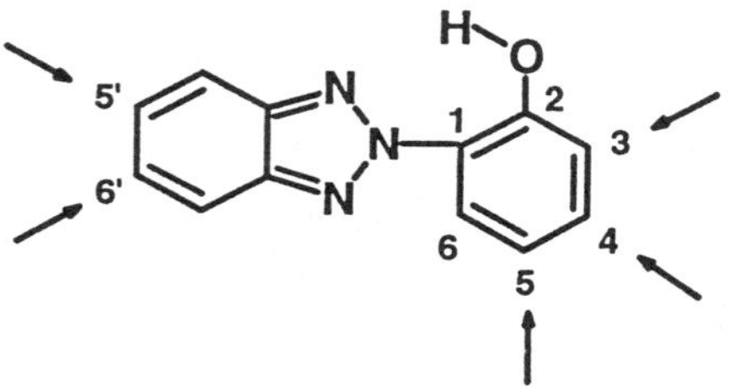

Scheme 4

vinylphenyl)2H-benzotriazole on thin film of low density polyethylene, high-density polyethylene and polypropylene was achieved. The grafting efficiency was highest for low density polyethylene, followed by high-density and polypropylene films. Weather-O-meter data suggested that the grafting had been efficient, and it was also shown that this type of stabilization was more effective than the incorporation of the corresponding low molecular weight UV stabilizers [37].

It was also our objective to incorporate 2(2-hydroxyphenyl)2H-benzotriazole with aliphatic hydroxyl groups or with phenolic hydroxyl groups into polymers, particularly into acrylic and styrenic polymers, by reacting the appropriate compounds with epoxy groups containing monomers [38], such as glycidyl methacrylate. The resulting monomers could be polymerized or copolymerized with the appropriate comonomers in the desirable proportions (Scheme 5).

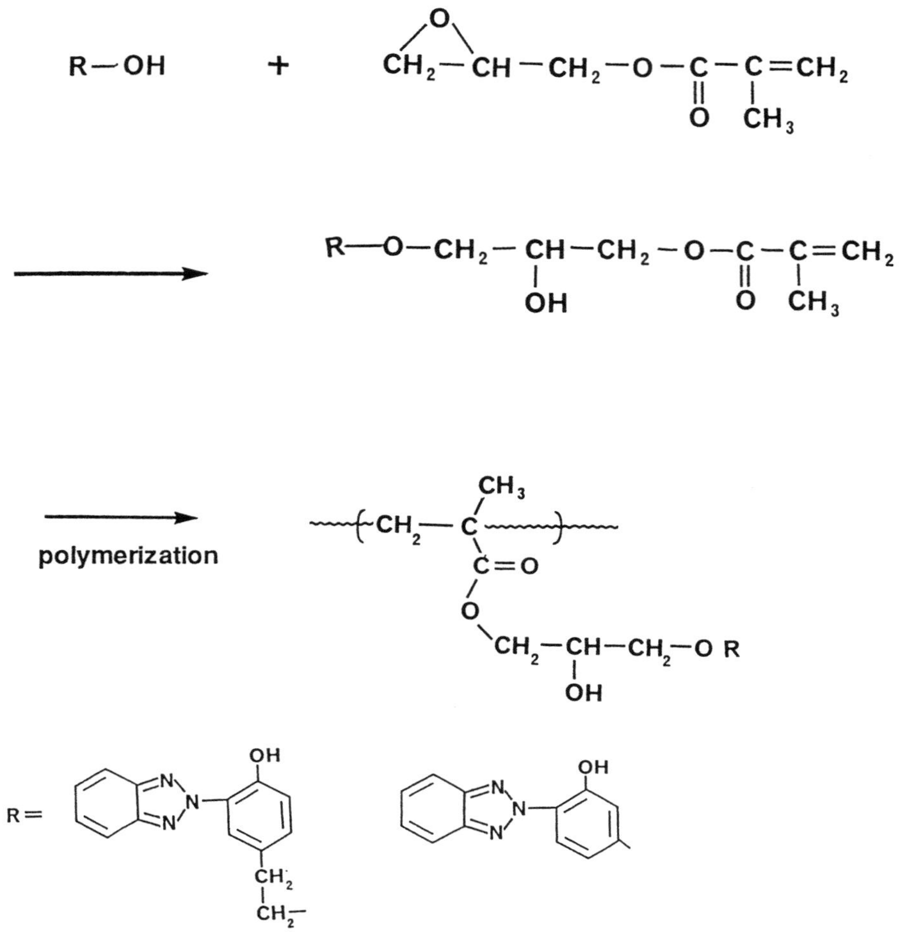

Scheme 5

Alternatively the hydroxyl containing 2(2-hydroxyphenyl)2H-benzotriazole could be reacted with a polymer that contained epoxy groups such as may be present in acrylic copolymers with glycidyl methacrylate groups as part of the polymer (38) (Scheme 6). Another alternative was to introduce 2(2-hydroxyphenyl)2H-benzotriazole derivatives with aliphatic hydroxyl groups such as 2(2-hydroxy-5-(2'-hydroxyethyl)-phenyl)2H-benzotriazole into methyl methacrylate polymers by transesterification.

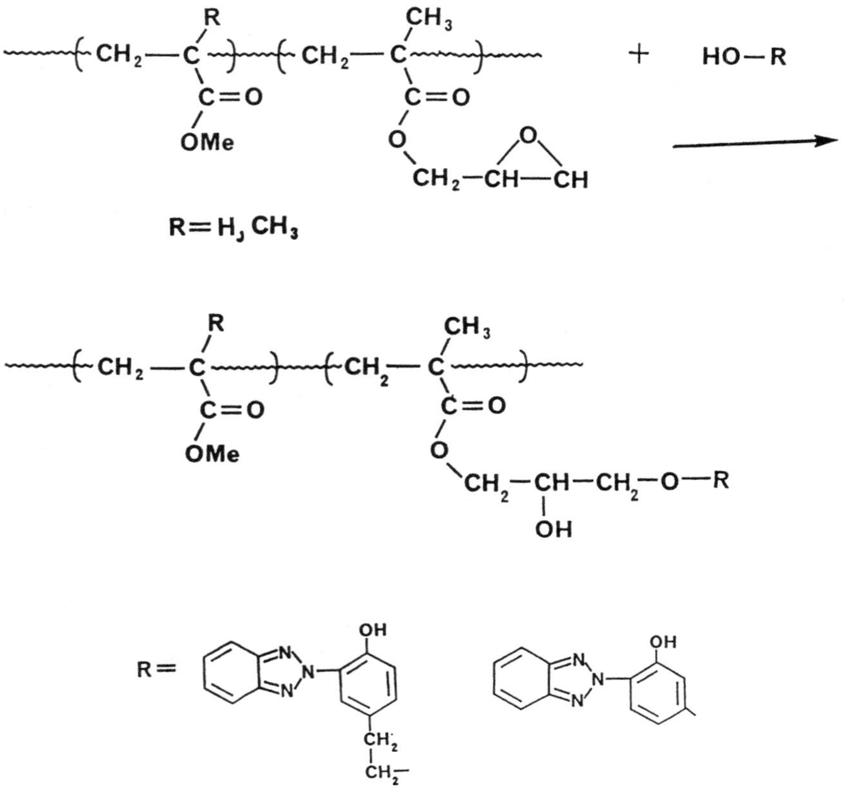

Scheme 6

Polymer bound UV stabilizers were of great interest to provide UV stabilized contact lenses. [39,40]. Three 2(2,4-dihydroxy-4-acryloxyphenyl)2H-benzotriazole derivatives, unsubstituted, 5'-methoxy and 5'-chloro derivatives were reacted with the epoxy group of glycidyl methacrylate. The three 2(2-hydroxyphenyl)2H-benzotriazole derivatives were also allowed to react with the epoxy groups of acrylic copolymers with glycidyl methacrylate concentrations of up to 20 mole % [41].

Over the years we have synthesized 2(2-hydroxyphenyl)2H-benzotriazoles with vinyl groups in the 5 and 5' position, with the isopropenyl group in the 5-position, with the meth)acryloxy group in the 4 position and the (meth)acrylamido and maleimido group in the 3 position. These monomers have been incorporated into acrylic and styrenic polymers - including unsaturated polyesters and ABS type resins. Some of these monomers have also been incorporated at the end of methacrylate polymers by group transfer polymerization [42] (Scheme 4).

For the use in condensation polymers, a number of disubstituted 2(2-hydroxyphenyl)2H-benzotriazole UV stabilizers have also been synthesized. They include the 5,5'-diydroxy derivatives, but also the 5',5''-dihydroxy derivative and the

5'5"diacetoxy derivative of dibenzotriazole. 5',5"-dicarbomethyl- and other 5,5"-disubstituted dibenzotriazoles have also been incorporated into polyesters, polycarbonates and polyamides [40] Scheme 7).

The tailor-making of 2(2-hydroxyphenyl)2H-benzotriazole derivatives has reached a new level of sophistication - the synthesized polymerizable UV stabilizers which, when incorporated into the polymers influence the surface polymer morphology (morphology engineering) [16,42]. 2(2-Hydroxyphenyl)2H-benzotriazoles with a long hydrocarbon, silicon and fluoro carbon group and - with a (meth)acrylamidomethyl group attached to the 3-position gave monomers which could readily be incorporated into acrylic polymers. When properly applied as coating, the hydrocarbon, fluorocarbon and silicon groups aggregate on the surface; this was determined by ESCA and by surface angle measurements. As a consequence the UV absorbing moiety is now placed right at the surface where it is most needed.

As another approach to the synthesis of polymerizable UV stabilizers we have also studied the behavior of 2(2-hydroxyphenyl)2H-benzotriazole compounds that are potential precursors for polymerizable UV stabilizers. Our objective was the modification of the UV spectra of 2(2-hydroxyphenyl) 2H-benzotriazoles toward a sharper cut-off at 385 nm, the beginning of the visible and the first sign of yellowness, or significant absorbency at 400 nm (for UV absorbing lens application). We have synthesized a number of 5'-substituted 2(2-hydroxyphenyl)2H-benzotriazoles with halogen methoxy and alkylamino substituents These compounds had usually also a hydroxyl group in the 4-position of the phenyl ring for easy reaction with glycidyl methacrylate - to produce a new family of polymerizable, hydrolysis stable 2(2-hydroxyphenyl)2H-benzotriazole stabilizers. We have also synthesized a number of 2(2-hydroxyaromatic)2H-benzotriazoles where the aromatic group is multinuclear [14].

Now to some practical aspects of the general synthesis of 2(2-hydroxyphenyl)2H-benzotriazoles. The reaction sequence starts with the

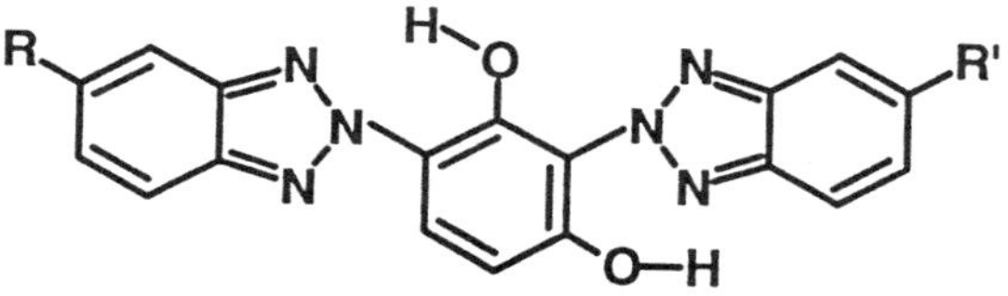

R = R' = H
OH
OCOCH$_3$
OCH$_3$
COOH
COOCH$_3$
Cl

Scheme 7

condensation of ortho-nitrodiazonium salts in aqueous solution with the phenol. If the phenol is not substituted or has no substitution in the 4-position, the electrophilic reaction will take place in the 4-position, otherwise in the 2-position of the phenol. Since it is necessary for an efficient UV stabilizer of the 2(2-hydroxyphenyl)2H-benzotriazole variety it is necessary to block the 4-position. Consequently, all the commercial stabilizers have either a methyl group (from condensation of o-cresol or 3(4-hydroxyphenyl)-propionic acid (made from phenol and acrylic acid) in 4-position of the phenyl ring or are derived from hydroquinone. A special case is the reaction of the o-nitrodiazonium compound with resorcinol which is much more reactive as regular phenols. As the 1,3-dihydroxybenzene, even the condensation in 4-position will give a 2-hydroxy compound. This reaction must be carried out in acidic medium, under basic conditions disubstitution will occur. As the next step the nitro group of the diazo-compound is then reduced with zinc powder which also causes cyclization to the benzotriazole ring. For reasons of solubility but also to block the o-position to the phenyl group, and to provide some electron donation which is desirable for the effective hydrogen bonding of the phenolic hydroxyl group to the triazine ring, we find usually a teritiary butyl group in that position (easily introduced by reaction of the 2(2-hydroxyphenyl)2H-benzotriazole with isobutylene) [43].

Oriental lacquer, the *urushi* component of the sap of the Japanese lacquer tree, when applied to wood, has a beautiful appearance and an excellent stability in the normal environment. The urushi components of the oriental lacquer are a mixture of catechol derivatives which have a n-C_{15} or n- C_{17} aliphatic side chain (with various degrees of unsaturation) in 3 position of the catechol ring [44,45]. If the triene content of the mixture approaches 60 % of the catechol mixture, the urushiol mixture is readily polymerized. Urushiol behaves similarly for oxidative polymerizations to the so-called *drying* oils and *tung* oils, which are glycerol esters of oleic, linoleic and linolenic acids The polymerizability increases with increasing double bond content. Esters with high linolenic acid (triene) content are particularly effective in this kind of polymerization. The same is also true for the triene unsaturation content in oriental lacquer compositions.

Poor photooxidative stability as a consequence of extensive crosslinking of the oriental lacquers on exposure to the environment and aging, limits the outdoor use. We have synthesized 2[2-hydroxy-(3'-hydroxypropyl)-phenyl]2H-benzotriazole esters of polymerizable groups such as oleic, linoleic and linolenic acid These long unsaturated aliphatic groups were subjected to photooxidative co-curing with the highly unsaturated aliphatic C_{15} side chains of the urushiol components [46]. The final product, when tested under simulated UV conditions, showed a significant increase in the photooxidative stability of the oriental lacquer. The UV stabilizers described here are also effective stabilizers for the stabilization of polymers derived from drying oils.

References

1.   B. Ranby and J.F. *Rabek, Photdegradation and Photostabilization of Polymers; Primnciples and Applications,* John Wiley and Sons. New York, 1975
2.   Plastics Additives Handbook, Hanser Publishers, 1987.
3.   N.S. Allen, *Development in Polymer Photocheistry,* Applied Science Publishers, 1981

310

4. G. Scott, *Developments in Polymer Photostabilization,* Applied Science Publishers, 1980
5. K. Elbs and W. Keiper, J. fuer Prakt. Chem., 67(2), 580 (1905)
6. K. Elbs, O. Hirschel, K. Himmler, W. Turk, A. Heimrichand E. Lehmann, J. fuer Prakt. Chem., 108, 209 (1924)
7. G.N. Gantz and W. Summer, Text. Res. J., 27, 244 (1957)
8. J. Milionis and Hardy, U.s. Patent 3,159,646 (1964)
9. L. Balaban, J. Borkovec and D. Rysavy, Czech. Patent 108,792   (1963)
10. .S.J. Li, A. Gupta and O. Vogl, Monatsh. Chem. 114, 937 (1983)
11. F. Xi, W. Bassett, Jr. and O. Vogl, Makromol. Chem., 185(12),   2497 (1984).
12.. S.J. Li, W. Bassett, Jr., A. Gupta and O. Vogl, J. Macromol. Sci.,   Chem., A20(3), 309 (1983).
13. O. Vogl and S.J. Li, U.S. Patent 4,745,194 (5/17/1988).
14. A. Sustic, Ph.D. Dissertation, Polytechnic University, Brooklyn, New York 11201, (1992) p. 220. ref 149
15. L. Stoeber, Ph.D. Dissertation, Polytechnic University, Brooklyn,   New York, (1995)
16. O. Vogl J. Macromol. Sci., Chem, A27(13&14), 1781 (1990)
17. D. Bailey and O. Vogl, J.Macromol. Sci., Macromolecular Reviews, C14(2), 267 (1976).
18. D. Bailey, D. Tirrell and O. Vogl, J. Macromol. Sci., Chem., A12(5), 661 (1978).
19. D. Bailey, D. Tirrell and O. Vogl, J. Polym. Sci., Polym. Chem. Ed., 20. 2725 (1976).
21. D. Bailey, D. Tirrell, C. Pinazzi and O. Vogl, Macromolecules, 11(2), 312 (1978).
22. Y. Sumida and O. Vogl, Polymer J. (Japan), 13(6), 521 (1981).
23. O. Vogl and S. Yoshida, Rev. Rou. de Chimie, 25(7), 1123 (1980).
24. Otto Vogl, A.C. Albertsson and Z. Janovic, P. Klemchuk Ed., ACS Symposium Series, 280, 197 (1985).
25. S. Yoshida and O. Vogl, Polymer Preprints, ACS Division of Polymer Chemistry, 21(1), 203 (1980).
26. S. Yoshida and O. Vogl, Makromol. Chem., 183, 259 (1982).
27. S.J. Li, A.C. Albertsson, A. Gupta, W. Bassett, Jr. and O. Vogl, Monatshefte Chem., 115, 853 (1984).
28. J. Bartus, P. Goman, A. Sustic and O. Vogl, Polymer Preprints,   ACS Division of Polymer Chemistry, 34 (2), 158 (1993)
29. O. Vogl and E. Borsig, U.S. Patent 4,523,008 (6/11/1985).
30. O. Vogl, J. Bartus and H. Horacek, Austrian Appl. No. AZ: 2117/93; October 19, 1993
31. P. Gomez and O. Vogl, Polymer J., 18(5), 429 (1986).
32. P.M. Gomez, L.P. Hu and O. Vogl, Polymer Bulletin, 15(2), 135   (1986).
33. S.K. Fu, S.J. Li and O. Vogl, Monatshefte Chem., 119, 1299 (1988).
34. E. Borsig, A. Karpatyova and O. Vogl, Collec. Czechoslov. Chem. Comm., 54, 96 (1989).
35. M. Kitayama and O. Vogl, J. Macromol. Sci., Chem., A19(3), 375 (1983).
36. M. Kitayama and O. Vogl, Polymer J. (Japan), 14(7), 537 (1982).
37. W. Pradellok, A. Gupta and O. Vogl, J. Polym. Sci., Polym. Chem. Ed., 19, 3307 (1981).
38. J. Lucki, J.F. Rabek, B. Ranby, B.J. Qu, A. Sustic and O. Vogl,   Polymer, 31, 1772 (1990).

39. A. Sustic, C.L. Zhang and O. Vogl, J. Macromol. Sci., Pure & Applied Chem., <u>A30(9/10),</u> 741 (1993)
40. S. Loshaek, U.S. Patent 4,304,895 (1981)
41. J. Rosevear and J.F.K. Wilshire, Austral. J. of Chem., <u>38,</u> 1163 (1985)
42. P.M. Gomez and H.H. Neidlinger, Polymer Preprints, ACS Division of Polymer Chemistry, 28(1), 209 (1987)
43. A. Sustic, L. Stoeber and O. Vogl, in preparation
44. O. Vogl, J. Bartus, M.F. Qin and J.D. Mitchell, in *Progress in Pacific Polymer Science 3*, K.P. Ghiggino ed., Springer Verlag. p.423 (1994)
45. J. Bartus, W.J. Simonsick jr. and O. Vogl, Polymer J., <u>25(7),</u> 703 (1995)
46. M.F. Qin and O. Vogl, Cellulose Chemistry and Technology, <u>29,</u> 533 (1995)

Chapter 20

Overview of Mechanical Property Changes During Coating Degradation

Loren W. Hill

Resins Division, Solutia Inc., 730 Worcester Street, Springfield, MA 01151

Dynamic mechanical analysis (DMA) and chemical stress relaxation determinations indicate that automotive topcoats undergo significant changes in glass transition temperature and crosslink density during exposure to UV light and moisture in weatherometers designed to mimic but accelerate the effects of outdoor exposure. The resulting changes in physical properties are believed to contribute to failure by clearcoat cracking or by delamination at various interfaces between paint layers in the auto coating system. Previous studies emphasized loss of gloss, but current pigmented basecoat/clearcoat (BC/CC)systems lose gloss very slowly. Topcoats are frequently still glossy when failure by some other mode occurs. Cracking and delamination occur abruptly whereas gloss loss, observed with older technologies, occurred gradually. The benefits of using service life prediction (SLP) methods are well documented, but application to auto coating durability will require consideration of abrupt versus gradual failure modes. Property change data may be important for understanding failure mechanisms, but more work is needed to effectively integrate property change data with SLP methodology.

Opportunities for advancement often occur when methods proven in one field are transferred to another field where a well-recognized need has been documented. Martin et al (*1*) state that service life prediction (SLP) methodology has been used in the electronics, medical, aeronautical and nuclear industries to reduce the time necessary for introduction of new products. The coatings industry, however, has continued to depend mainly on long-term in-service tests or long-term outdoor exposures in Florida or Arizona for qualification of new products. Use of SLP methodology is just beginning. Bauer (*2, 3*) and Dickie (*4*) have made a strong case for understanding weathering mechanisms in terms of chemical changes that take place during exposure. The combination of SLP methodology and chemical mechanistic information has been demonstrated for automobile coatings (*2*). Further advances may be possible if better integration can be achieved between three approaches to weathering study: changes in chemical composition (*2-4*), changes in physical properties (*5-9*), and use of SLP (*1*).

© 1999 American Chemical Society

Uses of Physical Property Determinations in Weathering Studies

Initial physical properties of coatings are purposefully optimized by selection of formulation components and specification of application and cure conditions. Material selection takes into account the need to maintain the optimized properties when the coated article is placed in service. The purpose of many determinations of physical properties during weathering is simply to confirm that the desired and optimized properties remain unchanged or change as little as possible during in-service exposure. From the point of view of a coating supplier, the goal often has been to show that a new, hopefully improved, coating is more resistant to changes in properties than the coating that is being replaced. Attempts to answer even this simple question (... is the new paint better or not?) may have under estimated the required sample size. SLP methodology is designed to deal with the high variability in performance among nominally identical specimens exposed at the same time to the same environment (*1*). As a minimum, SLP methodology teaches that a rather large number of samples must be tested if we want to have a high level of confidence that the new paint is better. Within the realm of R&D, this property discussion, so far, focuses on the "D". What about the "R"?

Physical property determinations are useful for understanding the mechanisms of coating failure. Nichols et al (*5*) used chemical stress relaxation to study bond forming and bond breaking processes in acrylic/melamine auto clearcoat systems that had previously been extensively characterized by chemical change methods such as hydroperoxide formation. A Q-U-V® weatherometer (Q-Panel Co.) was modified to house a small load frame for mounting of free film samples. Samples were placed under a predetermined strain (3%), and stress was measured over time of exposure to UV light at constant temperature (80°C) and humidity (dew point 28°C). Both continuous and intermittent relaxation experiments were performed. The acrylic/melamine system that was less resistant to photooxidation showed a rapid decrease in crosslink density beginning immediately on exposure. Crosslink density, expressed as moles of elastically effective network chains per cubic centimeter, decreased from 2.5×10^{-3} moles/cc before exposure to 1.0×10^{-3} moles/cc after about 400 hours exposure. Between 400 and 800 hours, crosslink density was quite constant, but after 800 hours it began to increase again. The system that was more resistant to photooxidation showed an initial increase in crosslink density (2.5×10^{-3} to 3.0×10^{-3} moles/cc) followed by a slow steady decrease. Studies of this type show promise of combining the chemical degradation and physical property information in elucidation of failure mechanisms, but the current state of knowledge is not sufficient to provide a clear linkage between failure modes and environmental exposure factors.

Dynamic mechanical analysis (DMA) before and after exposure of acrylic/melamine (*7, 8*) and acrylic/isocyanate (*6, 7*) clearcoats in laboratory accelerated tests has provided data related to failure by clearcoat cracking and delamination. For clearcoats alone, changes in T_g and crosslink density show remarkably strong dependence on the acrylic/ melamine ratio and on the presence or absence of stabilizers [ultraviolet absorbers (UVA) and free radical scavengers of the hindered amine light stabilizer type (HALS)] (*8*). For example, T_g changes reported at 250 hours Q-U-V® exposure with variation in acrylic/MF weight ratio are as follows: 60/40 (unstabilized) 33°C change, 75/25 (unstabilized) 15°C change, 60/40 (stabilized) 8°C change , and 75/25 (stabilized) 3°C change. Although these changes are for clearcoats alone, it is logical to expect cracking or delamination problems when such changes take place in complete (multi-layer) coating systems. Acrylic/isocyanate clearcoats undergo very large T_g changes during exposure but relatively small changes in crosslink density (6, 7).

Most of the property change studies were carried out on single layer coatings, but some work has been reported for BC/CC systems (7). BC/CC systems of the acrylic/MF type have been analyzed by DMA before and after 500 hours Q-U-V® exposure. The shape of the loss tangent vs. temperature plots depended on the presence or absence of HALS. Maximum values on such plots were taken as an indication of a glass transition. Films that contain UVA but not HALS, show one T_g before exposure and two T_g's after exposure. When both UVA and HALS are present, only one T_g is seen both before and after exposure. Studies of this type show promise of linking property change data with time-to-fail data, but the current state of understanding of fracture processes in coatings is not advanced enough to form the linkage quantitatively. Fracture mechanics (9) is a well developed field for engineering plastics and composites, but it has not been applied frequently to coatings.

Changes in Automobile Coating Technology

More stringent regulations of volatile organic compound (VOC) emissions and recent listing of some coating solvents as hazardous air pollutants (HAPS) have resulted in the need for more rapid approval of new coating products. In-service tests are much too slow to meet the need for changes mandated by regulations. For automotive topcoats, Bauer (2) noted that qualification of new products is complicated further by rapid and continuing changes in technology. The changes have included an increase in the solids content of primer/surfacers (P/S) and topcoats achieved mainly by reduction of molecular weights of the binder components. Appearance preferences by consumers have caused a continuing shift from solid color topcoats to metallic color topcoats. Initially metallic color topcoats were monocoat systems, but development of basecoat/clearcoat systems dramatically improved initial appearance and gloss retention (2, 3). Improved gloss retention also depended on wide spread use of stabilizers (10). Modification of basecoat/clearcoat systems to reduce VOC has continued, but reduction of solvent content in solvent-borne basecoats (SBC) is somewhat counter productive for metallics because the shrinkage associated with solvent loss is important for increasing metallic flake orientation. Waterborne basecoats (WBC) have been developed because they permit extensive volatile loss (mainly water) without exceeding VOC emission regulations (11). The change from SBC's to WBC's is proceeding steadily but rather slowly because the change is usually put off until the paint shop of the assembly plant is scheduled for equipment replacement. The clearcoat is still usually solvent-borne (SCC) so the majority of topcoat systems now are SBC/SCC, and the change is to WBC/SCC. Powder coating clearcoats, which would provide another major step in VOC reduction, are being studied by the Low Emission Paint Consortium made up of the major automobile companies and their OEM coating suppliers, but questions of performance and the timing of acceptance of powder clearcoats remain uncertain (12).

The development of cationic electrocoat primers in the late 1970s followed by increased use of galvanized steel more recently has essentially eliminated corrosion as a failure mode for auto coating systems (2-4).

Gloss Retention - Did We Stay With It Too Long?

As these technology changes were occurring, conventional weatherability testing continued. Gloss retention during outdoor exposures in Florida or Arizona was taken as the main determiner of acceptability of new systems (2). The effects of weathering were reported in terms of gloss loss, and for monocoat systems gloss loss started early and occurred gradually. The poor candidates could be screened out in one to two years. The BC/CC systems behaved very differently, but it took some time to fully realize the differences. For example, for BC/CC systems early gloss loss is very slow and high gloss levels are often maintained for five years or more of outdoor exposure. The long times needed for observing gloss loss outdoors resulted in greater dependence on results obtained by accelerated weathering in weatherometers, but the exposure times required to observe gloss loss in weatherometers also became inconveniently long. For example (10), gloss data supporting the effectiveness of a new HALS was reported for an exposure time of 8400 hours in a Q-U-V® Weatherometer (Q-Panel Company). Since there are 8760 hours in a year, this "accelerated" lab test was continued for 0.96 years. In terms of gloss retention, the BC/CC systems were phenomenally successful. Instead of acknowledging success and changing focus to other aspects of weathering, coating chemists continued to try to use gloss retention for assessing the durability of topcoats. Accelerated tests were modified to increase the severity of the exposure in hopes of seeing gloss loss in shorter times. In SLP terminology increasing the severity is called "increasing the intensities of weathering factors" (1). Lamps with greater intensity and a wider band of UV wavelengths were used. In some cases lamps were used that had significant emission of UV wavelengths not present in natural sunlight. The daily outdoor changes in temperature and moisture were also "accelerated" by both shortening the time of the day/night cycle and by increasing the magnitude of changes during the cycle. In retrospect it is not surprising that BC/CC films were forced to undergo chemical changes that do not occur in outdoor service. Chemical characterization (3) revealed "un-natural chemistry", and it was recommended that the coatings community return to more natural forms of accelerated testing. At this time, the more severe 'un-natural" tests were already lasting very long, and coating developers were reluctant to change to milder "more natural" tests that would likely take much longer. Of course, the real problem here was continuing to concentrate on gloss retention for many years after the gloss retention problem had essentially been solved by BC/CC systems. In product development, solution of one problem sometimes leads to or permits increased emphasis on other problems.

Other Failure Modes

For BC/CC systems other problems are cracking, delamination and environmental etching. Martin et al (1) recommend "fault tree analysis" for displaying the current state of knowledge about relationships between variables and failure modes. A skeleton form for fault tree analysis is given in Figure 1. The uppermost block names the failure mode. The next block provides some key words describing the "failure mechanism". The terms in parentheses are specific to the particular mode under analysis, but the next row of five

		(FAILURE MODE)		
		(Failure Mechanism)		
Application Considerations	Design Considerations	Materials Processing	Materials Properties	Environmental Factors
(surface preparation)	(substrate type)	(cure conditions)	(crosslink density)	(ultraviolet)

Figure 1. Form Used to Prepare a "Fault Tree Analysis" (*1*). Terms in Parenthesis Are Changed to Correspond to a Particular Failure Mode

blocks contains terms that are the same for all modes. These are called "root faults" (*1*). "Environmental Factors" for example, is a root fault. Under each root fault one lists mode-specific causes of failure called "basic faults". "Ultraviolet" for example, is a basic fault. One of the fault tree examples, provided in reference (*1*), is for gloss loss or chalking. The "failure mechanism" block for gloss loss contains the statement "photooxidation and photo enhanced hydrolysis". This comment identifies a cause of gloss loss, but the specific mechanism of gloss loss is likely generation of surface roughness. The distinction is important because photooxidation is also the cause of other failure modes such as cracking and delamination. Terminology that differentiates clearly between failure modes is preferable for the mechanism block. Figures 2, 3 and 4 represent an attempt at fault tree analysis for failure modes observed to occur for BC/CC systems.

Clearcoat Failure By Cracking

For cracking (Figure 2), the failure mechanism is given as hygrothermal stresses as concluded by Oosterbroek et al (*6*). A CoRI Stressmeter® (Braive Instruments) was used to determine the stress in films before and after lab accelerated weathering in Q-U-V® (Q-Panel Co.). Stress was determined as a function of temperature and relative humidity. The T_g and content of hydrophilic groups increased during weathering, and both of these changes resulted in increased hygrothermal stresses. Dynamic mechanical analysis and advanced IR techniques indicated that new crosslinks were formed during exposure. Frequent temperature and humidity changes associated with accelerating the day/night cycle forced the film to expand and contract frequently. Dynamic fatigue from this diurnal cycle is reported to eventually contribute to cracking (*6*). The diurnal cycle is included in the "environmental factors" column. Although cracking was a frequent problem in early BC/CC systems, improvements in resistance to cracking were made quite rapidly.

		CRACKING, CHECKING		
		Hygrothermal Stresses		
Application Considerations	Design Considerations	Materials Processing	Materials Properties	Environmental Factors
BC/CC - wet-on-wet	special effects-metallic, pearlescent	cure conditions	glass transition temperature, T_g	diurnal cycle
		storage stability		temperature
thickness uniformity	two tone vs. single color	total system-pretreatment, cationic EDP, primer/surfacer, basecoat, clearcoat	crosslink density	moisture
			moisture uptake	ultraviolet
overbake resistance	plastic substrates		photostability	
			cohesive strength	
two package vs. one package				

Figure 2. Fault Tree Analysis for Clearcoat Failure by Cracking of Checking

Since photooxidation is identified as a key occurrence for both gloss loss of pigmented coatings and for cracking of clearcoats, it is of interest to review the current accepted belief as to why different modes of failure result. For pigmented coatings, loss of binder around the pigment particles is believed to cause an uneven surface that scatters light rather than reflecting incident light at the specular angle. Surface roughness is believed to develop because pigment particles are freed from binder but remain on the surface or because pigment particles partially protect the binder over a small area so that high points occur where particles are located. For clearcoats, the degradation of binder still occurs, but there is no pigment to cause an uneven surface. It is believed that loss of gloss is slow because the clearcoat surface remains smooth despite degradation, not because degradation itself is slow. There are many chemical (2-4) and physical property (5-8) indications that clearcoats undergo significant degradation.

Automobile Coating Failure By Delamination

Delamination is considered in Figure 3. The "failure mechanism" block indicates boundary layer degradation. The layers where delamination has been observed to occur in service are between the cationic electrocoat and the topcoat or between the basecoat and the clearcoat. Electrocoat/topcoat delamination was only possible for a brief period during which the primer/surfacer (P/S), normally used between the cationic electrocoat and the topcoat, was eliminated. The idea was to use a thicker, smoother electrocoat layer so that P/S would not be necessary. The goal was to save the cost of an application and bake step.

DELAMINATION				
Boundary Layer Degradation				
Application Considerations	Design Considerations	Materials Processing	Materials Properties	Environmental Factors
BC/CC - wet-on-wet	special effects - metallic, pearlescent	thick EDP with no P/S vs. thin EDP with P/S	photostability-stabilizer type, permanence	ultraviolet
				temperature
	two tone vs. single color	water-borne vs. solvent-borne BC-catalyst interference	component migration	moisture
			boundary mixing	

Figure 3. Fault Tree Analysis for Auto Coating Failure by Delamination

It is now quite well established that electrocoat/topcoat delamination occurred because the topcoat, despite the use of UVA, did not prevent some UV penetration to the top surface of the electrocoat primer. Photooxidative degradation at the upper boundary of the electrocoat is believed to result in loss of intercoat adhesion and eventually delamination. The binder used in electrocoat is known to be very susceptible to UV degradation. The term "boundary layer degradation" is used to describe the mechanism in Figure 3. Recently it has been shown that UVA concentration is not as constant as previously believed, and considerable attention has been directed at understanding loss of UVA (2). Of course, loss of UVA may have contributed to transmission of uv light to the top of the electrocoat. Although the practice of omitting the P/S did not last long, numerous warranty claims have resulted from delamination at the electrocoat surface.

Delamination at the boundary layer between BC and CC has received more attention recently. The possibility of forming a weak boundary layer is increased when the BC is water-borne, because there is the possibility of migration of low molecular weight, surface active materials to the interfacial region. The low polarity end of such components will be in the solvent-borne CC and the high polarity end in the water-borne BC. Delamination of WBC/SBC systems has not been widely observed in service, possibly because OEM coating suppliers have used state-of-the-art instrumental methods to characterize the interfacial region between BC and CC. Systems that form weak boundary layers during exposure have mostly been avoided.

When the failure mechanism is cracking or delamination, failure occurs abruptly rather than gradually as is the case for gloss loss. For these mechanisms, failure has been described as "occurring without warning". One of the possible uses of mechanical property determinations during weathering is to provide some warning by establishing property changes that are leading up to an abrupt failure such as cracking or delamination. An abrupt failure mode without prior indication of a developing problem results in the need for a larger number samples. In SLP methodology (*1*), it is seldom practical to wait for all of the coated panels in a test to fail. Those that have not failed before a prespecified time (or number of failures) are said to be "censored". Defining failure in terms of a critical value for a continuously changing property, such as T_g or modulus, will reduce the number of censored panels and would likely permit service life prediction with fewer specimens.

Topcoat Failure By Environmental Etching

Another failure mechanism for automobile topcoats is environmental etching (*13*). This term is used to include visible damage in the form of spots caused by acid rain, bird droppings, insect remains or other materials deposited on cars outdoors. A fault tree analysis for acid etching is given in Figure 4. Acid rain droplets are the most frequent cause of damage. The "failure mechanism" comment indicates that acid catalyzed hydrolysis of the binder is thought to be the critical cause. Even if droplets of rain are initially only weakly acidic, evaporation of an undisturbed droplet will eventually produce a very small amount of very strong acid. Unlike water droplets, acid droplets do not evaporate to dryness but rather to an equilibrium acid concentration depending on the relative humidity. The strong acid is believed to etch away the topcoat surface by hydrolysis reactions resulting in an a spot of uneven depth, often called an etch spot. A less severe form of damage consists of a pale white ring or area that looks like a water spot, but such spots often cannot be removed by washing. They are shallow compared to etch spots and can sometimes be removed by polishing (*13*). Spots are more easily observed on dark metallic colors than on lighter colors.

		Environmental Etching		

		Binder Hydrolysis		
Application Considerations	Design Considerations	Materials Processing	Materials Properties	Environmental Factors
season painted- summer vs. winter	curvature of flat surfaces	weather conditions where newly painted cars are parked	hydrolytic resistance	acid rain
	color - dark vs. light	shipment method- rail or truck, removable PE film	glass transition temperature, T_g	temperature
				moisture
			moisture uptake	ultraviolet

Figure 4. Fault Tree Analysis for Failure by Environmental Etching

As indicated in the materials properties column of Figure 4, etch resistance is improved by increasing hydrolytic stability of crosslinks, increasing the T_g (to the extent possible without losing chip resistance), and selecting film binder components that result in low water uptake. Crosslinks formed in acrylic polyol/MF systems are reported to be very susceptible to hydrolysis, and alternative crosslinking systems of greater hydrolytic stability have been described (13).

An unusual feature of environmental etching is that other changes caused by weathering appear to prevent it. Ultraviolet is listed as an environmental factor in Figure 4, but in this case photooxidation is considered to be beneficial because it is believed to prevent etching. This idea is based on the observation that if etching does not occur early in the life of a car it is unlikely to occur at all. Most of the etching damage occurs during shipping or at car dealers rather than after sale of the car. This time course of failure is quite different from that of the other auto coating failure modes. In SLP terminology (1) etching is said to have "a decreasing hazard rate". This means that "old is better than new" or that resistance to etching improves with time.

A patented approach to improved etch resistance is exposure of a newly applied acrylic/melamine clearcoat (14) or an acrylosilane/melamine clearcoat (15) to a rather high dose of UV light in the presence of oxygen. It is believed that photooxidation occurs to form a thin surface layer of altered chemical composition that is resistant to etching. Attempts to analyze the modified surface layer have not been very successful, which supports the idea that the altered layer is very thin. The protection afforded by UV pre-treatment and the observed decreasing hazard rate in service are believed to come from a similar mechanism, but understanding remains sketchy. Decisions about possible commercial introduction are reported to be imminent at the time of this writing.

One of the major differences between the current weathering protocol and the proposed SLP protocol is the role of outdoor (Florida or Arizona) exposures. Traditionally coatings chemists have accepted outdoor results as the ultimate predictor of durability. Laboratory weatherometer tests were validated based on their ability to cause the same degradation chemistry as occurred outdoors. In SLP methodology lab tests are not validated by comparison with outdoor results. SLP holds that weather never repeats itself, that weathering factors are not reproducible from year to year and that there is no such thing as an "average Florida year".

This outdoor variability is more evident in tests of environmental etching than observed for other failure modes. Tests of the same potentially etch resistant clearcoat in Jacksonville on successive years can give wide ranging results. This variability is believed to be related to the episodic nature of etching. In other words etch spots do not build up uniformly over time, but rather a high fraction of the spots may be formed during a few days when conditions are the worst (high temperature, high humidity, smog, and a very brief rain shower). The number of such episodes are likely to vary from summer to summer in Jacksonville. Acid deposition is reported to be a regional phenomenon, but attempts to establish specific source-receptor relations have not been successful (16). In general, emissions from coal-burning power plants are considered to be the main source of compounds that are converted to acid rain. A rather elaborate environmental smog chamber has been used to produce etch spots and water spots that look very much like those observed in service (17).

Summary and Conclusions

Historically the two most frequent modes of failure of automobile coatings were corrosion (i.e. rust spots and even metal perforation) and gloss loss. Corrosion has been eliminated as an important failure mode by cationic electrocoat primers and increased use of galvanized steel. Gloss loss has been eliminated as an important failure mode by use of BC/CC systems and incorporation of effective UVA and HALS stabilizers. These are notable successes.

Despite previous success, there is no doubt that SLP has a lot to offer for improvement in weathering studies. SLP recognizes the high variability in time-to-fail of nominally identical specimens, and this variability has huge implications for sample sizes needed to draw conclusions with confidence. SLP stresses vigilance for various failure modes. When BC/CC auto coatings were developed, emphasis on gloss retention continued when focus should have shifted much sooner to other failure modes such as cracking, delamination and environmental etching. "Fault tree analysis" diagrams are presented for these three modes to indicate the current state of understanding of the relationship between variables and failure modes. This exercise makes it clear that our knowledge of property changes during weathering is limited. The possibility of complete mechanistic analysis of failure requires more work.

In current SLP work, time-to-fail distributions are often chosen empirically, i.e the Weibull distribution(1). For mechanism-based SLP, the distribution will be derived from the mechanism. Our current state of knowledge about failure mechanisms is insufficient to permit such derivation. Bauer (2) has used time-to-fail models based on degradation chemistry and the assumption that delamination occurs at a certain level of photooxidation. To eliminate the assumption, we need to know the linkage between chemical degradation and physcal property changes and the relationship between physical properties and delamination. Examples of studies directed at the linking the chemical changes and physical property changes have been reviewed here. Two acrylic/MF clearcoat systems, which previously had been shown to differ greatly in photooxidation rate, were analyzed for property changes during accelerated weathering by chemical stress relaxation. This method permitted determination of crosslink density (5). The pattern of change in crosslink density with exposure was very different for the two systems. Dynamic mechanical analysis has also been shown to be useful for determining changes in T_g and crosslink density of auto coatings during weathering (7, 8). Knowledge of property change is not sufficient, however. In addition, we need to establish the connection between physical properties and the details of the failure mechanisms for cracking or delamination. This is a subject for the field of fracture mechanics (9).

SLP places strong reliance on short-term laboratory tests, and therefore, shows promise of reducing the time required to introduce new coatings. The current mind set among coatings chemists that places so much emphasis on outdoor testing and on gloss retention will not be easily changed. The need for more rapid qualification of new coating systems, however, is a very strong driving force for change. Failure mechanism work will undoubtedly continue, but it is important to note that complete mechanistic understanding is not necessary before many of the benefits of SLP can be realized.

322

Literature Cited
1. Martin, J.W.; Saunders, S.C.; Floyd, F.L.; and Weinberg, J.P.; In *Methodologies for Predicting the Service lives of Coatings Systems*; Editors, D. Brezinski abd T.J. Miranda; Federation Series on Coatings Technology; Federation of Societies for Coatings Technology:Blue Bell, PA ., 1996, pp. 1-34.
2. Bauer, D.R.; *J. Coat. Technol.*, **1997**, *69*, No. 864, pp. 85-96.
3. Bauer, D.R.; *J. Coat. Technol.*, **1994**, *66*, No. 835, pp. 57-65.
4. Dickie, R.A.; *J. Coat. Technol.*, **1994**, *66*, No. 834, pp. 28-37.
5. Nichols, M.E.; Gerlock, J.L. and Smith, C.A.; *Poly. Degrad. And Stability*, **1997**, *56*, pp. 81-91.
6. Oosterbroek, M.; Lammers, R.J.; van der Ven, L.G.J. and Perera, D.Y.; *J. Coat. Technol.*, **1991**, *63*, No. 797, pp. 55-60.
7. Hill, L.W.; Korzeniowski, H.M.; Ojunga-Andrew, M. and Wilson, R.C.; *Prog. Org. Coat.*, **1994**, *24*, 147-173.
8. Hill, L.W.; Grande, J.S. and Kozlowski, K., *Proc. of the ACS Div. Polymeric Materials Sci. And Eng.*; **1990**, *63*, pp. 654-659.
9. Williams, J.G., *Fracture Mechanics of Polymers*; John Wiley & Sons: NY, 1984.
10. Bramer, D., Holt, M.S., Mar, A.; *Proc. of the ACS Div. Polymeric Materials Sci. And Eng.*; **1990**, *63*, pp. 647-653.
11. Backhouse, A.J.; *J. Coat. Technol.*, **1982**, *54*, No.693, pp. 83-90.
12. Reisch, M.S., Chemistry & Engineering News, **1997**, Oct. 27 Issue; pp. 43-54.
13. Gregorovich, B.V. and Hazan, I.; *Prog. Org. Coat.*, **1994**, *24*, pp 131-146.
14. Tyger, W.H.; Cornuet, Jr., R.F. and Johnston, B.K.; USP 5,106,651, April, 1992.
15. Nordstrom, J.D.; Omura, H.; Smith, A.E. and Thomson, D.M.; USP 5,532,027, July, 1996.
16. Schwartz, S.E.; *Science;* **1989**, *243*, pp 753-763.
17. White, D.F.; Fornes, R.E.; Gilbert, R.D. and Speer, J.A.; *J. App. Poly. Sc.*, **1993**, *50*, pp. 541-549.

Chapter 21

Stress Development and Weathering of Organic Coatings

Dan Y. Perera, Patrick Schutyser, and D. Vanden Eynde

Coatings Research Institute (CoRI), Ave. P. Holoffe, 1342 Limelette, Belgium

Regardless of the mechanism involved in the development of stress in organic coatings due to weathering, when high hygrothermal stresses are induced they can, alone or in combination with fatigue processes, cause coating degradation, such as cracking and/or detachment. It is suggested in this paper that hygroscopic stress participates in formation and/or enlargement of pathways in the coating, which enable the transport of the electrolyte to the metallic substrate provoking its corrosion.

Protection and decoration, two main functions of an organic coating, are directly dependent on the coating resistance to environmental degradation. Competitive and/or legislative pressures to produce environmentally and user friendly, durable coatings is generating abundant research. In attempting to develop such coatings it is important to be able to predict their service life. To reach this aim at least two conditions are necessary: (i) to have a methodology for predicting the long term weatherability (1), and (ii) to understand the main facets of coating degradation (2).

To reach this understanding a variety of techniques, such as spectroscopy and measurement of mechanical properties, gloss, transport of water, oxygen and ions (3,4), could be used. A number of studies (5,9) contributed significantly to elucidating the chemical changes occurring in certain coatings during weathering, and to the development of tests for monitoring the rate of photo-oxidation.

Physical aging, another process that can affect coating durability, is the relaxation process of a material in glassy state arising as a consequence of its non-equilibrium condition (10-12). In the coating's approach towards equilibrium, its density increases thus inducing important changes in its mechanical, thermal and electrical properties, some of which, e.g., increase in hardness, can decrease the coating's durability.

The importance of electrochemical aspect of corrosion in the comprehension of the degradation of organic coating used to protect metallic substrates has also to be mentioned (13,14).

If chemical changes in coating composition due to the action of UV radiation, moisture and temperature are unquestionable factors in the degradation of most of organic coatings, it is now accepted that stresses arising in a coating are also playing an

© 1999 American Chemical Society

important role in this process (15,16). There is also evidence that the stress accelerates the photochemical and thermo-oxidation degradations (17).

This paper focuses mainly on the role of stress in the failure of organic coatings exposed to weathering and wet conditions.

Stress phenomena

If pure mechanical effects, e.g., stone chipping, are neglected, the main causes inducing stress in a coating are (18) film formation, and temperature (T) and relative humidity (RH) variations, producing, respectively, internal (S^F), thermal (S^T), and hygroscopic (S^H) stresses. In practice, these stresses interact resulting in a low or high total stress (S_{tot}):

$$S_{tot} = S^F \pm S^T \pm S^H \qquad (1)$$

The positive and the negative signs, arbitrarily chosen, indicate the coating tendency to contract (tensile stress) and expand (compressive stress), respectively. Since during film formation the coating is practically always shrinking, S^F is positive.

Equation (1) indicates the possibility of a high tensile stress at low T and RH, and a high compressive stress at high T and RH. Other situations are discussed in detail in references 18 and 19. It could be shown that in an identical environment, the stress arising in a coating is dependent on the coating previous history. For example, the stress in a coating cured and kept at room climatic conditions (e.g. 21°C and 50% RH) will be different from that in a coating brought to a temperature above the glass transition temperature (T_g) for sufficient time. In the first case $S_{tot} = S^F - S^H$ and in the latter one $S_{tot} = S^T - S^H$. This is so because at $T > T_g$, S^F could relax, and the cooling develops S^T.

It is important to add that a high stress reduces adhesion (20-22), and favors the formation and propagation of cracks in a coating. This is especially true for any coating in the glassy region where the stress cannot easily relax. The interdependence between stress and adhesion (21) is expressed by:

$$\beta \cong c\, S\, \epsilon \qquad (2)$$

where β, c, S and ϵ are the elastic energy acting against adhesion, coating thickness, stress and strain, respectively. Therefore, the higher the stress magnitude (S) arising in the coating, the higher is the coating tendency to detach from the substrate.

Failure of organic coatings by loss of cohesion can occur at relatively insignificant stress levels, far below the tensile strength. One of the reasons for this is the presence of stress concentrations in the film due to coating heterogeneity that can initiate fissuring and/or cracking.

Experimental

Stress. The thermal tensile and the maximum hygroscopic compressive stresses measurements were made with CoRI-Stressmeter, Braive Instruments, Liège, Belgium (18,19), and an experimental device that allows stress measurement under liquid water (23,24). The thermal tensile stress was measured by cooling the samples at a rate of 0.4 °C/min, under dry conditions, from a temperature above T_g to one below it (i.e., free of physical aging).The maximum hygroscopic compressive stress was determined by exposing the samples to various RH or water. In this study the samples consisted of coated, pre-calibrated stainless steel substrates.

Materials. The materials used in this study were an acrylic/polyurethane, an acrylic/melamine, a polyester/melamine, a polyester/TGIC powder coating, a polyisobutyl methacrylate (Plexigum), a high T_g acrylic latex, a fast degrading epoxy, and a polyurethane.

Accelerated weathering tests. These tests were carried out in a QUV apparatus by submitting the samples to continuous exposure of UV-A or UV-B at 70°C, or alternating dry/wet cycles of 4 hours of UV-B at 60°C, and 4 hours condensation at 40° or 50°C.

Hygrothermal stress and weathering

If during weathering a coating undergoes photo-initiated oxidation, hydrolysis, and/or thermal degradation, it is also exposed to an increasing hygrothermal stress that often results in coating cracking and/or delamination. Figure 1 shows the stress (S) dependence on T and RH of a 40 μm thick acrylic/melamine coating exposed for 7 days,

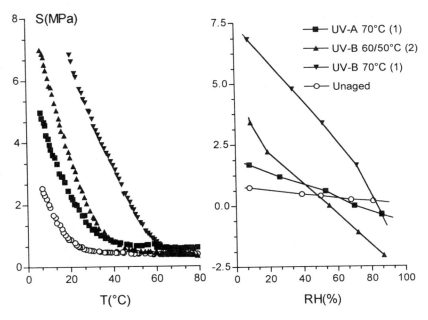

Figure 1. Stress (S) dependence on temperature (T) at 0 %RH (left), and on relative humidity (RH) at 21°C (right) for a 40 μm thick acrylic/melamine coating weathered in a QUV apparatus during 7 days. (1) = continuous UV-A or UV-B exposure; (2) = alternating cycles (4 hours UV-B at 60°C, and 4 hours condensation at 50°C).

in a QUV apparatus, to different weathering conditions. The results indicate not only that, as expected, the type of weathering affects the magnitude of stress as a result of different chemical reactions, but also the great sensitivity of the stress method to detect changes in organic coatings. For many organic coatings important changes in the S = f(T) and S = f(RH) are observed even after a few hours of weathering.

Figure 2, describing S = f(T) and S = f(RH) for an acrylic/polyurethane coating about 90 μm thick after different periods of weathering, demonstrates that weathering strongly affects the stress development of this coating. The displacement of the curves S = f(T) to higher temperatures, and the increase in the slope of the curves S = f(RH) indicates an increase of T_g and of the coating hydrophilicity, respectively, with the duration of weathering (15,16). These properties induce a high tensile stress under dry conditions, and a high compressive stress under wet conditions which, in combination with the fatigue process due to alternation in T and RH occurring during weathering, eventually causes the cracking of the coating.

The increase of T_g with the duration of weathering observed with many coatings can have one or more causes acting separately or, more likely, simultaneously. Such causes could be loss of residual solvents, cure completion (25), and above all, transformation of flexible segments into less flexible ones (15,16). This last cause results from a chain scission due to the photo-oxidation reactions occurring during weathering. This is confirmed by the change of the slope of the curves describing S = f(RH) (see the right side of Figure 2), an indication of an increase in polar functional groups.

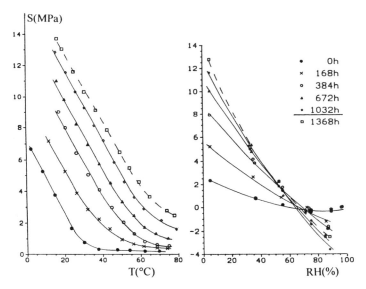

Figure 2. Stress (S) dependence on temperature (T) at 0 %RH (left), and on relative humidity (RH) at 21°C (right) for a 90 μm thick acrylic/polyurethane coating after different periods of exposure (h, hours) in a QUV apparatus. Weathering conditions: alternating cycles (4h, 60°C, UV-B and 4h, 40°C, condensation). Cracking of the coating occurs between 1032 and 1368 h.

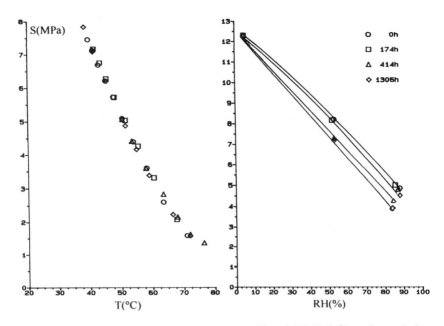

Figure 3. Stress (S) dependence on temperature (T) at 0 %RH (left), and on relative humidity (RH) at 21°C (right) for an about 75 μm thick polyester/TGIC powder coating after different periods of exposure (h, hours) in a QUV apparatus. No cracking of the coating occurs during 1306 hours of exposure.

It is important to add that there were also cases where the coating is cracking without a significant increase of T_g. For these cases an important vertical upward displacement of curves S = f(T) to high tensile stress values takes place. One possible explanation of this process would be that weathering induces the formation of new, photo-oxidized products with higher cross-link density.

Figure 3 illustrates the case of an organic coating with good weathering resistance, i.e., a polyester/TGIC powder coating. Contrary to acrylic/melamine coatings used in this study, this powder coating is practically unaffected by the accelerated weathering for the period of time investigated. Indeed, for this coating S = f(T) and S = f(RH) are not changing or changing very little with weathering. The results obtained here are in agreement with the behavior of these coatings under tropical conditions. While the acrylic/polyurethane coatings investigated are cracking after a few years, the powder coating is still undamaged after the same period of time.

It is important to note that in Figures 1-3 the stress was calculated by considering the total film thickness, i.e., as if the whole thickness (40, 75 or 90 μm) is submitted to the UV aggressive action. In reality, the weathered layer is certainly much thinner, at least during the first stages of aging. This implies that the coating can be considered at least as a bi-layer system. The mathematical equations developed to calculate the stress in each layer of a multicoat system (26) can also be used to determine the stress in the weathered layer more accurately [see equation (3) for a bi-layer system]. Such calculations indicate that in most cases the stress developed in the weathered layer is much higher during the first period of weathering than those presented in Figures 1-3, and references 15 and 16.

$$S_2 = \frac{1}{c_2(t+2c_1+c_2)} \frac{E_S\, t^3}{6(1-v_S)} \left[\frac{1}{R_2} - \frac{1}{R_1}\right]$$

(3)

where:

t, c_1, c_2:   thickness of the substrate, layer 1 and layer 2, respectively;

R_1, R_2:   radius of curvature of the bent coated substrate with one and two layers, respectively;

E_S, v_S: elastic modulus and the Poisson's ratio of the substrate, respectively

Hygroscopic stress and failure of coating/metal systems

A case often encountered in practice (i.e., in natural and accelerated weathering) is the stress arising during immersion of a coated substrate in water or its exposure to a high RH. Previous studies (18,23) indicate that for a great number organic coatings the stress development as a function of time can be represented schematically by the curves shown in Figure 4. These curves indicate that the immersion of a coated substrate in water induces a hygroscopic compressive stress (arbitrarily chosen as negative values, i.e., curve 1) that, after a period of time, starts to decrease (curve 2).

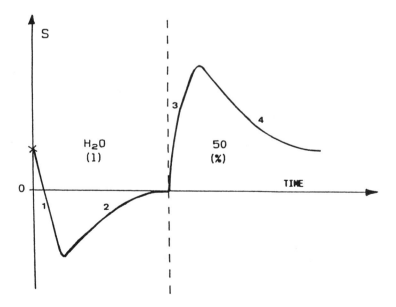

Figure 4. Schematic representation of the stress dependence on time during immersion of a coating in water and exposure to 50 %RH, at 21°C; X = initial stress

The exposure of the coatings to 50% RH induces first the development of a hygroscopic tensile stress (curve 3) that also eventually decreases (curve 4). If no delamination occurs, these decreases (curves 2 and 4) are due to stress relaxation, the relaxation rate being related to the real T_g of the coating immersed in water or exposed to 50% RH, respectively. The lower the T_g the faster is the relaxation process.

The determination of the maximum hygroscopic compressive and tensile stresses, as well as the time necessary to reach them, is useful for a theoretical interpretation of the stress results and from practical point of view. For example, the consideration of these times in programming the wet/dry cycles, with/without UV, will accelerate the deterioration of coatings, thus reducing the time necessary for their selection.

Another important fact is the role of the hygroscopic stress in the failure of coating/ metal systems exposed to electrolyte or water (27). Examples of the hygroscopic compressive stress development in five organic coatings of different nature are shown in figure 5.

Alternative Current (AC)-impedance, wet adhesion, water uptake and thermal analysis measurements were also carried out. The examination of the data collected indicated only one good correlation, namely between the hygroscopic stress and AC-impedance measurements. The coating which provided the best protection against corrosion, in this case the powder coating, developed the lowest hygroscopic compressive stress and had the highest impedance moduli at low frequency.

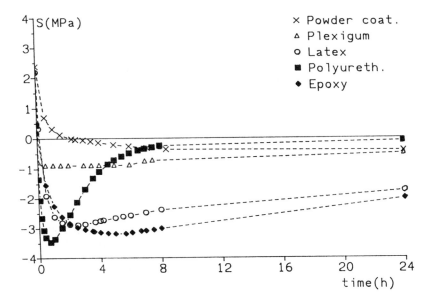

Figure 5. Stress (S) as a function of time (hour, h) at 21°C for five coatings immersed in water.

It is important to add that a coating which develops a relatively high hygroscopic compressive stress, still has a satisfactory behavior if the stress relaxes relatively fast.

The good correlation between the values of hygroscopic stress and AC-impedance signifies that the stress contribution to coating degradation can be significant. The heterogeneous nature of coatings with regions of various degree of hydrophilicity induces local stresses when exposed to a high RH. This is a consequence of a difference in the swelling between the hydrophilic and hydrophobic regions. We suggest that these hygroscopic local stresses contribute to the enlargement of pathways providing the passage for the electrolyte to reach the substrate, thus causing corrosion in agreement with the model discussed in the reference 13.

Conclusions

Chemical changes occurring during weathering can induce high hygrothermal stresses that, alone or in combination with fatigue processes due to variation in temperature and relative humidity, can provoke coating cracking and/or loss of adhesion. It is suggested that the hygroscopic stress participates in formation and/or enlargement of pathways in the coating, which enables the transport of the electrolyte to the substrate provoking its corrosion. The consideration of the time necessary to reach the maximum tensile and compressive stresses should result in efficiency improvements in the accelerated weathering tests. The measurement of stress as a function of temperature and relative humidity is also a sensitive way to detect early changes in an organic coating due to weathering.

Acknowledgement

The active participation of Dr. Tinh Nguyen from National Institute of Standards and Technology, NIST (USA), in part of the project is acknowledged.

Literature cited

1. Martin, J.W., Saunders, S.C., Floyd, F.L. and Wineburg, J.P., *NIST Building Science Series 172*, "Methodologies for Predicting the Service Lives of Coating Systems", U.S. Department of Commerce, Washington, DC 20402-9325, 1994.
2. Bauer, D.R., *J. Coat. Technol.*, **1997**, 69, N° 864, 85.
3. *Prog. Org. Coat.*, **1987**, 15, No.3, 8 papers.
4. Symposium on "Durability of Coatings", **1993**, *ACS (PMSE Div.)*, 68, 26 papers.
5. Gerlock, J.L. and Bauer, D.R., *J. Polym. Sci. Polym. Lett.*, **1984**, 22, 447.
6. Bauer, D.R., Briggs, L.M. and Gerlock, J.L., *Polym. Sci.Polym. Ed.*, **1986**, 24, 1651.
7. Bauer, D.R., Mielewski, D.F. and Gerlock, J.L., *Polym. Deg. Stab.*, **1992**, 38, 57.
8. Bauer, D.R., *ACS (PMSE Div.)*, **1993**, 68, 62.
9. van der Ven, L.G.J. and Hofman, L.H., *ACS (PMSE Div.)*, **1993**, 68, 64.
10. Struik, L.C.E., *Physical Aging in Amorphous Polymers and Other Materials;* Elsevier, Amsterdam, Holland, 1978.
11. Perera, D.Y. and Schutyser, P., *22th FATIPEC Congress*, Budapest, Hungary, **1994**, 1,25.
12. Perera, D.Y. and Schutyser, P., *Prog. Org. Coat.*, **1994**, 24, 299.
13. Nguyen, T., Hubbard, J.B. and Pommersheim, J.M., *J. Coat. Technol.*, **1996**, 68, No.855, 45.

14. Piens, M. and Verbist, R., in Leidheiser, H. Jr., (ed.), *Corrosion Control by Organic Coatings*, Natl. Assoc. Corros. Eng., Houston, TX, 1981 (p. 32).

15. Oosterbroek, M., Lammers, R.J., van der Ven, L.G.J. and Perera, D.Y.., *J.Coat.Technol.*, 1991, 63, No.797, 55.

16. Perera, D.Y. and Oosterbroek, M., *J.Coat.Technol.*, **1994**, 66, No.833, 83.

17. White, J.R. and Rapoport, N.Y., *Trends in Polym. Sci.*, **1994**, 2, No.6, 197.

18. Perera, D.Y., in Koleske, J.V.(ed.), *Paint and Coating Testing Manual, 14th edition of the Gardner-Sward Handbook*, ASTM (MNL 17), 1995 (p. 585).

19. Perera; D.Y. and Vanden Eynde, D., *J.Coat.Technol.*, **1987**, 59, No.748, 55.

20. Croll,S.G., in Mittal, K.L. (ed.), Adhesion Aspects of Polymeric Coatings, Plenum, 1983, (p.107).

21. Perera, D.Y., *Prog. Org. Coat.*, **1996**, 28, 21.

22. De Deurwaerder, H.L., *23nd FATIPEC Congress*, Brussels, Belgium, **1996**, vol.A, 1.

23. Perera; D.Y. and Vanden Eynde, *20th FATIPEC Congress*, Nice, France, **1990**, 125.

24. Perera; D.Y. and Vanden Eynde, D., *J.Coat.Technol.*, **1981**, 53, No.677, 39.

25. Hill, L.W., Kerzeniowski, H.M., Ojunga-Andrew, M. and Wilson, R.C., *19th Intern. Conf. Org. Coat. Sci. and Technol.*, Athens, Greece, **1993**, 225.

26. Boerman, A.E. and Perera, D.Y. to be published in *J.Coat.Technol.*.
    Perera, D.Y. and Nguyen, T., *EUROCOAT 96*, Genova, Italy, **1996**, vol. 1, 2 or 3 (p.1); or *Double Liaison*, No; 489, **1996**, 22 and 66.

Chapter 22

The Effect of Weathering on the Stress Distribution and Mechanical Performance of Automotive Paint Systems

M. E. Nichols and C. A. Darr

Ford Motor Company, P.O. Box 2053, MD 3182 SRL, Dearborn, MI 48121

The sources of stress in complete automotive paint systems have been identified and measured as a function of weathering. The main sources of stress are thermal expansion coefficient mismatch, humidity expansion mismatch, and densification of the clearcoat. Stresses generally increase during weathering due to a slow densification of the clearcoat and increasing water absorption and desorption stresses. Finite element analysis (FEA) was used to compute the stress distribution in full paint systems. Stresses are typically in-plane and highest in the primer and clearcoat. Stresses approaching those required to propagate cracks can be attained in weathered paint systems. The presence of flaws, either cracks or incipient delaminations, will lead to large stress concentrations that can give rise to peeling forces not present in coatings without cracks.

Modern automotive paint systems are highly complex, multilayer structures where each layer performs multiple functions. For example, the primary role of the clearcoat is to enhance appearance by maintaining a high level of gloss. However, the clearcoat must also screen the underlying layers from harmful ultraviolet radiation, which is accomplished by doping the clearcoat with ultraviolet light absorbers. While each individual layer is engineered to possess certain chemical and physical properties, many aspects of paint performance depend on system properties and interactions between layers. Hindered amine light stabilizers (HALS), for example, are often added to the clearcoat, but during cure the HALS can migrate to the basecoat thereby distributing the additive and coupling the basecoat performance to the composition of the clearcoat (1). Similarly chip resistance is related to the brittleness of the clearcoat, the stiffness of the underlying layers, and the geometry of the vehicle structure (2,3).

While potential weathering-induced failures on vehicles usually have their origin in chemical composition changes brought on by photooxidation (4-10), the failures that may finally be observed are typically mechanical in appearance: cracking of the clearcoat or delamination of one layer from another. These failure

© 1999 American Chemical Society

modes are driven by two separate events, the changing toughness of the material or interface where the crack may grow and the stresses the material or interface experiences during exposure (see Figure 1). While quantification of the fracture toughness or fracture energy of structural materials and interfaces is becoming common, only recently has the fracture behavior of thin polymer films, and in particular automotive coatings, been explored (11,12). This has been due to the difficulty in handling and preparing samples and also to the lack of theoretical understanding of the fracture mechanics of thin films. However, these problems have been recently addressed (13-16).

The question of the stress distribution in thin films is equally important when considering their mechanical failure, be they single or multilayer systems. The most brittle material or film will not crack without a stress to drive the crack tip forward. Stresses can arise in coatings from a number of sources. A mismatch in the thermal expansion coefficients between the coating and substrate (or between coating layers) will cause stresses to arise when the temperature is changed. Changes in the humidity of the environment can cause swelling and plastisization of a coating leading to changes in it's stiffness and dimensions. These changes can lead to stresses in coatings adhering to substrates as can a slow increase in the density of the coating.

For a simple single layer coating on a substrate, the stresses produced by any of the above phenomena can be simply calculated given the appropriate material constants for both the substrate and the coating, i.e. the coefficient of thermal expansion, the density change, or the amount of swelling and the elastic modulus of the coating and substrate. A number of techniques also exist to experimentally measure the stress in coatings on substrates. The classical method of applying a coating to a thin metal shim has been used widely, and a commercial instrument is available (17-19). For small deflections the use of Corcoran's equation is quite satisfactory (20). In addition, the use of interference fringes (21), Raman scattering (22), and fluorescent probes (23) can be used to measure the stresses in coatings. A limited number of reports have been published regarding the changes in stress as a function of weathering. Perera et. al. monitored the stresses as a function of humidity and weathering and found the stresses in most coating systems to increase with weathering and decrease with humidity (24,25). Perera ascribed most of the stress increase to an increase in T_g. Both Perera and Shiga have noted that stress relaxation, that is the decay of stress over time, is important in quantifying the stress response to any environmental perturbation (23,25).

For multilayer systems the problem of deducing the stress in individual layers is much more difficult. Knowledge of the stress in an individual layer from data taken in multilayer systems is essentially impossible using beam deflection techniques because only information on the overall or mean stress in the coating system is accessible. The fluorescent probe technique is an exception and appears promising, as it can probe the stresses in each layer of a coating system (23). However, the stress range over which current probes are effective is limited. By measuring the relevant material properties of isolated, individual layers, the stresses in a multilayer system can be calculated. However, because no closed form

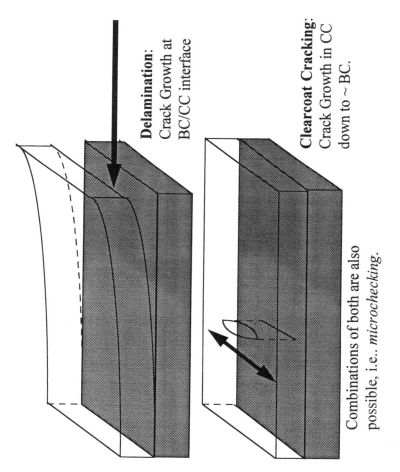

Delamination: Crack Growth at BC/CC interface

Clearcoat Cracking: Crack Growth in CC down to ~ BC.

Combinations of both are also possible, i.e.. *microchecking*.

Figure 1. Two possible modes of crack propagation in basecoat/clearcoat paint systems. Delamination (upper), where the clearcoat peel off of the basecoat. Cracking (lower), where small cracks form in the clearcoat and propagate down to the basecoat and spread laterally in the clearcoat.

analytical solution for the stresses in each layer of a multilayer system with arbitrary geometry exists, one must resort to numerical methods to calculate stresses in individual layers.   A number of techniques exist for doing this including finite element analysis (FEA) and boundary element methods (BEM) (26,27). The main hurdle to applying such methods is the time involved in generating the material properties for each layer of a multilayer system.

By knowing the stresses in various layers of a modern automotive paint system one could more accurately comment on the likelihood of potential failures in any given layer or at any interface. Also, by knowing which material properties play the most important role in determining the stresses, better coating systems could be designed which could more effectively manage environmental inputs. In this paper we report on our initial attempts at quantifying the stresses in various layers of a modern automotive paint system as function of weathering and comment on how these stresses relate to the material properties that likely govern potential failure mechanisms.

Experimental

Materials. The thermoelastic constants of three different clearcoats (clearcoats B, C, and D), two basecoats, and one monocoat (Monocoat A) were measured. All of these coatings were acrylic/melamine based. In addition an epoxy based e-coat and polyester based primer coat were studied. For humidity-stress testing an additional clearcoat (E) based on acrylic/silane chemistry was also examined. All of the coatings were prepared as isolated films. No multilayer laminates were made. The coatings were applied to substrates, either tin-plated steel, poly(tetraflouroethylene), or aluminum, with a Byrd applicator and cured for the recommended times and the recommended temperatures. Free films were made by removing the coating from the substrate by peeling or my forming an amalgam of the tin with mercury. Free films were die cut into strips for testing.

Accelerated Weathering. All clearcoats and basecoats were weathered for various times in a QUV weathering chamber (QUV Co.). FS-340 nm bulbs were used as the light source. The temperature was held constant at 40°C and the dew point at 25°C. These conditions correspond to the "near ambient" exposure used by Bauer et. al (4).

Dynamic Mechanical Analysis (DMA) and Thermomechanical Analysis (TMA)
The elastic modulus as a function of temperature was measured in tension for all the free coating films using a Polymer Laboratories DMTA. The coefficient of thermal expansion (CTE) was measured using the same instrument in TMA mode. In this configuration, the film is held in tension under a negligible load as the temperature is changed. The instrument adjusts the length of the film to maintain zero load. The CTE is the change in length divided by both the change in temperature and the original film length. CTE measurements were made as a function of weathering on

each basecoat and clearcoat. CTE measurements were also made on unweathered primer and e-coat.

Density. The density of each clearcoat was measured as a function of weathering using the method of Guy-Lussac. Small pieces of clearcoat were immersed in a beaker containing an aqueous solution of ZnCl. The concentration of the solution was adjusted until the piece of clearcoat was perfectly suspended in the middle of the solution, indicating a match in densities of the solution and clearcoat. The solution was then transferred to a Guy-Lussac flask (the exact volume and mass of which had been previously determined). The filled flask was weighed and the mass and then density of the solution calculated. The density of polymers can be measured accurately to ±0.002 g/cm^3 using this technique.

Humidity Stresses. Changes in the internal stress of coatings produced by changes in humidity were measured using a custom built load frame. The small load frame was equipped with a sensitive load cell and signal conditioner (Transducer Techniques, Temecula CA) as shown schematically in Fig. 2. The load frame was placed inside a UV2 weathering chamber (Atlas Electric Co.) with the load cell and strain adjustment screw outside the chamber. The chamber temperature was maintained at a constant 40°C. A free clearcoat film was subjected to a small strain, by adjusting the screw at the top of the load frame. The load as a function of time was recorded with a strip chart recorder. A dry atmosphere (0°C dew point) in the chamber was produced by purging the chamber with nitrogen gas. A dew point of approximately 25°C was produced by filling the bottom of the chamber with water. The stress in the coating as a function of dew point was recorded. All stresses were allowed to relax to relatively steady-state values before environmental conditions were changed. Weathering was accomplished by taking the free films and mounting them with tape on an aluminum panel. These films were then exposed in the QUV and removed periodically for the stress measurements described above. For a given clearcoat, the exact same specimen was used for each measurement at different weathering times. In this manner, the humidity induced stresses could be measured as a function of weathering.

Modeling of Stresses in multilayer Coatings. A number of methods exist to approximate the stresses that arise in thin multilayered structures due to changes in temperature. Vilms and Kerps recognized previous attempts to solve these problems were plagued by inconvenient calculation schemes and cluttered nomenclature (28). Their analysis addresses these issues and is improved upon by Townsend et. al (29). Suhir has furthered the work by extending the calculations to real structures with finite dimensions, taking into account the nontrivial edge effects (30). Suresh has accounted for plasticity in one or more of the layers (31).

All of the methods used to calculate the stresses in each layer rely on the same fundamental principle: the different thermal expansion coefficients of each material will cause each layer to seek a new equilibrium size when the temperature is changed. If the layer is constrained by adhering to another layer(s), a stress will

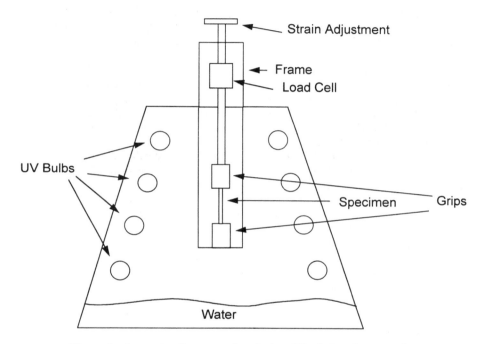

Figure 2. Apparatus for measuring the humidity induced stresses in clearcoats.

arise in the film. If the substrate is not infinitely thick compared to the films on top of it, a curvature of the substrate/film structure will ensue as the forces and moments on the composite structure must sum to zero. The approach is general. Differential thermal expansion is just the most common subset of circumstances under which stresses arise in multilayer thin films. This can easily be extended to account for changes in dimensions due to moisture induced swelling or the gradual change in specific volume of the various layers due to physical or chemical aging. However, summing the effect from moisture, thermal expansion and densification proved cumbersome using analytical techniques. In addition, the analytical solutions are restricted to simple planar geometries, and the introduction of flaws cannot be accounted for.

For these reasons, the stress distributions were determined using finite element analysis (FEA). Commercial FEA software (ANSYS, Ansys Corp.) was used. 8-node quadrilateral elements were employed for the analysis. Unlike analytical solutions which can only deal with simple geometries, FEA analysis can solve structural mechanics equations for complex geometries with complex constitutive behavior. This is accomplished by breaking down complex geometries into smaller "elements" on which a computer can solve the standard set of differential equations for displacements and stresses. In addition material nonlinearities can be accounted for as well as the summation of various stress sources. However, a number of simplifying assumptions were made in the FEA analysis presented in this paper to prevent the details of the analysis from obscuring the general conclusions. It was assumed that (1) the thermal expansion (or humidity expansion) coefficient is constant over the relevant temperature range, (2) the elastic modulus remains constant over the same temperature range, and (3) the effects of relaxations can be accounted for by simply using an appropriately smaller modulus. The validity of these assumptions will be addressed later in this paper. It should be emphasized that the FEA technique does not require these assumptions, and that the added complexity of the required analysis would not dramatically change the results, but would significantly increase the analysis time.

Results

Thermoelastic constants of Individual Layers. Table I shows the measured thermal expansion coefficients and elastic modulus for each of the clearcoats and basecoats, along with the primer and e-coat that were common to each system. Thermal expansion coefficients were generally higher in the clearcoat and basecoat and lower in the primer, e-coat and the monocoat topcoat. The modulus of the primer was higher than the other coatings because of its substantially higher pigment concentration. Table 1 also shows the changes in CTE and modulus for each coating layer as a function of weathering.

Humidity Stresses in Single Layers. Stresses will result in coatings when the ambient humidity is changed. The coatings will take up water and swell when the

Table I. Thermoelastic constants of coatings. Note: All Moduli (E) are in units of GPa. All CTE (α) are in units of $°C^{-1}x10^{-6}$. Subscripts refer to weathering time in hours.

Material	E_o	α_o	E_{1500}	α_{1500}	E_{3000}	α_{3000}
CC B	1.6	84	1.8	74	2.1	109
BC B	1.4	95	1.1	70	1.2	74
CC C	2.1	109	2.2	101	2.5	108
CC D	1.9	123	2.5	102	2.3	102
BC D	1.9	106	2.0	92	2.1	92
Monocoat A	2.1	77	2.6	70	2.3	77
Primer	3.2	82	---------	---------	---------	---------
E-coat	0.8	63	---------	---------	---------	---------
Steel	205	10	---------	---------	---------	---------

humidity is raised and will shrink and stiffen when the humidity is lowered. The humidity response of a typical clearcoat is shown in Figure 3, where the stress as function of time is shown. The dew points at various times throughout the test are shown on the graph. The measurements were made at a constant dry bulb air temperature of 40°C. The stress in the clearcoat initially relaxes due to viscoelastic effects. At point A the dew point is lowered to 0°C and the stress goes up. When the dew point is quickly raised (point B), the stress quickly drops. A plateau level is eventually attained (C). As the dew point slowly decreases, the stress slowly rises (D to E) until at point E it reaches its former level just before point A. We have defined the stress response to humidity as the difference in stress between points E and C. By normalizing the stress difference (E to C) by the change in dew point (in this case 25°C) the humidity response can be quantified for arbitrary changes in dew point. This is shown for three different clearcoats in Figure 4. The results have been normalized to changes in stress per one °C change in dew point. Clearcoat E is quite sensitive to changes in humidity. The fully stabilized version of clearcoat D (contains HALS and UVA) is quite insensitive to changes in humidity while the unstabilized version (no HALS or UVA) is more sensitive.

Densification Stresses. The density of all the clearcoats changes with weathering time. Table II shows the density of the clearcoats as a function of weathering time. Because most of the chemical composition changes occur in the clearcoat, the density of the other layers was not monitored.

Multilayer Stresses. All of the stress profiles for multilayer systems were done using the same thickness numbers for the coatings: 25 μm e-coat, 25 μm primer, 25 μm basecoat, and 50 μm clearcoat. The analysis assumes a biaxial stress state in the

340

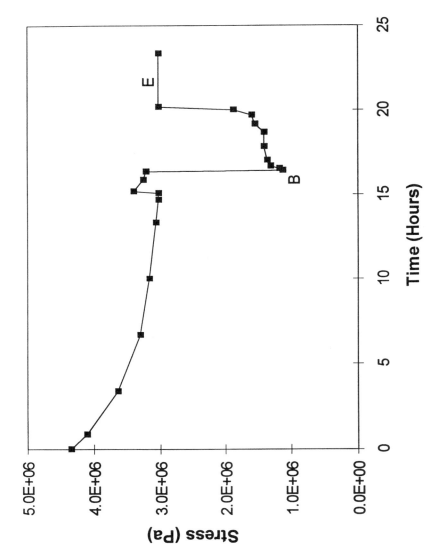

Figure 3. Typical stress response of clearcoat to changing humidity.
Dewpoints: Before A, 5°C; A, 0°C; B, 25°C; C, 25°C; D, 25-0°C; E, 0°C.

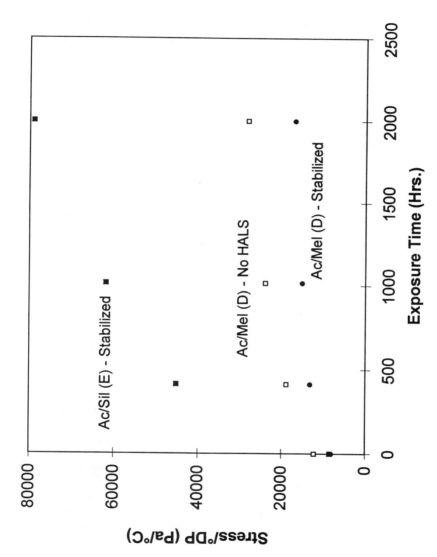

Figure 4. Change in humidity/stress response with weathering for three clearcoats.

Table II. Density (g/cm^3) of three acrylic/melamine clearcoats as a function of accelerated weathering time.

Clearcoat	Density(t=0)	Density(t=1500 hrs)	Density(t=3000 hrs)
Monocoat A	1.127	1.131	1.132
Clearcoat B	1.148	1.155	1.162
Clearcoat C	1.20	1.196	1.205
Clearcoat D	1.165	1.172	1.174

Table III. Maximum calculated stresses in each layer of a complete unweathered and weathered paint system. Stresses do not account for relaxation effects, which would reduce values by roughly 50%. Moduli and CTE values determined at 10°C dew point.

Coating	Stress Max. t=0 hrs (MPa)	Stress Max. t=3000 hrs. (MPa)
Clearcoat	8.8	15.5
Basecoat	8.8	5.7
Primer	17.2	17.2
E-coat	3.2	3.2

coating system, as would be the case on a large fraction of the vehicle. This enhances the stress over the values observed in a simple beam deflection experiment by a factor of 1/(1-v) due to Poisson's effects.

Figure 5 shows the stress distribution in a BC/CC paint system after the temperature has dropped 50°C to 25°C. The temperature is uniform throughout the paint system and the different grey-scales signify different contours of stress. The top of the clearcoat is at the top of the figure and the analysis extends down approximately 350 μm into the paint and steel. The thermoelastic constants chosen represent the unweathered paint system B as in Table I. The stresses range from 17 MPa in the primer to only 3 MPa in the e-coat. The stresses shown are the in-plane stresses. The normal stresses (not shown) are comparatively uniform throughout the coating system and approach zero. Stresses are highest in the primer due to its high modulus. Figure 6 shows the in-plane stress distribution for a system whose thermoelastic constants correspond to weathered paint system B. Changes occur mostly in the clearcoat and basecoat regions. Table III lists the maximum stress in each layer of the coating for the two weathered and unweathered coating systems.

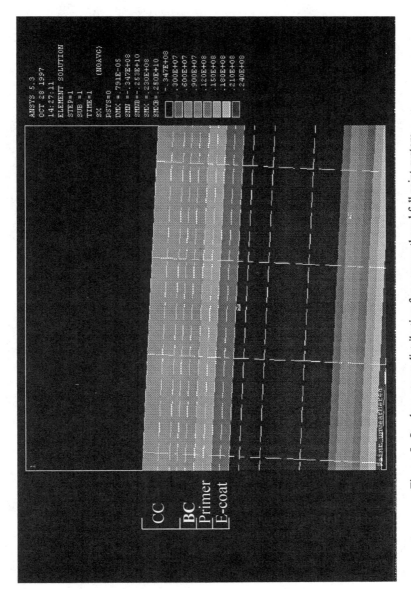

Figure 5. In-plane stress distribution for a unweathered full paint system on cooling from 70°C to 20°C. Grey-scales correspond to different stress levels. Stresses are as in Table 3.

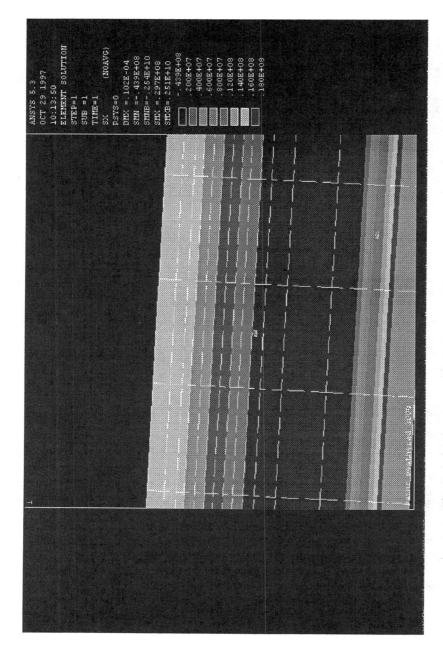

Figure 6. In-plane stress distribution for a full paint system weathered 3000 Hrs and cooled from 70°C to 20°C. Stresses as are in Table 3.

Discussion

The combination of increasing stresses and changing mechanical properties lay at the root of most mechanical failures of paint systems. The changing mechanical properties are largely driven by the weathering induced chemical composition changes of the various coating layers. The details of the chemical changes play a key role in determining the mode and location of any possible mechanical failures (8). However, even the most brittle materials and interfaces will not fail in the absence of stress. At least three sources of stress have been identified: thermal expansion mismatch, humidity expansion mismatch, and densification.

The coefficient of thermal expansion results indicate that, depending on the formulation of the clearcoat, the CTE can vary between clearcoats by up to 50%. Larger CTEs will always lead to higher stresses given the same modulus, but typically, coatings with high CTEs will have lower elastic moduli, as the two properties are not fully independent on a molecular scale. The thermal expansion coefficient first decreases and then increases with weathering for all three clearcoats studied. This mirrors changes that have been observed for other properties, such as crosslink density. Similar results were observed for the basecoats, with first a decrease then increase in CTE. Much smaller changes were observed in the room temperature modulus of the coatings as a function of weathering. This is not surprising as well below T_g the modulus of most polymers is quite constant and similar to each other. Changes in modulus at elevated temperatures, where changes in crosslink density have occurred, are much more pronounced (32).

The changes in density exhibited by the clearcoats can vary from near zero to almost 1.5% after 3000 hours of weathering. Assuming all of the density change is due to a collapse of volume and not an increase in mass, a coating constrained on a substrate will experience a tensile stress as the density increases. The change in any linear dimension is the cubed root of the change in the density so a change of 1.5% in the volume will lead to a roughly 0.5% change in the linear dimension. For clearcoats with typical elastic moduli this will lead to a strain of almost 10 MPa in the unrelaxed state. However, this strain will relax in the same manner as the thermal strain.

The exact origin of this densification has been studied extensively in many polymers and recently for a number of coatings. If the densification is reversible it is termed physical aging and is due to the amorphous polymer slowly progressing towards its equilibrium specific volume (33-35). This densification proceeds more quickly at temperatures close to, but below, the polymer's glass transition temperature (T_g). When the polymer is exposed to a temperature above its T_g it will recover to a higher specific volume, thus the process is reversible. Chemical aging can also cause densification and occurs when the polymer undergoes chemical composition changes, due to degradation or continued curing, over the course of time. These changes are not reversible as they result from the rupture and formation of chemical bonds. In many cases these chemical changes can lead to a decrease in the specific volume of a polymer. This is because most curing reactions are volume reducing (with the exception of expanding monomers) and most degradation reactions will increase the polarity of the network increasing the likelihood of hydrogen bonding. While volume decrease effects probably dominate the density increase phenomena, photooxidation can lead to the incorporation of oxygen which can initially lead to an increase in density. This is likely offset in the latter stages of photooxidation as small molecules such as CO_2, methanol, and formaldehyde are produced (36). If the density increase is indeed due to volume reduction, not mass increase, tensile stresses will occur. However, this assumption has not been proven. Most coatings as they weather will decrease in thickness due to the loss of the most degraded material near the surface. The density of the material left behind is the crucial parameter. Assessing its density is easy; confirming that it is indeed due to volume collapse and not mass increase is difficult due to the concomitant thickness loss.

The complications from stress relaxation are common to both the density and thermal stress measurements and are due to the viscoelastic nature of polymers. Because the elastic modulus is a function of frequency or time, a given strain will produce a constantly decreasing stress with time. For crosslinked polymers this stress will eventually plateau. For coatings well below their T_g this may take an experimentally unrealizable time, but for coatings near T_g the relaxation processes can occur quite rapidly, making it unlikely that large stresses can be supported by coatings near their T_g.

Some estimate must be made for the fraction of stress on a coating that will remain after reasonably long periods of time. We accomplish this by comparing the residual stress in coatings after curing to the stress values calculated from measured moduli and CTE values. For clearcoat B, for example, the residual stress measured using the shim bending technique was 2.0 MPa. From CTE and moduli values for brass and the clearcoat and assuming stress does not build up until after the polymer is cooled below its T_g (55°C). The stress can be calculated using

$$\sigma = E\Delta T\Delta\alpha \tag{1}$$

where E is the elastic modulus of the clearcoat, ΔT is the temperature change, and $\Delta\alpha$ is the difference in thermal expansion coefficients between brass and the clearcoat. The calculated value is approximately 4.0 MPa. The ratio between the two being approximately 1:2. This ratio appears to hold for most clearcoats. Therefore, as an estimate, we use a factor of two as the amount by which the stress will relax during normal service conditions. Analytically this can be affected by using a modulus value one half of the measured value. Of course in-service vehicles see constantly changing temperatures and humidities. During hot cycles the stress will be much lower and during cold cycles the stress can be higher. These temperature extremes, where the material properties could be quite different from the room temperature properties used in this analysis, can be accounted for using FEA techniques. In addition, the transient stresses set up when a paint system transitions from one environment to another will be highly dependent on diffusion and on the viscoelastic nature of the polymer coatings. If required, these effects could also be accounted for in the analysis. In addition, the increase in T_g can lead to higher stresses on cooling a coating from above T_g, due to the larger difference between T_g (below which stresses can be supported) and the ambient temperature (19).

Weathering has a much more pronounced effect on the humidity induced stresses. The majority of chemical changes that can occur in clearcoats during weathering produce polar or hydrophilic groups in the network. The production of these groups increases the clearcoat's tendency to absorb moisture. This is also the main reason the T_g of coatings tend to increase as weathering progresses. Some coatings are much more susceptible to these chemical changes than other coatings. For example, in Figure 4 the acrylic/silane clearcoat becomes quite sensitive to moisture as weathering progresses, while the conventional, stabilized acrylic/melamine clearcoat changes comparatively little. However, the unstabilized version of the clearcoat D shows a greater humidity/stress response as weathering proceeds, in accordance with the greater amount of photooxidation occurring in the clearcoat and a greater proportion of polar species being produced. The acrylic/silane clearcoat likely shows an increasing sensitivity to moisture as weathering progresses due to the enhanced network formation these coatings undergo after exposure to moisture.

The changes in dew point that are experienced either in a weatherometer or outdoors rarely exceed 25°C. Thus, the humidity induced stresses in a clearcoat can range from 0.25 MPa to up to 2 MPa. These stresses are the fully relaxed stresses as the experiments were performed after waiting for the clearcoats to come to stress plateaus. These values would be enhanced by transient effects. Perera has shown that large stress overshoots can occur when coatings are taken from wet to dry environments very quickly (37). Also in multilayer coatings, moisture will diffuse from the surface first leaving the underlying layers swollen, giving rise to enhanced surface stresses. In Fig. 3 the effects of these transient can be seen as a stress undershoot and then reverse relaxation to a higher stress when the humidity is quickly increased. While not intuitive, this reverse relaxation is viscoelastically permitted and is related to different rates of stress and volume relaxation occurring in the polymer (38).

After measuring the stress in individual layers, the stresses from each source can be summed (assuming simple additivity) and then computed for the entire paint system. The harshest conditions would be cold dry conditions after some weathering has occurred, so as to induce a stress due to densification also. Weathering is assumed to effect the modulus and CTE of the clearcoat and basecoat only and densification of the clearcoat only. The assumption used, that the modulus and CTE of the coatings are independent of temperature, is obviously incorrect, but since we are only trying to calculate stresses at the temperature at which the CTE and modulus were measured, we have taken them to be constant (after dividing the modulus by 2 to correct for relaxation effects). Figure 7 shows the stress distribution that would be brought about by a 50°C drop in temperature and a 25°C drop in dew point, along with a 1% increase in density. Stresses are highest in the cleacoat. These stresses are the fully relaxed stresses, unlike those in Figs. 5 and 6 where the stresses are the initial stresses. These stresses will be sustained for a significant amount of time providing the opportunity for mechanical damage. The

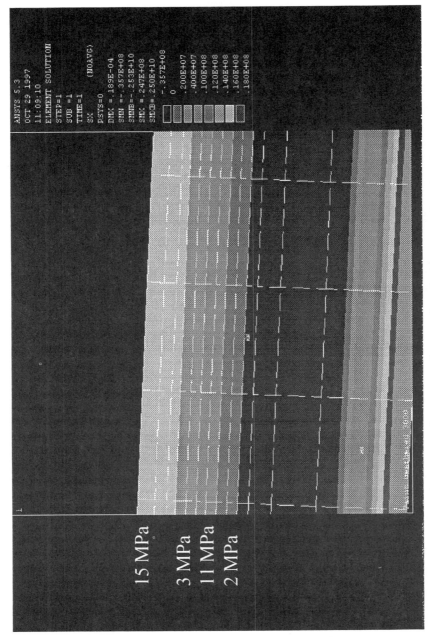

Figure 7. In-plane stresses for a weathered paint system after relaxation.
See text for environmental conditions. Stresses shown on plot.

initial stresses in the clearcoat would be approximately twice as high, approaching 40 MPa for some coatings. These stresses would also be magnified due to the inhomogeneous nature of the degradation. Because UV light absorbers are used in all commercial clearcoats, a gradient of photooxidation exists in the clearcoat. This gradient in photooxidation will set up a gradient in the thermomechanical properties also, with changes being greatest at the surface. This gradient will only increase the surface stresses over what has been calculated using average layer properties here.

As noted earlier, the stresses that exist in intact paint systems are almost exclusively biaxial in the plane of the paint system. Only near the edges and near very sharp radii of curvature are any significant normal or shear stresses incurred. However, paint can delaminate. Because delamination stresses do not typically occur for intact systems, the question must be asked where do delamination stresses arise from? Certainly absorption of moisture, particularly if it is trapped at interfaces can give rise to capillary pressure and some normal stresses. More likely however, flaws in the clearcoat will provide the opportunity for normal stresses. It has been shown analytically that the normal and shear stresses drop away exponentially to zero from a crack or edge, but that very close to these flaws the normal and shear stresses are quite high (30). Figure 8 shows some initial FEA results on the normal stress distribution in a paint system around a 10μm crack in a clearcoat that has been cooled 50°C. For an uncracked clearcoat the normal stresses approach zero. However, the stresses are quite high in this cracked clearcoat (depending on the radius of curvature of the tip) - in this case up to 30 MPa. The shear stresses are likewise high in the tip region. In addition, if a small delaminated zone is introduced near the crack tip (Figure 9), the peeling forces are clearly seen, as is the deflection of the clearcoat away from the surface. In this case the delaminated clearcoat initially laid with no adhesion on the surface of the basecoat for three crack tip radii away from the normal crack. As the temperature was reduced the clearcoat peeled away in the delaminated region and large stresses at the tip of the delamination were produced which would drive the delamination outward.

Knowledge of the stress distribution in paint systems is crucial to designing accelerated weathering tests and understanding the differences between failures in outdoor exposures and accelerated weathering. To more closely mimic outdoor exposure (with Florida as the standard comparator) accelerated test conditions must not greatly distort the weathering chemistry and the test must produce similar mechanical failures as are observed outdoors. Current weatherometers can closely mimic outdoor chemistry by using borosilicate inner and outer filtered xenon arc light, but the mechanical failures are often incorrect because the stresses produced in weatherometers are different than those produced outdoors. It is not uncommon for test panels to remain intact inside a weatherometer but crack when they are removed due to the rapid drying and cooling which sets up large thermal and humidity stresses gradients. By understanding the stress distribution in paint systems - and how heat, moisture, and age effect the stress distribution - better accelerated weathering protocols should be realized.

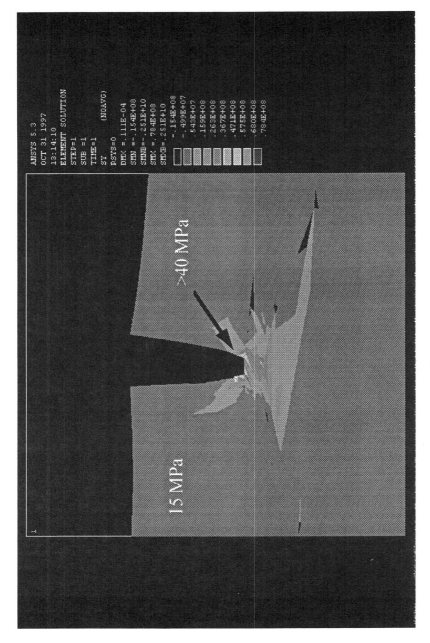

Figure 8. Normal stress distribution around 10mm crack in clearcoat. Note high stresses (~50 MPa) near crack tip. Stresses shown on plot.

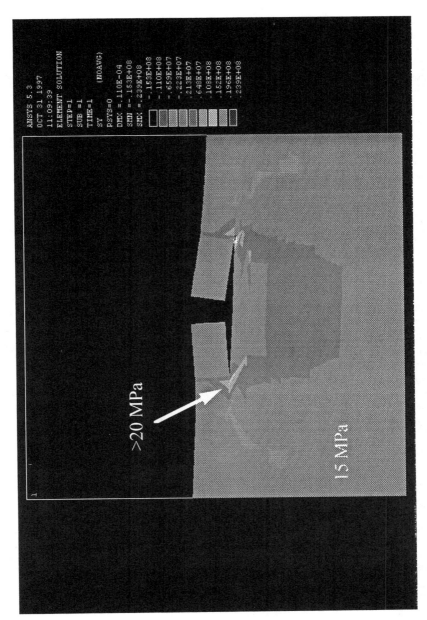

Figure 9. Normal stress distribution near incipient delamination. Note high stresses ahead of delamination front (~20 MPa). Stresses shown on plot.

Final remarks must be made regarding the interaction between stresses and relevant material properties. Because most possible failure mechanisms in clearcoat/basecoat systems involve either crack propagation in the clearcoat or at the clearcoat/basecoat interface, the fracture energy of these materials and interfaces is important. The fracture energy is the amount of mechanical energy required to propagate a crack in a material or at an interface and is a direct measure of the brittleness of a material. For example, the fracture energy of silica glass and mild steel are is approximately 6 J/m^2 and 12,000 J/m^2 respectively. For typical unweathered automotive clearcoats the fracture energy can range from 15 - 350 J/m^2, which would lead to failure stresses on the order of 20-40 MPa for these same clearcoats at nominal film thicknesses (11). Weathering typically embrittles materials lowering their fracture energy significantly, by over half in some cases. The failure stresses required for weathered clearcoats can be easily reached through a combination of the stresses outlined in this paper. This increasing embrittlement coupled with increasing stresses can lead to mechanical failure.

Conclusions

Aside from externally applied impacts and loads, the main sources of stress in complete automotive paint systems have been identified to be: thermal expansion mismatch, humidity expansion mismatch, and densification. These stresses have been measured for individual layers as a function of accelerated weathering time. In general, stresses increase with weathering, due mainly to continued densification of the clearcoat and increased stress sensitivity to moisture. Thermal expansion mismatch stresses, while large, change relatively little with weathering. The stress distribution in complete paint systems, as calculated by finite element analysis, shows uniform in-plane stresses in uncracked coatings. Stresses are often highest in the primer or cleacoat due to their thermoelastic properties. Cracks in the clearcoat will lead to large normal stresses that can lead to delamination. These normal stresses are typically not present in uncracked coating systems. The stresses present in weathered paint systems approach those that will propagate cracks in many of these systems.

References

1. Haacke, G., Andrawes, F. F., and Campbell, B. H., *J. Coat. Tech.*, **1996**, *68*, 57.
2. Ryntz, R. A., Ramamurthy, A. C., and Holubka, J. W., *J. Coat. Tech.*, **1995**, *67*, 23.
3. Oosterbroek, M., *Proc. XVth International Conference on Organic Coatings Science and Technology*, **1989**, Athens.
4. Bauer, D. R., Mielewski, D. F., and Gerlock, J. L., *Polym. Deg. and Stab.*, **1992**, *38*, 57.
5. Bauer, D. R., Gerlock, J. L., and Mielewski, D. F., *Polym. Deg. and Stab.*, **1993**, *41*, 9.
6. Dickie, R. A., *J. Coat. Tech.*, **1994**, *66*, 29.
7. Wypich, G., Handbook of Material Weathering, ChemTec Pub., Toronto, **1995**.
8. Gerlock, J. L., Prater, T. J., Kaberline, S. L., and deVries, J. E., *Polym. Deg. and Stab.*, **1995**, *47*, 405.

9. Gerlock, J. L, Smith, C. A., Nunez, E. M., Cooper, V. A., Liscombe, P., Cummings, D. R., and Dusibiber, T. G., in Polymer Durability, eds. R. L. Clough, N. C. Billingham, and K. T. Gillen, ACS Advances in Chemisty Series 249, Wash. D. C., **1996**, p 335.

10. Bauer, D. R., *J. Coat. Tech.*, **1994**, *66*, 57.

11. Nichols, M. E., Darr, C. A., Smith, C. A., Thouless, M. D., and Fischer, E. R., *Polym. Deg. and Stab.*, in press.

12. Hashemi, S., *J. Mat. Sci.*, **1997**, *32*, 1563.

13. Thouless, M. D., Olsson, E., and Gupta, A, *Acta Metall. Mater.*, **1992**, *40*, 1287.

14. Beuth, J. L., *Int. J. Solids Struct.*, **1992**, *29*, 1657,.

15. Hu, M. S., Thouless, M. D., and Evans, A. G., *Acta Metall.*, **1988**, *36*, 1301.

16. Hutchinson, J. W. and Suo, Z., in Advances in Applied Mechanics, Academic Press, New York, **1992.**

17. Sato, K., *Prog. Org. Coat.*, **1980**, *8*, 143.

18. Perera, D. Y., in Paint and Coating Testing Manual, J. V. Koleske, ed., ASTM Manual Series, **1995.**

19. Perera, D. Y. and Eynde, D. V., *J. Coat. Tech.*, **1987**, *59*, 55.

20. Corcoran, E. M., *J. Paint. Tech.*, **1969**, *41*, 635.

21. Hetenyi, M., Handbook of Experimental Stress Analysis, John Wiley and Sons, New York, **1950.**

22. Sato, N., Takahashi, H., and Kurauchi, T., *J. Mater. Sci. Lett.*, **1992**, *11*, 365.

23. Shiga, T., Narita, T., Tachi, K., Okada, A., Takahashi, H., and Kurauchi, T., *Polym. Engin. Sci.*, **1997**, *37*, 24.

24. Oosterbroek, M., Lammers, R. J., van der Ven, L. G. J., and Perera, D. Y., *J. Coat. Tech.*, **1991**, *63*, 55.

25. Perera, D. Y. and Oosterbroek, M., *J. Coat. Tech.*, **1994**, *66*, 83.

26. Logan, D. L., A First Course in the Finite Element Method, PWS Engineering, Boston, MA, **1986.**

27. Brebbia, C. A., Boundary Element Method for Engineers, Pentech Press, London, **1984.**

28. Vilms, J. and Kerps, D., *J. Appl. Phys.*, **1982**, *53*, 1536.

29. Townsend, P. H., Barnett, D. M., and Brunner, T. A., *J. Appl. Phys.*, **1987**, *62*.

30. Suhir, E., *J. Appl. Mech.*, **1988**, *55*, 143.

31. Shen, Y-L. and Suresh, S., *J. Mater. Res.*, **1995**, *10*, 1200.

32. Hill, L. W., Korzeniowski, H. M., Ojunga-Andrews, M., and Wilson, R. C., *Prog. Org. Coat.*, **1994**, *24*, 147.

33. Struik, L. C. E., Physical Aging in Amorphous Polymers and Other Materials, Elsevier, Amsterdam, **1978.**

34. Perera, D. Y. and Schutyser, P., *Prog. Org. Coat.*, **1994**, *24*, 299.

35. Perera, D. Y. and Schutyser, P., *Proc. of 22nd FATIPEC Congress*, **1994**, Budapest.

36. Killgoar, P. C. and van Oene, H., in ACS Symposium Series num. 25, Ultraviolet Light Induced Reactions in Polymers, S. S. Labana, ed., ACS.

37. Perera, D. Y. and Eynde, V., *Proc. of 20th FATIPEC Congress*, Nice, **(1990)**.

38. Ferry, J. D., Viscoelastic Properties of Polymers, John Wiley and Sons, New York, **1976.**

Chapter 23

Improved Service Life of Coated Metals by Engineering the Polymer–Metal Interface

W. J. van Ooij

Department of Materials Science and Engineering, University of Cincinnati, Cincinnati, OH 45221–0012

An important factor in coated metal systems is the polymer-metal interface. This paper focuses on this interface. It is discussed how it controls the overall corrosion rate, and how it can be engineered to improve the service life of the entire coated system. Examples are given of coated cold-rolled steel, galvanized steel and Galvalume, where the interface was modified and the performance of the system increased. We have studied two methods for interface modification, viz., a) by depositing a thin film of a plasma-polymerized organic monomer, b) by depositing thin films of organofunctional silanes. EIS and accelerated corrosion tests measured the corrosion rates of the coated systems. The main conclusions of this work are that pretreatments of metals based on plasma or silane treatments could be developed that would improve the service life of the coated metal systems.

In many industries metals are commonly pretreated, prior to painting, by solvent cleaning, alkaline cleaning and optionally phosphating or chromating. These pretreatments involve many different treatment and rinse steps. In general, such pretreatments work satisfactorily. Therefore, relatively little research has been published in recent years on the mechanisms of these treatments. No universal alternatives have been proposed, although a worldwide search for chromate replacement is in effect *(1)*. Chromate is such an effective inhibitor that, to date, automotive, aerospace or other industries have adopted none of these alternative systems.

It is generally assumed that the paint system on a metal is decisive in determining the overall corrosion protection hence the service life of the metal. Prior to paint application, the metal should be thoroughly cleaned, in order to avoid delamination and blistering. The purpose of the phosphate coating is to anchor the paint better, *i.e.*, to improve paint adhesion. The phosphate coating is not believed to

© 1999 American Chemical Society

improve the corrosion per sé. Indeed, exposure of phosphated metals directly to a corrosive environment without paint confirms this conclusion (2). The function of the chromate final rinse of the phosphate coating is believed to be to seal the existing pores in the phosphate coating. In summary, other than providing adhesion and/or anchoring of paint films, the interface is not believed to contribute much to the overall corrosion performance of a painted metal.

In our research programs we have studied alternative metal pretreatments of various metals that could possibly replace the phosphating pretreatment. Our goal was to develop processes that would comprise fewer steps than conventional systems. Our assumption was further that the service life of the entire system consisting of metal, pretreatment processes and organic coating, could be improved if the new interface process would improve the adhesion and the corrosion (passivation) of the metal at the same time. Thus, the new processes had to meet the following requirements:

i)   Be cost effective in terms of cost of chemicals and process steps;
ii) Be environmentally acceptable;
iii) Offer corrosion protection at least equal to that offered by phosphates and/or chromates;
iv) Provide excellent paint adhesion both in wet and dry condition;
v)   Passivate the metal, i.e., also provide corrosion protection without paint.

We report here on the status of two approaches we have taken. One is to deposit a thin organic film on the metal by the process of plasma polymerization. In this process an organic monomer is polymerized in a plasma and directly deposited on a metal surface. Such a process is attractive as it is solvent-free, requires low energy for deposition, and allows thorough cleaning of the metal by a reactive plasma just prior to film deposition (3,4). We will demonstrate here that the corrosion protection provided by these films can be outstanding. The limitations of this process are the elaborate equipment and the vacuum requirements.

The other approach is to deposit very thin films of organofunctional silanes on the metal surface from an aqueous solution (5,6). The assumption was here that corrosion protection provided by the coating could be improved by the presence of covalent metallo-siloxane bonds of the type -Me-O-Si- that would passivate the metal underneath the coating. In the case of the plasma-polymerized films, the assumption was that the films would act as dense well-adhering barrier layers below the organic coating. Thus, the two approaches contrasted each other in mechanisms by which they could improve overall performance, namely by barrier protection and by covalent bonding, respectively.

In developing the new treatments, our assumptions were that the rate of corrosion of a painted metal under the paint or at a defect is minimized if the interface is optimized. By optimization we mean:

i)   The metal is properly cleaned. All water-soluble material as well as all hydrocarbon contaminants must be thoroughly removed; the metal is thus receptive to interface modifiers such as the silanes or plasma coatings described here;
ii) If the surface is cleaned properly, covalent bonds can be formed between the metal or its oxide and the interface modifier (4,7);
iii) These covalent bonds remove the hydroxyl groups at the metal oxide. Such OH

groups initiate corrosion processes under the paint by attracting water through hydrogen bond formation;

iv) Such bonds must be stable in water, *i.e.*, should not hydrolyze easily in a wide range of pH conditions;

v) Further improvement, in terms of resistance against corrosion, can be achieved if this interface can be made hydrophobic, for instance during the cure of the paint;

vi) The interface modifier has certain functionalities that react chemically with the polymer during the cure of the paint.

It is one of the objectives of this paper to demonstrate the validity of these assumptions.

Experimental

A. Materials and Silane Film Deposition. All metal test panels were of $10 \times 15 \times 0.05$ cm^3 dimensions. Galvalume panels of 7 μm coating thickness were obtained from the production line at SSAB Tunnplått in Borlänge, Sweden. Prior to silane rinse, they were cleaned in either Ytex 4234 (pickling) or Ytex 4345 (non-pickling, *i.e.*, inhibited) from Ytteknik, AB, Sweden. The first one resulted in a zinc-rich surface, the second cleaner left the Al-rich oxide intact. Cleaning conditions were 5 min. at 60°C. CRS panels of 1010 quality and HDG of 20 μm coating thickness were obtained from ACT in Hillsdale, Michigan. Al 3003 and 2024-T3 panels were purchased from Q-Panel. CRS, HDG and the Al alloys were all cleaned in Parker 338 at 55°C for 1-5 min.

The silanes bis-1,2-(triethoxysilyl)ethane (BTSE), γ-aminotriethoxysilane (γ-APS), styrylaminoethylaminoproplytrimethoxysilane (SAAPS), and vinyltrimethoxysilane (VS) were all obtained from Witco Osi Inc. They were used without further purification. Their purities were listed as 95-98% except SAAPS which came as a 40% solution in methanol. All silanes were mixed with demineralized water and either acetic acid or NaOH to adjust the pH. These solutions (about 10 vol.-% silane) were hydrolyzed by stirring for 1-2 hours. They were then applied to the metals by dipping the panels into 2-5% solutions of pH ranging from 4 (BTSE), to 6 (VS) and 8 (γ-APS). Only freshly hydrolyzed solutions were used to deposit the films. Dipping times were 2 min. at room temperature in all cases. The film was then blown dry with compressed air and optionally dipped into a second silane solution. The films were stored in ambient conditions and were not heated. Painting was usually done within one day after dipping.

Pyrrole and hexamethyldisiloxane (HMDS) for plasma polymerization were obtained from Aldrich Chemical Inc. in the highest available purity (around 99%) and were used without further purification.

Polyester (VP-317-A) and polyurethane UFB 550 S9) powder paints were commercial products from Ferro and O'Brien, respectively. They were cured at 190°C/15 min. (PE) or 205°C/10 min. (PU), respectively. The cured film thicknesses were 60 ± 10 μm.

B. Plasma Polymerization. Plasma polymerization was carried out in a three-electrode parallel plate reactor which has been described *(4)*. The plasmas were generated by DC power which could be pulsed. The substrates were CRS panels of 20 × 20 cm² which were first plasma-cleaned in the same reactor by an Ar⁺ plasma of 25 W for 20 min., unless specified otherwise. The conditions for pyrrole and HMDS depositions typically ranged from 2-20 W power and 5-65 Pa pressure. Monomer flow rates were always of the order of 4 standard cm³. No carrier gases were mixed with the monomers.

C. Corrosion Testing. The painted panels were scribed with a silicon carbide tool (7 cm, 45°, 0.5 mm width) and then exposed in the ASTM B-117 neutral salt spray test or in the GM 9540P Scab test. In the case of Galvalume panels, three of the four edges were sealed with paraffin and the fourth (long) edge was abraded with medium silicon carbide paper. The edge creep was evaluated on that edge only. Following corrosion test exposure, the panels were rinsed, brushed and then tape-tested. The creep from the scribe or edge was measured and averaged for five panels.

D. Electrochemical Measurements. DC linear polarization and EIS were performed on a CSM 100 system from Gamry instruments, Inc. using a three-electrode standard test cell (flat cell). The electrolyte was aerated 3% NaCl in all cases. The frequency range in EIS was 10^5 to 10^{-2} Hz. The AC voltage was 200 mV and 5 points per decade were collected.

It should be noted that EIS does not primarily measure the corrosion performance of a coated metal. What is actually determined are the changes in the system during continuous exposure to an electrolyte. When these changes lead to the onset of corrosion under the paint coating, its rate can be estimated from the charge transfer resistance R_p which is inversely proportional to the corrosion current. While there is generally a good agreement between EIS data and field performance of coated metals, in terms of ranking of systems, the correlation is far from perfect because the conditions in an EIS experiment differ greatly from those in field exposure. Examples of these differences are the degree of wet time, interfacial pH, defects in the coating and the presence of the radiation.

Results and Discussion

A. Plasma-Polymerized Films

A-1. Insulating Films. In Figure 1 we show EIS results obtained after 24 hours immersion in 3% NaCl of films of plasma-polymerized films of hexamethyl-disiloxane (PP-HMDS) deposited on cold-rolled steel substrates (CRS) *(8)*. The thickness of the films was of the order of 50-100 nm. In this experiment EIS was measured of the film only, *i.e.*, the panel was not painted. The variables in this experiment were the pretreatment of the steel panels prior to the deposition of the plasma films. It was verified by XPS and IR analysis that the plasma polymer films themselves did not differ between the panels *(8)*. It is thus clearly observed that thorough solvent cleaning of CRS (panel labeled 'no pretreatment') is insufficient. The total resistance of the system at low frequency is 1½ orders of magnitude lower than the system which has been thoroughly plasma-cleaned by a combination of oxygen and argon/hydrogen treatment. These data indicate that the adhesion of the plasma polymer film to steel has to be optimized. It has been suggested that only

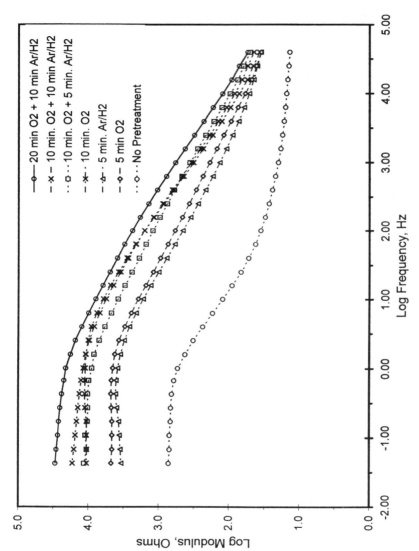

Fig. 1. EIS Bode plots of CRS samples coated with 100 nm PP-HMDS coatings as a function of metal pretreatment prior to plasma polymer deposition. Immersion time 24 hours in 3% NaCl.

after a very specific combination of these two plasmas the steel surface is sufficiently low in carbon that covalent FeSi or Fe-C bonds can be formed when the plasma coating is deposited *(9)*. The strong effect of metal cleaning procedure prior to plasma film deposition has also been noted by others *(10)*.

Figure 2 shows the results of a similar experiment, but now the panels were additionally coated with a cathodic electrocoat system. CRS with and without a standard zinc phosphate system are also compared here. The results now demonstrate that only the sample with a plasma coating deposited on a solvent-cleaned CRS shows a poor performance. In this case the plasma film actually deteriorates the performance. All other panels still show a completely capacitive behavior after 24 hours immersion. Apparently, the plasma film does not adhere well to the CRS panel that has been solvent-cleaned only, but the E-coat does adhere to that substrate.

Figure 3 presents the change in impedance during continuous immersion for a plasma-coated CRS with the optimum pretreatment and film and, for comparison, the behavior of a standard fine-grained automotive zinc phosphate. The thickness of the E-coat in the two systems was approximately equal (30 μm), as was verified by SEM of cross sections. This conclusion can also be drawn from the regions of the curves where the behavior is completely capacitive. The moduli of these curves at $f = 0.16$ Hz are practically identical. Thus, we arrive at the conclusion that a plasma-polymerized films of no more than 50-100 nm can outperform a modern zinc phosphate, at least in EIS experiments. The interface in a painted metal can thus be manipulated to improve the performance of the entire system.

Since there was no defect in the E-coated panels that were tested by EIS, these panels were also tested in an accelerated cyclic corrosion test after scribing them with a sharp tool into the base metal. Table I shows the underfilm corrosion from the scribe after 4 weeks in the GM scab test. The conclusions drawn from the EIS data are completely confirmed here: i) the plasma coating deposited on the CRS substrate with the optimum pretreatment performs as well as the zinc phosphate and, ii) the performance of the plasma film varies strongly with the pretreatment of the steel and only a very specific combination of plasma pretreatments gives optimum results.

Table I. Scribe Creep in E-coated CRS Panels after 4 weeks in GM Scab Test

E-coat-CRS Interface	Scribe Creep*, mm
Standard zinc phosphate	2 ± 0.5
PP-HMDS film after 5 min. Ar/H$_2$ Cleaning	13 ± 2
PP-HMDS film after 10 min. Ar/H$_2$ Cleaning	8 ± 2
PP-HMDS film after 10 min. O$_2$ Plasma followed by 10 min. Ar/H$_2$	2 ± 1

*averaged for 5 readings and 5 panels

A-2. Conductive Films. In the experiments with PP-HMDS interface modifiers, the composition, structure and thickness of the plasma films were not

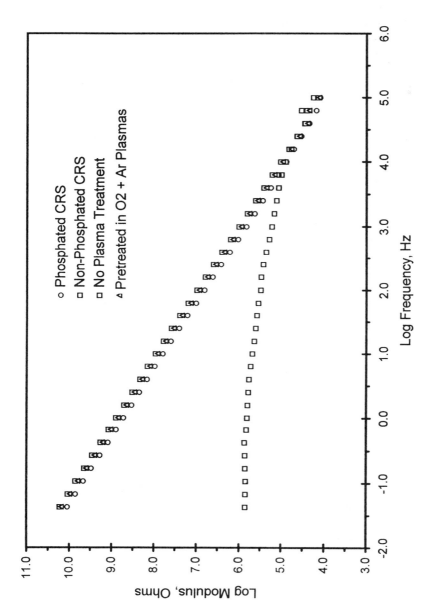

Fig. 2. EIS Bode plots after 24 hours immersion in 3% NaCl of CRS samples phosphated or coated with 100 nm PP-HMDS coatings and a cathodic E-coat.

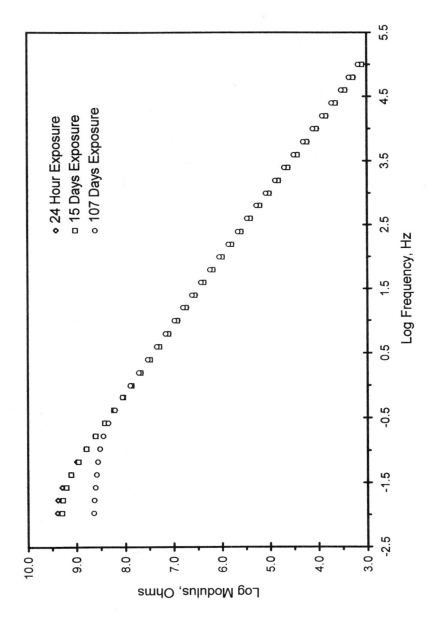

Fig. 3a. EIS Bode plots of CRS coated with 100 nm PP-HMDS coatings and a cathodic E-coat and immersed in 3% NaCl for 107 days.

362

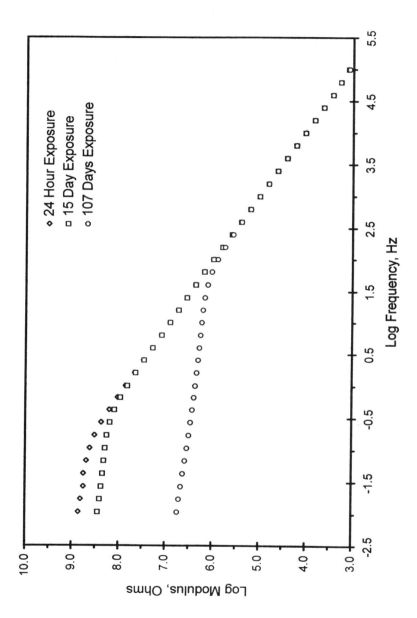

Fig. 3b. EIS Bode plots of zinc-phosphated CRS coated with a cathodic E-coat and immersed in 3% NaCl for 107 days.

varied. In other experiments two variables were introduced, namely the conductivity of the plasma film and the structure of the film. Pyrrole was used as the monomer in the experiments summarized in Table II. Films of plasma-polymerized pyrrole (PPy) have a fairly high electrical conductivity of up to 10^{-1} S/cm *(11)*. The conductivity of the PP-HMDS films was 10^{-9} S/cm. The structure of the films was varied by means of the power and pressure. We have demonstrated that in DC plasmas, the combination of low power and high pressure (LW/HP) results in soft films of low crosslink density. Films deposited at HW/LP conditions are harder and more crosslinked *(12)*.

Table II presents EIS data for these films after deposition on CRS panels which had been precleaned with the conditions that were the best of Table I. The paint was a powder paint here of 60 μm thickness. Shown is the polarization resistance of the painted panels, which were not scribed here. The R_p data were determined from a simple equivalent circuit which we have published *(8)*. The other components of the circuit did not show a strong change after 14 days. The phosphate was the same zinc phosphate as in Figure 2 and 3 and Table I. It is apparent from these data that the PPy film deposited under HW/LP conditions is superior to the phosphate coating. Another conclusion is that the softer films (LW/HP) have failed after 14 days. A third conclusion is that an interfacial film between a paint and CRS does not have to be an insulator.

Table II. EIS Data for Polyurethane-Powder-Painted CRS* in Aerated 3%NaCl*

CRS Pretreatment	Time, Days	$R_p, \Omega \times 10^{-6}$
Solvent-Cleaned only	1	334
	14	9
Zinc Phosphated	1	998
	14	677
PPy, LW/HP	1	2000
	14	46
PPy, HW/LP	1	4177
	14	2500

*The PPy panels were pretreated by 10 min. O_2 plasma followed by 10 min. Ar/H_2 plasma;
LW/HP = 2 W, 65 Pa; HW/LP = 20 W, 5 Pa

The mechanism of protection by these PPy films is most likely also the result of the formation of covalent bonds with the iron oxide as the marked effect of the type of precleaning was also observed here. The effect of the crosslink density of the PPy films is attributed to a strong swelling of the film with low crosslink density. The EIS data suggested that delamination occurred between the E-coat and the PPy film in this case as an increase of C_{dl}, the double-layer capacitance, was observed.

In Figure 4 our model for the mechanism by which interfacial plasma-

364

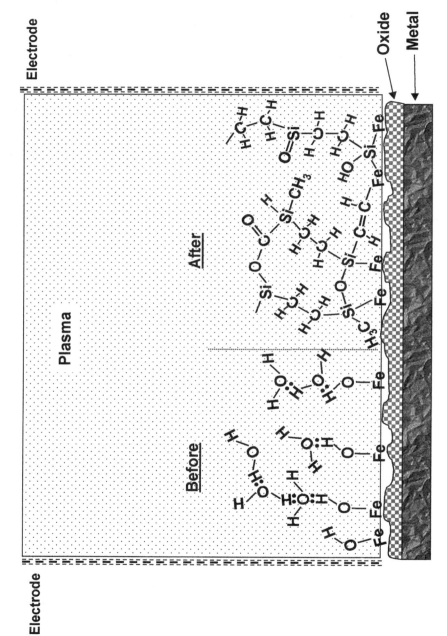

Fig. 4. Steel surface modification in a plasma of HMDS.

polymerized films enhance the corrosion performance of a painted metal is summarized. It is based on the passivation of the metal as a result of the formation of covalent bonds between the metal and the plasma film.

B. Films of Organofunctional and Non-Organofunctional Silanes. Organofunctional silanes of the type $Y-C_nH_m-Si(OX_3)$ are primarily known to improve the adhesion between ceramic (*e.g.*, glass or silica) or metallic oxide surfaces and polymers. They are not known as corrosion inhibitors. The mechanism by which they function is still under debate, but a widely accepted mechanism is shown in Figure 5. The silane is assumed to form a monomolecular film at the interface and is covalently bonded to the polymer via the Y groups and to the ceramic surface via the silanol groups, which are formed upon hydrolysis of the OX ester groups.

We have studied the adsorption of a wide range of silanes on metal surfaces in the past several years *(13)*. The purpose of these investigations was to develop novel interfaces between metals and paints that not only provided good adhesion (as suggested by Figure 4) but corrosion resistance as well. In these studies it turned out that an interface as sketched in Figure 5 can indeed be obtained by rinsing the thoroughly cleaned metal briefly in a dilute silane solution which has been freshly hydrolyzed in water to yield $Y-C_nH_m-Si(OH_3)$. However, although excellent adhesion can be obtained in this way, corrosion performance of a painted metal is poor. One monomolecular layer of silane molecules is not sufficient. Excellent corrosion performance of a painted metal with or without scribe or even without paint can be obtained by thicker silane films. In summary, our findings have indicated that an ideal silane film can be described as follows:

1. The film is solidly anchored to the metal via metallo-siloxane bonds of the type -Me-O-Si- formed from hydroxyl groups and silanol groups. This implies hat the silane has to be hydrolyzed first and then adsorbed. Adsorption of non-hydrolyzed silane ester molecules followed by hydrolysis at the surface (by surface-bound water or water from the atmosphere) does not result in films with good corrosion resistance *(14)*;

2. The metallo-siloxane bond -Me-O-Si- must be hydrolytically stable. Metals that form very stable bonds are Fe and Al. Much less stable is the bond formed by Zn. Metals that do not have basic hydroxyl groups cannot easily be treated by silanes. An example is Cu. Also, inverted adsorption, with the functional group down and adsorbed on the metal via hydrogen bonding, as indicated in Figure 4 (left), reduces the corrosion resistance markedly, as these hydrogen bonds attract water molecules and will hydrolyze;

3. The optimum thickness of the silane film is of the order of 50-100 nm. Thinner films are difficult to deposit homogeneously and they do not mix well with the polymer, *i.e.*, an Interpenetrating Network is not formed (see below). Thicker films lack mechanical strength and may become brittle;

4. The films have to be homogeneous and pore-free. This can only be achieved if the metal is very thoroughly cleaned by alkaline cleaners until the metal surface is water break-free. Also, the silane concentration needs to be optimized for the deposition of pore-free films that are within the optimum thickness range;

5. The orientation of the first layers of surface molecules has to be extremely uniform and regular. In this way the free silanol groups that have not reacted with

366

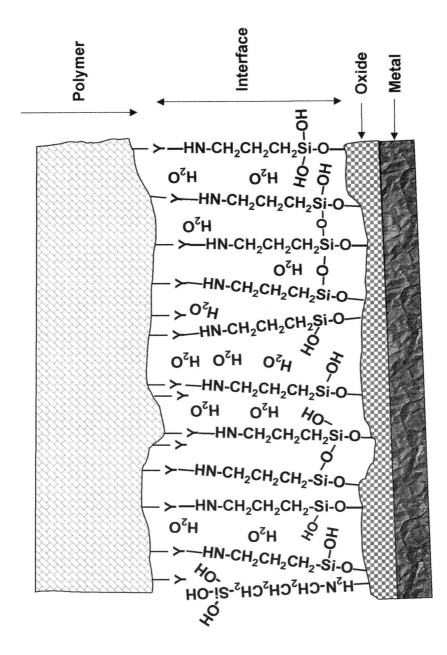

Fig. 5. Polymer-metal interface modified by a monomolecular layer of a functional silane.

the metal hydroxyls can react with each other and form and extended siloxane network at the interface. This network reduces the diffusion of water molecules to the metal surface. However, it is important that the film does not completely crosslink and lose water before the polymer is applied. If this would happen, the polymer cannot penetrate into the silane film and may not even wet the film properly. As a result, the adhesion between the polymer and the substrate becomes poor;

6. The film can have functional groups that are reactive to certain functionalities in the polymer. If this happens, the metal-polymer bond strength will be enhanced further. However, it is a major finding of our previous work that excellent adhesion (*i.e.*, exceeding the cohesive strength of the polymer) can be obtained even if the silane film does *not* contain functional groups that are reactive to the polymer.

These findings have led to the following model for the mechanism of adhesion and corrosion by silanes. It is schematically shown in Figure 6. The silane whose film is shown in the figure is 1,2-bis-(triethoxysilyl) ethane (BTSE). It is a non-organofunctional, but bis-silylfunctional, silane that can be used for adhesion promotion of metals by itself or in combination with a second, organofunctional silane. For corrosion protection, its use as a first treatment step is essential for steel and aluminum. It is, however, not very reactive to zinc substrates.

The film is bonded to the metal by Me-O-Si units. The remaining -SiOH groups have largely crosslinked along the surface and in the film as well. This crosslinking has occurred after the polymer has been applied over the freshly deposited, non-crosslinked BTSE film. If there is sufficient compatibility of the BTSE film with the polymer, the liquid polymer will penetrate the partially crosslinked BSTE film. Upon curing, both polymers then crosslink and the situation as sketched in the figure will develop. In this model the adhesion is solely due to the formation of an Interpenetrating Network (IPN). However, some –SiOH groups may also anchor to the polymer, depending on the functional groups in the polymer. The silanol groups can be expected to be reactive to other hydroxyl groups (epoxies) and to isocyanato groups in polyurethanes, for instance.

If the silane film contains an additional silane with functional groups that are reactive to the polymer, the adhesion will, in principle, be enhanced. However, this effect will not be noticeable as the interface as sketched in Figure 6 will already result in bond strengths higher than the cohesive strength of the polymer. Nonetheless, the addition of a second silane to the BTSE film of Figure 6 may improve the adhesion and/or corrosion performance because it improves the compatibility (miscibility) between the liquid polymer and the silane film. This mechanism offers an explanation for the observation that sometimes excellent adhesion is obtained by silanes, although no apparent chemical reactivity exists between functional groups in the silane and the polymer. The film obtained in this so-called two-step process is schematically shown in Figure 7 for the process consisting of a rinse in a BTSE solution followed immediately by a rinse in solution of vinylsilane.

The advantages of the 2-step process are that the stability of the film is much increased. This is illustrated in Figure 8 for a comparison between γ-APS, BTSE

368

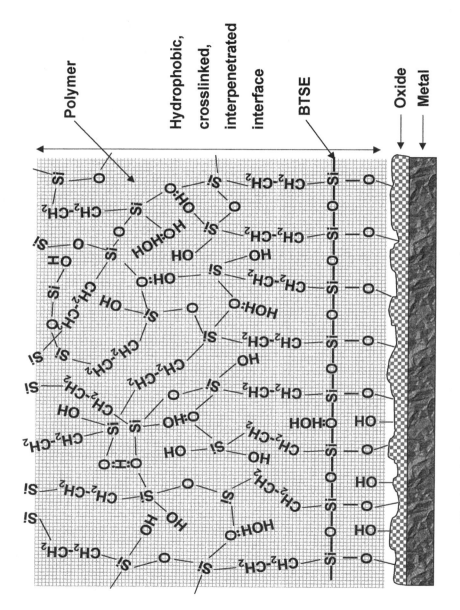

Fig. 6. Model for the bonding mechanism of BTSE films to polymers.

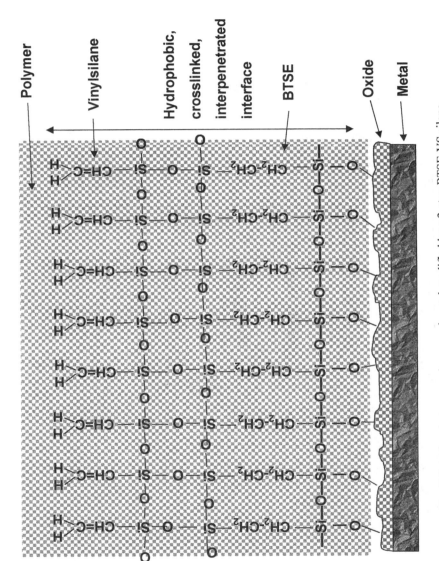

Fig. 7. Interface between polymer and metal modified by a 2-step BTSE-VS silane treatment.

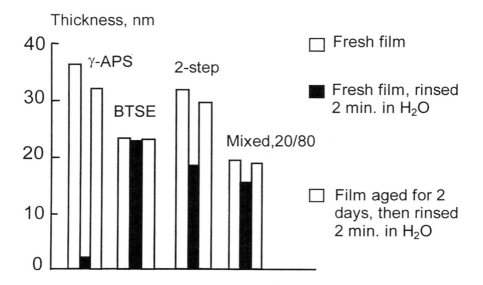

Fig. 8. Stability of silane films on Al 3003.

mixed and 2-step films. Only those films containing BTSE are stable immediately, *i.e.*, they form covalent bonds more readily than the functional silanes. It can be argued on the basis of inductive effects that silanols in BTSE are stronger acids than in γ-APS. As a result, BTSE will react with weaker basic hydroxyl groups and also form stronger bonds that are less sensitive to hydrolysis.

C. Corrosion Performance of the Two-Step Silane Process. The new silane processes have thus far been observed to work well on cold-rolled steel, electrogalvanized steel, hot-dip galvanized steel, Galvalume, and on aluminum and some of its alloys *(13)*. The paint systems that have been found to work very well on silane-treated metals include various powder paints (polyester, polyurethane, epoxies), solvent-based appliance coatings, coil paint systems, and cathodic electrocoat systems *(15)*. Some treatments also perform very well without paint when exposed to the atmosphere. Especially BTSE can passivate mild steel and aluminum surprisingly well. It is however, not very effective on zinc substrates. On that metal certain vinyl silanes or amino silanes perform better. The corrosion tests and exposures where the treatments have exhibited good performance comprise various accelerated laboratory test, filiform tests, salt immersion tests, condensation tests, stack tests, and in outdoor exposure in marine and industrial environments. For painted Galvalume both scribe creep and edge creep have been shown to benefit from a silane rinse prior to painting *(16)*.

In the following some selected results are presented of painted metals where the silane rinses have shown excellent performance. It should be noted that there is no universal silane rinse that works with all metals and all paints. The variables that have to be optimized are, in general, the use of one or two silanes, the type of silane, the silane concentration, and the pH of the application. The contact time is usually not a variable. The mode of application is not critical either. Dipping, wiping, spraying, brushing and other methods have all been used successfully *(17)*. Further, all applications that we have investigated so far can be covered by a total of no more than five different silanes, one non-functional and four organofunctional ones.

Table III. Scribe Creep in Painted CRS*

No.	Treatment	Scribe Creep, mm
1	Alkaline cleaned only	60 ± 20
2	Iron phosphated	22 ± 7
3	Iron phosphated + chromate rinse	12 ± 4
4	5% γ-APS; pH = 6; 2 min.	65 ± 20
5	Iron phosphated + 1% γ-APS	10 ± 2
6	2% BTSE + 5% γ-APS	3 ± 0.5

*Polyurethane powder paint; 60 μm; 12 min. 200°C; iron phosphate: ChemCote 3029 + ChemSeal 3603; 7 cm scribe, 0.5 mm width, 45°; humidity test, 60°C, 85% r.h.; average of 5 panels; silane-treated panels aged 30 days in air; silane rinses 30 s in each solution followed by blow drying.

In Table III scribe creep results are shown for powder-painted CRS. They are compared with phosphated controls and with a system in which the silane is applied over the phosphate as a replacement of the chromate final rinse. In this and all other Tables, only some selected data are shown in order to highlight the performance of the optimized process. In all cases at least a dozen different treatments were compared with most of them showing no performance better than the non-silane treated alkaline-cleaned control.

The results show that a treatment by an organofunctional silane alone (no. 4) does not work, as has also been found by others (18). In this experiment many other single-silane rinses were tested with similar results (15). However, the final chromate rinse can be replaced successfully with a γ-APS rinse. By far the best results were consistently obtained by a two-step treatment consisting of BTSE and γ-APS at the concentrations shown in the Table. Varying the conentrations had a minor effect on the performance.

In Table IV results are shown for Galvalume panels painted with two types of powder paints. Both scribe and edge creep were measured here. The paints did not contain the usual chromate anticorrosive pigment. Two silanes were found to work here, γ-APS and SAAPS, for the PET and PU paints, respectively. There is also a slight difference between panels treated by an etching or non-etching cleaner. This is related to the surface composition after the cleaning process, which is Al-rich for non-etching and Zn-rich for an etching process (19). Two steps perform better than one-step processes and in all cases a significant improvement of the edge creep over untreated or even over chromate-treated panels is observed. A slight improvement of the scribe creep is also observed.

Table IV. Performance of Silane Treatments on Galvalume*

| | | SST | | | | GM SCAB TEST | | | |
| | | PET | | PU | | PET | | PU | |
Cleaning	Silane	Scribe	Edge	Scribe	Edge	Scribe	Edge	Scribe	Edge
Non-etch	None	3.7	11.4	3.0	4.8	13.0	17.9	4.1	19.5
Non-etch	Chromate	0.2	0.2	0.9	0.4	12.4	7.3	6.3	2.7
Etch	BTSE/γ-APS	1.8	0.5	2.0	1.1	8.9	4.7	5.0	2.0
Etch	BTSE/SAAPS	3.2	0.5	3.4	0.8	8.9	5.7	4.3	3.3
Non-etch	BTSE/γ-APS	0.4	0.1	0.6	0.4	11.2	4.3	3.8	1.9
Non-etch	BTSE/SAAPS	1.0	0.5	2.0	0.3	9.2	5.3	3.7	1.5

*Creep in mm; averages for 5 panels of 10x15 cm; SST = ASTM B-117; 2 weeks (1 week for controls); GM Scab = GM 9540P; 4 weeks (2 weeks for controls); SAAPS = styrylaminoethylaminopropyltrimethoxysilane; PET = polyester powder paint, 50 μm; PU = polyurethane powder paint, 50 μm; Non-etch = non-etching cleaner, Al-rich surface; Etch = etching cleaner, Zn-rich surface

Spectacular results were obtained for hot-dip galvanized steel panels that were powder-painted and tested in two accelerated tests, as shown in Table V. The scribe creep is reduced to practically zero by the silane process. This is an example of a metal where a one-step simple rinse in a silane works well. The optimum silane for zinc was found to be vinyl silane. There is some evidence that the interaction between the metal and the zinc surface occurs via the vinyl groups and much less via the silanol groups. This is in agreement with the finding that BTSE does not form very stable films on zinc. Apparently, the zinc surface – after industrial cleaning processes – is too acidic to react with the acid BTSE silanol groups. It does adsorb the basic vinyl groups present in VS and SAAPS, however. The metals Fe and Al have a more basic oxide hence they adsorb BTSE strongly.

Table V. Scribe Creep in Painted HDG*

No.	Treatment	Salt Spray Test, 4wks		GM Scab Test, 4 wks	
		PET	PU	PET	PU
1	None	52 ± 20	57 ± 12	23 ± 12	10 ± 2.0
2	Zn Phosphate	4.0 ± 1.2	12 ± 4.0	11 ± 6.0	19 ± 10
3	5% VS	1.1 ± 0.6	1.3 ± 2.0	1.1 ± 1.0	1.3 ± 0.8

*In mm ± standard deviation; none = solvent cleaning + alkaline cleaning; polyester and polyurethane powder paints; 60 μm; 12 min. 200°C; 7 cm scribe, 45°, 0.5 mm width; GM 9540P Test; average of 16-24 readings

In Table VI we present data for the improvement of filiform corrosion by silane treatments of an aluminum alloy. Since these alloys do not develop scribe creep in a painted state, a filiform corrosion test was chosen. The conclusion that can be drawn from these data is that single silane rinses provide some improvement, but it is only a balanced two-step treatment consisting of BTSE followed by a functional silane (here γ-APS, but the optimum is paint-dependent). Since it is known that the mechanism of filiform corrosion involves the propagation of anodic front which is highly acidic, this performance of BTSE is related to the higher stability of the BTSE film at lower pH *(20)*. Films of γ-APS, for instance, dissolve at low and high pH, but BTSE films on Al are stable in the pH range 1–12 *(21)*. Remarkable is the performance of BTSE alone. In addition to a good corrosion protection it also gave good adhesion in this system.

Only one typical result is presented here of corrosion protection of a metal without paint. It serves to illustrate that the BTSE film actually passivates the metal after appropriate heating, for instance during paint cure. Table VII shows salt immersion and electrochemical corrosion data – obtained by linear polarization – of Al 2024-T3 alloy samples treated with BTSE. The effects of double dipping the sample and of curing the film at moderate temperature are also shown. The heating conditions were mild so as not to affect the mechanical properties of this temper-rolled alloy.

It is seen that the double-dipped cured film (which was around 100 nm thick) outperforms the standard chromate film in salt immersion. The corrosion rate in mpy as determined from a Tafel analysis is about equal to that of the chromate and gives a 1000-fold improvement over the non-treated metal. The corrosion potentials of the various samples indicate that the chromate is mainly a cathodic inhibitor, whereas the silane film is both an anodic and cathodic inhibitor.

Table VI. Filiform Corrosion of Silane-Treated Al-3003*

No.	Treatment	Performance in Filiform Test	
		PET	PU
1	None (solvent only)	4	3
2	Chromate	2	1
3	2% BTSE + 5% γ-APS	5	5
4	5% γ-APS (high pH)	3	4
5	5% γ-APS (low pH)	2	1
6	5% BTSE	1	1
7	2% BTSE + 5% VS	1	0

*Polyester and polyurethane powder paints; 65 μm; 12 min. 200°C; scribed panels 1 hour over 33% HCl; then 4 weeks in 40°C, 80% r.h.; filiform evaluation: 0 = no filiform; 5 = high filiform density with lengths up to 10 mm; no underfilm corrosion visible

Table VII. Performance of BTSE On Al 2024*

No.	Treatment	Salt Immersion Hours	Corrosion Rate mpy	E_{corr}, mV
1	None	< 10	5.3	−445
2	Chromate	170	0.004	−700
3	BTSE, 1 dip	80	0.2	−450
4	BTSE, 2 dips	200	0.01	−500
5	BTSE, 2 dips, cured (2×)	>250	0.004	−500

*Continuous immersion in aerated 3% NaCl; hours = time when corrosion became visible at waterline; cure of BTSE: 15 min. at 125°C; BTSE: 2%, 1 min. dip, pH = 4; mpy = corrosion rate in 3% NaCl milimeters per year determined from polarization data; E_{corr} = corrosion potential of sample vs. Ag/AgCl in 3% NaCl

The DC polarization studies with BTSE-treated Al 2024 were repeated with NaCl solutions of different pH values. The underlying idea was that corrosion processes under paints could process at pH values ranging from 2 (filiform) to 12 (salts fog exposure). The results are plotted in Figure 9. They demonstrate that the film is essentially stable in the pH range 2-11. In the range 2-4 pitting corrosion begins to occur. The BTSE film reduces the rate of pitting corrosion as a result of the lower cathodic activity of the film adjacent to the pit. However, once a pit has formed the BTSE cannot stop further growth, as the film does not have the defect healing capabilities that chromate films have.

Conclusions

It has been demonstrated in this paper that an interface between a paint system and a metal can be manipulated by depositing thin films of dense, crosslinked plasma polymer films or by silane rinses. Such films exert a strong effect on the overall corrosion performance of the system in a wide range of accelerated corrosion tests.

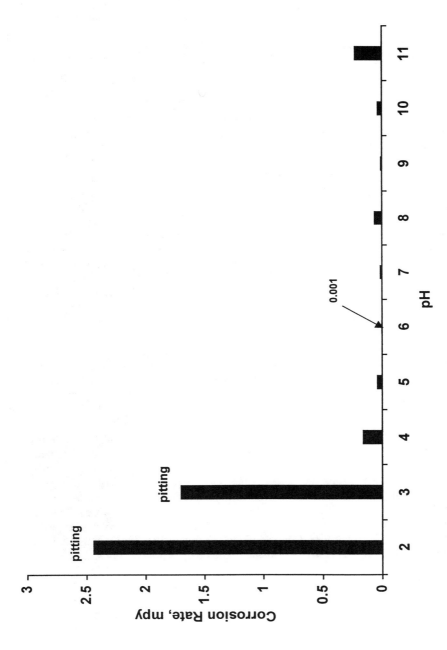

Fig. 9. Corrosion rate of BTSE-treated Al 2024-T3 in 3% NaCl at various pH values.

Even films as thin as 50 nm can outperform currently used systems such as phosphates and chromates. Thus, it can be concluded that there is ample room for improving the service life of coated metals by developing the systems discussed here to industrial scale processes. Such systems, and especially those based on silanes, are also attractive from an environmental point of view. The silane rinses discussed here do not produce any toxic waste. Further, they comprise fewer steps *viz.*, cleaning, dipping or spraying, superficial drying, optionally dipping in a second silane, and again superficial drying. Heat treatment is not required, but could be implemented if quick drying and curing of the film would be desirable. The plasma treatments are also attractive as they are solvent-free and require very little energy for conversion of the monomer to the polymer. The limitation of these processes is, of course, the vacuum requirement.

The results presented here show that, in order to obtain good corrosion performance and a stable interface which is not sensitive to moisture, the formation of covalent bonds should be stimulated. Such bonds should not be sensitive to water (hydrolysis) and further improvement can be gained if the interfacial layer becomes hydrophobic after the cure of the interface. The silane rinses which have been presented here meet all these requirements, at least for steel, zinc and aluminum. As a consequence of the requirement that covalent bonds be formed, the cleaning processes used prior to treatment by plasma films or silanes are of great importance.

Both processes presented here are very flexible in that they can be optimized for a particular metal and/or paint. For plasma deposition, this is achieved through a variation of the monomer. Its structure can further be varied within wide limits by varying the deposition conditions. For silanes, the variables that can be used in optimization studies are the type of silane, the number of steps, the pH of the silane solution and the concentration. These parameters allow great control over the properties of the films, so that many metals can be treated, as was demonstrated here.

Acknowledgments

The US Environmental Protection Agency through its Office of Research and Development partially funded the work described herein under the Assistance Agreement No. CR822989 to the University of Cincinnati. George Górecki's help was instrumental in getting the panels painted and tested in the ASTM B-117 test at Brent America Inc. The author is also grateful to Terry Craycraft for his design and construction of the plasma reactor. The following students provided input in terms of experimental results which were part of their Ph.D. thesis research project at the University of Cincinnati: Kirk Conners, Stefan Eufinger, Bin Zhang, Jun Q. Zhang, Nie Tang, Jun Song and Vijay Subramanian. Their help is greatly appreciated.

Literature Cited

1. Proceedings 3rd Annual Advanced Techniques for Replacing Chromium, Champion, PA, November 4-6, 1996.
2. W.J. van Ooij and A. Sabata, J. Coatings Technol., 61, 51 (1989).
3. R. d'Agostino, editor, *"Plasma Deposition, Treatment and Etching of Polymers"*, Acad. Press, San Diego, 1990, and references therein.

4. S. Eufinger, W.J. van Ooij and K.D. Conners, Surf. Interface Anal., 24, 841 (1996).
5. W. Yuan and W.J. van Ooij, J. Coll. & Int. Sci., 185, 197 (1997).
6. W.J. van Ooij, B.C. Zhang, K.D. Conners and S-E. Hörnström, in "Organic Coatings, AIP Conference Proceedings 354", American Institute of Physics, Woodbury, New York, 1996, p.305.
7. W.J. van Ooij and A. Sabata, Surf. Interface Anal., 20, 475 (1993).
8. W.J. van Ooij, N. Tang, K.D. Conners and P.J. Barto, Proc. Int. Adhesion Symp., Yokohama, Japan, Nov. 6-10, 1994, Gordon and Breach, Amsterdam, 1997, p. 111.
9. W.J. van Ooij, P.J. Barto, S. Eufinger, K.D. Conners and N. Tang, ", in K.L. Mittal and K-W. Lee, editors, "Polymer Surfaces and Interfaces: Characterization, Modification and Application", VSP, Zeist, The Netherlands, 1997, pp. 319-343.
10. H.K. Yasuda, T.F. Wang, D.L. Cho, T.J. Lin and J.A. Antonelli, Progr. Org. Coat., 30, 31 (1997).
11. S. Eufinger, W.J. van Ooij, and T.H. Ridgway, Journal of Appl. Pol. Sci., 61, 1503 (1996).
12. W.J. van Ooij, S. Eufinger and T.H. Ridgway, Plasma and Polymers, 1, 231 (1996).
13. W.J. van Ooij and T.F. Child, Fourth International Forum and Business Development Conference in Surface Modification, Couplants and Adhesion Promoters, Boston, September 22-24, 1997; to be published in CHEMTEC.
14. V. Subramanian and W.J. van Ooij, CORROSION, in press.
15. Chunbin Zhang, Ph.D. thesis, University of Cincinnati, 1997.
16. S-E Hörnström, J. Karlsson, W.J. van Ooij, N. Tang, and H. Klang, J. Adhesion Sci. Technol., 10, 883 (1996).
17. W.J. van Ooij, C.P.J. van der Aar and A. Bantjes, International Conference on Rubbers, Calcutta, India, December 12-14, 1997.
18. J. Marsh, J.D. Scantlebury and S.B. Lyon in Proceedings of "Advances in Corrosion Protection by Organic Coatings", J.D. Scantlebury and M.W. Kendig, editors, The Electrochemical Society, Proc. Vol. 95-13, Pennington, NJ, 1995, p.243.
19. S.E. Hörnström, E.G. Hedlund, H. Klang, J.-O. Nilsson and M. Backlund, Surf. Interface Anal., 20, 427 (1993).
20. Zhengcai Pu, W.J. van Ooij and J.E. Mark, J. Adhesion Sci. Technol., 11, 29 (1997).
21. W.J. van Ooij, J. Song and V. Subramanian, ATB Metallurgie, 37, 137 (1997).

Chapter 24

Application of Failure Models for Predicting Weatherability in Automotive Coatings

D. R. Bauer

Ford Motor Company, Research Laboratory MD#3182,
P.O. Box 2053, Dearborn, MI 48121

In large part, paint and coatings determine the esthetic appeal of a vehicle. Long-term customer satisfaction with paint is determined by how well the paint protects the body and by how well the paint maintains its overall appearance. The average age of vehicles has been steadily increasing. Customers expect reasonable maintenance of paint appearance for 10 years or longer. Catastrophic changes in appearance (peeling, cracking, etc.) can cause significant loss of satisfaction with the entire vehicle. A critical goal of weatherability testing is to be able to predict the risk of catastrophic paint failure in service. Paint performance is a function of the intrinsic capability of the particular paint system and the environment into which it is placed. It is important to note that the in-service time-to-failure has to be described by a distribution function. To predict this distribution function, it is necessary to determine the distribution functions for both the environment harshness and the paint system capability. This paper describes how to develop paint failure models that are capable of predicting the distribution of in-service failure rates. By using mechanistic failure models, test time can be greatly reduced.

The ability to predict accurately the long-term weatherability performance of paints and coatings is essential for both the coatings industry and for those industries that coat their products. Failure to anticipate in-service failures leads to high warranty costs and dissatisfied customers. Over-engineering a paint system can lead to significantly higher cost that does not provide value to the customer. An "ideal" paint system is one that exceeds customer expectations for long-term performance at minimal cost. Performance is typically measured in terms of specific engineering metrics (e.g., gloss retention). Customer expectations are much more subjective. Both customer expectation and paint system performance are functions of time. Given the

© 1999 American Chemical Society

difficulty of translating customer expectations into engineering metrics, the time-to-failure is usually defined as the time at which a particular performance characteristic drops below a specified value rather than by the time when performance drops below customer expectations. The shapes of the performance functions depend on the nature of the failure mode. For example, gloss loss in basecoat/clearcoats tends to be gradual unless some catastrophic failure occurs. As shown in Figure 1a, for gradual changes, long-term performance can be readily predicted from short-term measurements. Of course, in-service, there will be a distribution of slopes (determined by environmental load and other material and processing variables) that will lead to a distribution of times-to-failure. Distribution functions are critical to risk assessment since no commercially viable coating system ever fails 100% in-service. The application of distribution functions to interpret weathering data has been discussed (1,2).

Other failure modes (delamination and cracking) tend to be abrupt. As is clear from Figure 1b, short-term testing is not a good predictor of long-term performance. Again, the performance (time-to-failure) has to be described in terms of distribution functions of environmental load as well as material and process variables. Catastrophic failures are harder to anticipate, require longer exposure times, and often have a bigger impact on customer satisfaction. The long outdoor exposure times necessary to induce failure have caused an increased reliance on accelerated exposures. The main problem with the use of accelerated exposures is that they can change the balance of degradation and stabilization chemistries as well as the physics of failure. The effect can vary from coating to coating leading to acceleration factors that depend on the coating, on the failure mode, as well as on the test (3-6). Recent analytical studies have focused on developing chemical methodologies to better understand the factors that control catastrophic failure (7,8). In essence, this work involves translating those analytical studies into failure models whereby measurements of critical coating attributes and their distributions can be used to anticipate long-term in-service performance. This paper extends the analysis presented in a previous paper (2) hereafter referred to as I. In the Sections that follow, the development and application of different coating failure models are described.

Development and Application of Failure Models

In order to develop failure models, it is necessary to describe all possible failure modes of a basecoat-clearcoat paint system. Such failure modes that can be related to weathering induced photooxidation include gloss loss, color change, clearcoat cracking, delamination of clearcoat from the basecoat, and delamination of the basecoat or primer from a photosensitive substrate. In principle, it is necessary to develop specific failure models for each of these failure modes. Before doing that, a general model that can be used to describe any failure mode controlled by photooxidation is presented.

Photooxidation Distribution Function: A General Model for Coating Weathering Failures. As noted above, a failure model must include information about both the coating performance and the environmental harshness. In the absence

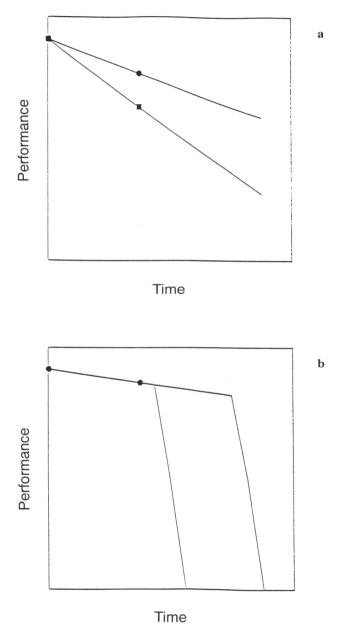

Figure 1. Performance versus time. Changes can be either gradual (a) or abrupt (b). Abrupt changes tend to be more noticeable and more difficult to predict from short-term measurements.

of specific mechanistic information concerning failure modes, it is still possible to account for the variation in environment. The time-to-failure, t_f, can be written as the product of a coating constant, C, and an exposure variable, EX:

$$t_f = C \times EX \tag{1}$$

Following the methodology used in I, EX is defined as the number of years of exposure on a particular vehicle that is equal to one year of exposure of a panel on a test rack in Florida. Thus, the value of C is the exposure time (years) in Florida required to fail the particular coating system. The value of EX is inversely related to the harshness of the exposure. EX will depend on the geographic location, customer parking habits, and on the specific dependence of coating photooxidation on environmental variables such as light intensity, temperature, humidity, etc. Based on mechanistic arguments and empirical data (9-12), it was proposed in I that the rate of coating photooxidation is proportional to accumulated light dose times an Arrenhius dependence on temperature (activation energy = ~7 kcal/mole).

Seasonal intensity and mean daily high temperatures were used to calculate the relative rate of photooxidation at different geographical locations in the continental US (13). An arbitrary correction factor for humidity was included to correct for differences between prediction and observation of weathering in Arizona and Florida. The distribution function is determined by counting the number of vehicles that reside in locations with specific harshness. This distribution function (Figure 1 of I) assumes that vehicles are exposed 100% of the time (i.e., as if they were exposure panels). In fact, some vehicles are parked mostly outside while others are mostly parked in garages or parking structures. A "parking" distribution function was estimated and combined with the photooxidation distribution to estimate an "in-service vehicle" distribution function for EX in the continental US. This distribution function is shown in Figure 2. The distribution function is fairly broad. Around 90% of the vehicles accumulate the equivalent of one year exposure in Florida between 1.2-5 years.

The distribution in Figure 2 can be recast to answer the following question: What is the % expected failure after 10 years in service for a paint system that fails after a particular number of years in Florida. This relationship is shown in Figure 3. The critical point of Figure 3 is that very long exposure times (~8 years) are required to reduce the in-service failure rate to below 5% after 10 years (a typical average customer expectation for paint life). A plot of % failure with time in service is shown in Figure 4 for a system that fails after 5 years in Florida. The onset of failure is about 6 years rising rapidly to a total failure rate of almost 40% after 10 years in service. Another critical point is that this analysis assumes that all paints of this technology will fail at exactly the same time in Florida, independent of material and process variability. This is highly unlikely. In the absence of a more detailed mechanistic failure models, evaluation of the effects of process and material variability on performance will require the exposure of a wide variety of samples prepared under all possible conditions. This, together with the very long exposure times required to ensure performance

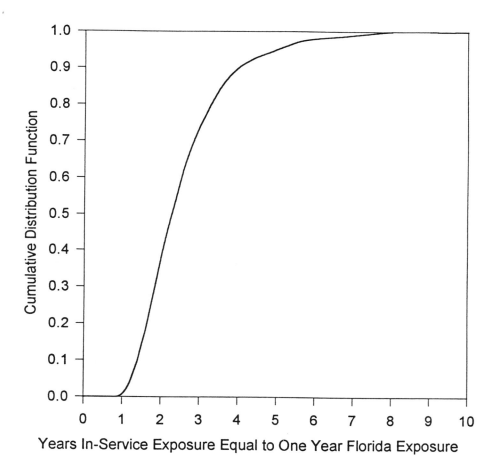

Figure 2. Cumulative distribution function for photooxidative harshness (see I for derivation). The abscissa is the number of years in service required to equal one year continuous exposure in Florida. The ordinate is the fraction of vehicles. For example, roughly 20% of the vehicles in service receive a exposure equal to a year in Florida in 1.6 years or less.

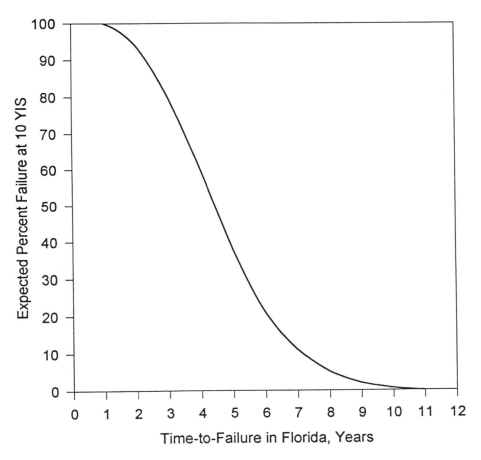

Figure 3. Percent failure on vehicles at 10 YIS versus time-to-failure in Florida for a given paint system.

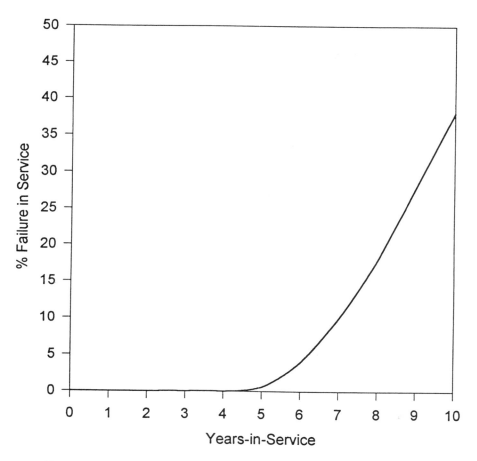

Figure 4. Percent failure in service versus time in service for a coating system that fails at 5 years in Florida.

against catastrophic failure, provides strong motivation to understand specific mechanisms of failure.

In the previous paper, I, two different mechanistic failure models were developed based on different hypothetical failure modes. The purpose of that development was to illustrate how failure models might be developed and applied. The failure models were based on two limiting cases for UV light induced delamination. In the first case, it was assumed that light transmitted through a black pigmented coating resulted in degradation and ultimate delamination of a photosensitive substrate. This case was used to illustrate the potentially large sensitivity of failure rates to process variations. In this case, performance was related to film thickness in an exponential fashion. Under some test protocols, field failures would be observed long before samples failed in Florida. In such cases, it is usually necessary to develop materials or processes that avoid this sensitivity. In the second case, it was assumed that basecoat-clearcoat delamination could be caused by a slow loss of ultraviolet light absorber (UVA) that is added to clearcoats. Once the light absorber is lost, UV light induces degradation of the basecoat ultimately leading to delamination. A simple model was derived based on this hypothesis. The model was used to illustrate how rapid measurement of specific parameters could be used to predict long term performance. In the next section, this model is evaluated in more detail. As a result of further experimental work, the model has been modified to more correctly represent the actual failure mode. The predictions of the old and new model are compared for the purposes of discussing the precision required to make successful predictions.

Model for Basecoat-Clearcoat Delamination. In the initial stages of the development of basecoat-clearcoat automotive paint systems, it was recognized that it is necessary to protect the basecoat from UV radiation (14). UVAs were added to clearcoats to achieve this protection. Samples where the UVA was deliberately left out exhibited delamination of the clearcoat from the basecoat after an exposure time ranging from 2 to >4 years in Florida. UVAs were developed with excellent physical retention (no loss due to volatility or extraction). It was assumed that the UVAs were basically permanent in the coating. Over the past several years, the work of Pickett and Moore, Decker and Zahouily, and Gerlock et al. demonstrated that UVAs in fact were slowly consumed in polymer systems (7,15,16). The rate of consumption depended on the UVA , the polymer host and the exposure condition. The exact mechanism(s) of loss are still not completely understood. There appear to be two basic processes: attack on the excited state of the UVA by oxidation products generated in the polymer matrix; and a photolytic process that does not depend on free radicals produced by the polymer matrix (15,17). The quantum efficiency is very low ($<10^{-6}$). Despite the existence of multiple mechanisms, the observed kinetic behavior is relatively simple. Pickett and Moore have shown that the rate of loss by UVA is given by the following equation (15):

$$\frac{d[\text{UVA}]}{dt} = -k < I > [\text{UVA}] \qquad (2)$$

where,

$$< I > = \frac{1 - 10^{-A}}{2.303A} \qquad (3)$$

Since UVA concentration is proportional to absorbance, Equation 2 can be rewritten in terms of absorbance loss:

$$\frac{dA}{dt} = -k(1 - 10^{-A}) \qquad (4)$$

The variation in absorbance with time can be determined by integrating Equation. 4. Using this dependence of absorbance with time, the % transmission to the basecoat can be integrated to yield the total dose of UV light hitting the basecoat as a function of time. If it is assumed that the interface delaminates after a particular dose is achieved in Florida, t_{BC}, then the time-to-failure in Florida as derived in I is given by,

$$t_{Fl} = t_{BC} + \frac{A_o^*}{k} L \qquad (5)$$

where L is the clearcoat thickness, A_o^* is the initial absorbance per unit thickness and k is the loss rate in absorbance/year as measured in Florida. The ratio A_o^* L/k essentially defines the length of time in Florida that the UVA in the clearcoat protects the basecoat.

In principle, the value of t_{BC} can be measured by exposing basecoat/clearcoat systems that do not contain UVA. In practice, relatively little is known about the factors that control this quantity. Unprotected, sensitive basecoat/clearcoat systems have been observed to fail after around 2 years of Florida exposure. Other unprotected basecoat/clearcoat systems have not delaminated after 4 years exposure. Values of A_o^* and L can be measured on the initial samples. In most cases, A_o^* is a material constant ranging from 1.5 - 3 absorbance units/mil. The only relevant processing variable is bake history. The concentration of a few UVAs are reduced by overbaking. The clearcoat film thickness, L, is a process variable whose distribution has to be determined for a particular part and processing setup. Since the system delaminates at the thinnest spot on a horizontal surface first, the most relevant distribution is the minimum horizontal clearcoat thickness. It was shown in I that to a reasonable approximation, it is permissible to use the average value of this quantity to predict performance. Typical average minimum clearcoat thicknesses are around 1.5

mils. As long as the total clearcoat absorbance is greater than 1, the initial rate of absorbance loss is exactly the rate constant k. Thus, it can be measured accurately by relatively short-term exposures. Florida exposures and xenon arc (boro-boro filters) have been used to obtain reliable values for k. Typical values for k range from 0.4 - 1 absorbance units/year, Florida exposure. Although the value of k is primarily determined by clearcoat and UVA chemistry, it may also, in some cases, depend on basecoat color and composition. Thus, it is necessary to determine both t_{BC} and k for every color.

Combining these measurements, it is possible to assess the degree of protection provided by the clearcoat ($A_o^* L/k$) for different basecoat/clearcoat systems. Typical values range from 5-9 years Florida exposure for some currently available systems. Measurement errors limit the accuracy of the determination to about $\pm$ 1 year in Florida.

By adding the degree of protection of the clearcoat with the value of t_{BC} it is possible to make predictions about the Florida resistance of basecoat/clearcoat systems to delamination based on relatively short-term experiments. In-service performance can then be predicted if the dependence of t_{BC} and k on exposure condition is known. In I, it was assumed that both quantities were functions of the coating photooxidation rate. Thus, the exposure distribution function of Figure 2 could be used to predict in-service performance. The validity of those predictions depends on the validity of that assumption as well as the validity of other assumptions used to derive Equation 5. For example, it is assumed that light of wavelength longer than the UVA cutoff (340-375 nm depending on UVA type) does not contribute to the degradation of the basecoat. It is recognized that UVAs can migrate from the clearcoat to the basecoat during application. This essentially changes the value of A_o^* in the total system versus the value for the clearcoat only. It is assumed, however, that there is no further diffusion of UVA during exposure. Migration of UVA out of the basecoat back into the clearcoat would tend to offset loss of UVA from the clearcoat. It is also assumed that the basecoat/clearcoat interface is clearly defined and that the delamination process is controlled solely by photooxidation of the basecoat. Only under these conditions can the value of t_{BC} be defined by the UV dose at the basecoat. Finally, it is assumed that HALS added to reduce photooxidation rates in the clearcoat and basecoat have constant effectiveness over the duration of the exposure. In the paragraphs that follow these assumptions are evaluated and modification to the failure model are made.

The compositions of the basecoat materials would suggest that their dependence of photooxidation on intensity, temperature, and humidity should be similar to clearcoats. In absence of any data to the contrary, the assumption that the exposure function in Figure 2 can be used to relate t_{BC} results in Florida to in-service results seems reasonable. The fact that basecoats contain pigments and colorants which absorb visible light suggests the possibility that they may have a different dependence on wavelength from clearcoats. This effect will be discussed shortly.

The second assumption used in I is that the dependence of UVA loss rate, k, on light intensity temperature and humidity is the same as that for overall coating photooxidation. This would be expected to be the case if the UVA loss was dominated by attack by products induced by clearcoat photooxidation. Under these circumstances, the time-to-failure in-service distribution function can be given by,

$$t_f = EX(t_{BC} + \frac{A_o^*}{k}L)$$ (6)

where EX is the exposure distribution function defined in Figure 2. However, for polymers with very low photooxidation rates, UVA loss appears to be dominated by a mechanism that is independent of polymer photooxidation. In this case, the data of Pickett and Moore (15) suggests that k may be proportional to light intensity but may to be independent of temperature and humidity. The distribution function for in-service harshness (EX') where the loss rate is solely determined by UV light intensity is given in Figure 5. Comparison of Figure 2 and 5 reveals that a significantly higher fraction of in-service vehicles see exposures as harsh as Florida for the light intensity only distribution function relative to the distribution of Figure 2. This is a result of the higher light intensities in the dry regions of the Western US. If this mechanism is valid for UVA loss, then the distribution function for in-service failure is given by,

$$t_f = EX\, t_{BC} + EX'\frac{A_o^*}{k}L$$ (7)

Equations 6 and 7 provide lower and upper estimates for % in-service failure for systems with particular values of the parameters. For example, if $t_{BC} = 3$ years, Florida and if $A_o^*L/k = 5.3$ years, Florida, then use of Equation 6 would lead to a prediction of 4.5% failure in service after 10 years while Equation 7 would predict a 9% failure rate. While this difference seems to be large, it is of the same order as the effect caused by variability of the weather over long periods of time. It is also similar in magnitude to the effect of measurement errors discussed above. The behavior of k in current basecoat/clearcoat systems will likely be controlled by a weighted average of these two mechanisms. In the absence of specific mechanistic data, it would seem prudent to err on the side of caution and use Equation 7 to assess failure rates.

Another critical assumption is that light beyond the UVA cutoff does not contribute to basecoat photooxidation. O^{18} TOFSIMS studies of basecoat photooxidation behind UV cutoff filters has recently shown that substantial photoooxidation can occur for some colors at wavelengths beyond typical UVA cutoff values (18). To account for this affect, we assume that light beyond the UVA cutoff causes photooxidation in the basecoat at a rate that can be expressed as a ratio, r, of the total induced in the absence of any UVA. Under these circumstances, it is easy to show that the time-to-failure is given by,

$$t_f = EX\, t_{BC} + EX'\,(1-r)\frac{A_o^*}{k}L$$ (8)

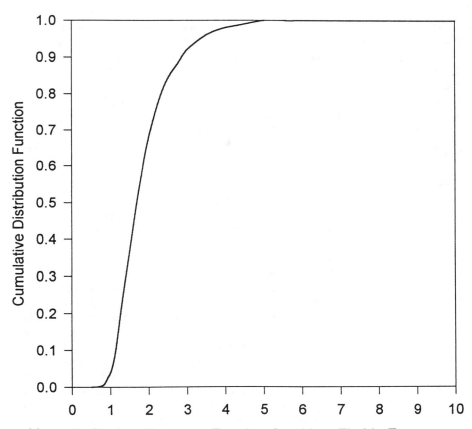

Figure 5. Cumulative distribution function for harshness that is solely a function of light intensity. Compare to Figure 2. Roughly 20% of the vehicles in service receive a light dosage equal to a year in Florida in 1.27 years or less.

or by,

$$t_f = \frac{EX\, t_{BC}}{r} \qquad (9)$$

whichever is smaller. In most cases, the value of r will be small enough that Equation 8 will apply. The fraction r will depend on basecoat color and chemistry as well as on the UVA cutoff. In principle, the value of r can be measured by measuring the extent of photooxidation in the basecoat with and without UVA (before the UVA concentration is significantly reduced). Alternatively, it is possible to expose basecoats behind a UV cutoff filter of the appropriate wavelength. The TOFSIMS experiments suggest that values of r for most basecoats are less than 0.2 for benzotriazoles, the UVA most commonly used now. For oxanilide UVAs, the cutoff wavelength (340 nm) is much shorter than the value for benzotriazoles (370 nm) and values of r as high as 0.5 have been observed. As shown in Figure 6, larger values of r lead to a dramatic increase in the in-service failure rate. This, together with a somewhat higher loss rate typically observed for oxanilides, probably accounts for the relatively poorer performance of basecoat/clearcoat systems fortified with oxanilide UVAs relative to those fortified with benzotriazoles. It should be clear that it is essential to determine the quantity r for each basecoat color and adjust the UVA content appropriately.

The assumption that UVAs do not migrate very much from the basecoat into the clearcoat during exposure seems to be confirmed based on UV microscopy of sections of basecoat/clearcoats exposed in various locations (19). The assumption that there is a clear boundary between the basecoat and clearcoat and that the delamination is controlled by basecoat photooxidation is less obvious. There is clear mixing of resin between the basecoat and clearcoat and the interface is better described as an interphase region. The O^{18} TOFSIMS experiments do suggest that the photooxidation that ultimately leads to delamination initiates in the basecoat/clearcoat interphase region after the UVA is depleted. The region of high photooxidation is relatively broad (5-10 microns) and grows as the degradation proceeds to eventually encompass most of the clearcoat. This may account for the observation that in many cases the degradation seems to be as much a mechanical failure of the clearcoat as it is a failure of the basecoat-clearcoat interface. It is unclear at this point how to use this knowledge to modify the failure model. It may be possible that t_{BC} depends on clearcoat thickness or other process parameters. Measurements of time-to-failure for systems with no UVA will be required to resolve this question.

The last assumption that is important to the validity of the model is the assumption that HALS are effective throughout the total exposure. If HALS were lost before the UVA was completely consumed, then it is likely that both k and t_{BC} would change from values determined in the presence of HALS. Such changes would drastically reduce the time-to-failure in Florida from that predicted causing significantly higher failure rates in the field. Techniques are being developed to quantify HALS longevity. Fortunately, it does appear that in most systems used currently, HALS are effective in coatings for very long times.

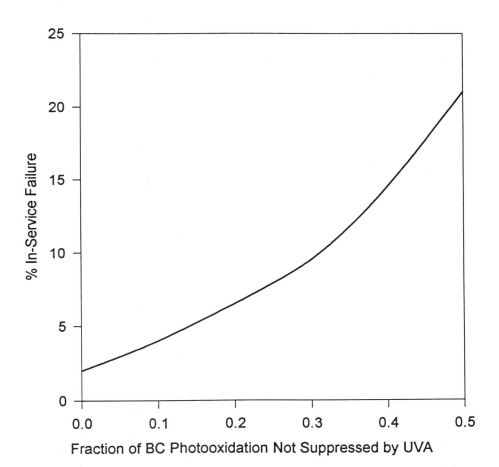

Figure 6. Percent failure after 10 years in service versus r, the ratio of basecoat photooxidation with and without UVA, for a particular basecoat/clearcoat system ($t_{BC} = 3$, $A_o^* L/k = 6$).

While Equation 8 is somewhat more complex than Equation 6, the amount of experimental work necessary to determine parameters for Equation 8 is not excessive in either effort or time. Substantial effort has been made in developing experimental data necessary to understand basecoat-clearcoat delamination. While further work (along with a reliable accelerated test) is necessary to understand the material and processing factors that control t_{BC}, it can be concluded that a reasonable failure model exists that can assist in the setting of requirements on particular parameters to prevent in-service failures. Evaluation of currently used materials suggests that systems are available that can prevent delamination out to 10 years-in-service. The status of models for other known failure modes is less advanced. In the sections that follow, we discuss the status of models for gloss loss, color shift, and cracking and make recommendations for further work.

Failure Models for Gloss Loss. Gloss loss due to weathering is by far the most widely studied coating appearance characteristic. For monocoats, gloss loss has been attributed to polymer photooxidation and erosion leading to surface roughness due to the presence of pigment particles near the surface. Specific pigment photochemistry has also been studied (particularly various forms of TiO_2). Clearcoats do not contain pigments and gloss retention of clearcoats is much better than for monocoats. Clearcoat gloss might drop from 95% to 80% in two years exposure in Florida. Gloss in a typical monocoat after two years in Florida might be anywhere from 20 - 60%. The factors that control gloss loss in clearcoats is not well understood. Surface roughness could still be caused by polymer photooxidation and erosion with the roughness occurring as a result of heterogeneous composition (presence of microgel additives, etc.). It has also been reported that acidic depositions lead to a loss of gloss in addition to causing localized etching (20). The relative contribution of these processes to gloss loss in-service is not known. Further work is required before pH can be considered along with intensity, temperature and humidity as an exposure factors for gloss loss. The effect of process variabilities on gloss retention is also not known. Initial state of cure does have a significant effect on acid etch resistance. On the other hand, no obvious sensitivity to process has been identified from Florida gloss exposures. However, it should be noted that gloss measurements are only reproducible to about 5%. Considering the small changes in clearcoat gloss with weathering, the variability due to measurement error might mask variations due to process. Gloss retention does depend on color. There are two possible explanations for this. First, dark colors will be hotter leading to higher rates of photooxidation and acid attack. Second, color specific dispersion resins and additives in the basecoat may migrate into the clearcoat changing its photostability. Customer expectations for long-term gloss retention in clearcoats are not understood in detail. Mild gloss loss due either to photooxidation or etching can be "repaired" by waxing and polishing. Thus, if a customer is dissatisfied with this particular appearance characteristic, he can restore the appearance at relatively minimal cost. At this point the only model that can be constructed for clearcoat failure due to gloss loss is to use Equation 1 with C being estimated for each basecoat/clearcoat using initial gloss retention data in Florida.

Failure Model for Color Change. Color change in basecoat/clearcoat systems is usually a result of photodegradation of a specific pigment or pigments in the basecoat. Reds tend to be the most sensitive to color shifts. Very dark colors such as blacks tend to lighten. Clearcoat yellowing may also contribute to long-term color changes, however, this appears to be relatively small in most cases. The amount of color change appears to be proportional to the dose of light. The effects of temperature, humidity, and other environmental variables are not known. That is, it is not known whether the EX, EX', or some other exposure distribution function should be used. The conservative approach would be to use the harsher EX' distribution function. The rate of color shift in the basecoat should increase when the UVA in the clearcoat is depleted. The color shifts can be measured in short-term Florida tests for basecoat/clearcoat systems with and without UVA. The ratio, r', of the color shift with and without UVA is analogous but not necessarily equal to the ratio, r, of the basecoat photooxidation with and without UVA. Color shifts will be most noticeable (and thus, most objectionable) between horizontal and vertical panels and between components with different paint systems that degrade at different rates. We assume that there is a value for color shift, ΔE, that can be used to establish a failure criterion. The measured color shift per year of Florida exposure without UVA, ΔE_{Fl}, can be used together with the value of r' and UVA retention data to yield an expression for the time-to-failure for color shift,

$$t_f = EX' \left(\frac{\Delta E}{\Delta E_{Fl}} + (1-r') \frac{A_o^* L}{k} \right) \tag{10}$$

or by,

$$t_f = EX' \frac{\Delta E}{\Delta E_{Fl} r'} \tag{11}$$

whichever is smaller. These expressions are analogous to Equations 8 and 9 for delamination. It should be noted that this model assumes that there are no colorants in the clearcoat. Tinted clears have recently been introduced. The presence of colorants in the clearcoat significantly changes the degradation process and new models will have to be developed for clearcoat gloss retention and color shift.

Failure Models for Clearcoat Cracking. Prevention of clearcoat cracking is probably the next most important long term weatherability issue after the prevention of delamination. Cracking, like delamination, is an abrupt failure mode that only be repaired by repainting. At this point there is no comprehensive mechanistic model that can predict in-service crack formation as a function of time. Long-term outdoor exposures are the only reliable source of performance information. Useful knowledge is being generated that could lead to such a model. Perara et al. have studied the change in internal stress in coatings as a function of exposure condition and weathering time (21). Nichols et al. have studied how to measure fracture toughness in coatings (22). By combining and extending these studies it should be possible to develop a rapid predictive model for cracking

Conclusion

This paper presents the development and application of several models aimed at predicting long-term, in-service weatherability performance of automotive coatings. Of critical importance is the need to treat failure parameters as distributions. Materials, processes, and environment are all variable leading to a broad distribution of failure times for any particular system. Two different environmental distribution functions have been generated. The first incorporates light intensity and temperature to model polymer photooxidation in coatings. An arbitrary correction factor for humidity is also included. This model seems to accurately reflect geographical differences in photooxidation induced coating weathering over the continental US. The second function only involves light dosage and may be applicable in some cases to UVA loss and color shift. Using these functions and mechanistic insights into different failure modes (delamination, gloss loss, color shift, and cracking) models have been developed which incorporate measurable material and processing parameters and distributions. The model for basecoat/clearcoat delamination is the most complete, allowing for the prediction of in-service failure rates as a function of such variables as UVA concentration, loss rate, clearcoat film thickness distribution, and basecoat sensitivity to light of different wavelengths. Models for the other failure modes are much less complete. The model to predict clearcoat cracking is in particular need of significant mechanistic study. Another area where further work is required is in the extension of the environmental exposure distributions to other areas of the world. Factors that were ignored in the US model such as variation in wavelength distribution with latitude and a better understanding of the role of humidity and rainfall will have to be incorporated. Finally, it is necessary to do a better job in translating customer expectations into engineering performance terms. When this is accomplished, it should be possible to specify coating performance to achieve superior customer satisfaction in specific markets.

References

1. Martin, J. W., Saunders, S. C., Floyd, F. L., and Wineburg, J. P., "Methodologies for Predicting the Service Lives of Coating Systems", NIST Building Science Series 172 (1994).
2. Bauer, D. R., J. Coat. Technol., 69 (864), 85 (1997).
3. Simms, J., J. Coat. Technol., 59, 45 (1987).
4. Bauer, D. R., Paputa Peck, M. C., and Carter III, R. O., J. Coat. Technol., 59 (755), 103-109, (1987).
5. Bauer, D. R., Gerlock, J. L., and Mielewski, D. F., Polym. Deg. Stab., 36, 9-15 (1992).
6. Bauer, D. R., Prog. Org. Coat., 23, 105-114 (1993).

7. Gerlock, J. L., Smith, C. A., Nunez, E. M., Cooper, V. A., Liscombe, P., Cummings, D. R., and Dusbiber, T. G., in "Polymer Durability", R. L. Clough, N. C. Billingham, and K. T. Gillen, Eds, ACS Advances in Chemistry Series, No. 249, p. 335 (1996).

8. Gerlock, J. L., Prater, T. J., Kaberline, S. L., and de Vries, J. E., Polym. Deg. Stab., 47, 405 (1995).

9. Mielewski, D. F., Bauer, D. R., and Gerlock, J. L., Polym. Deg. Stab., 41, 323 (1993).

10. Bauer, D. R., Polym. Deg. Stab., 48, 259 (1995).

11. Gerlock, J. L., Bauer, D. R., Briggs, L. M., and Hudgens, J. K., Prog. Org. Coat., 15, 197 (1987).

12. Bauer, D. R. and Mielewski, D. F., Polym. Deg. Stab., 40 349 (1993).

13. See for example, "National Atlas of the United States of America:, U.S. Dept. of the Interior, Geological Survey (1970).

14. Puglisi, J. and Schirmann, P. J., Org. Coat. Plast. Prep., 47, 620 (1982).

15. Pickett, J. E., and Moore, J. E., Polym. Deg. Stab., 42, 231 (1993).

16. Decker, C. and Zahouily, K., Polym. Mat. Sci. Eng. Proc., 68, 70 (1993).

17. Gerlock, J. L., Tang, W., Dearth, M. A., Korniski, T. J., Polym. Deg. Stab., 48, 121 (1995).

18. Dupuie, J. L., Gerlock, J. L., Prater, T. J., Kaberline, S. L., and Bauer, D. R., in preparation.

19. Gerlock, J. L. and Smith, C. A., private communication.

20. Wernstalhl, K. M. and Carlsson, B., J. Coat. Technol., 69 (865), 69 (1997).

21. Oosterbroek, M., Lammers, R. J., van der Ven, L. G., and Perera, D. Y., Coat. Technol., 63, 55 (1991).

22. Nichols, M. E., Darr, C. A., and Smith, C. A., Polym. Deg. Stab., in press.

Chapter 25

Computerizing Complex Data: Data Quality Issues and the Development of Standard Formats for Representing Materials Properties

J. R. Rumble, Jr. and C. P. Sturrock

Standard Reference Data, National Institute of Standards and Technology, Gaithersburg, MD 20899

The development of the computer models necessary to support the prediction of service life of complicated products is proceeding rapidly. Not only is the available computer power growing continuously, but also our understanding of the behavior of engineering materials on a nano- and micro-scale has increased significantly. However, the robustness of prediction is critically affected by the quality of data used in modeling. Many of these data are quite complex and depend on a large number of independent variables. Data must be drawn from many different sources, often incompatible in terms of storage formats, nomenclature, and ancillary information. Prediction of the service life of organic coatings is just now becoming an area of great interest and in the future will rely considerably on modeling. This paper addresses issues related to capturing and maintaining high quality materials properties data and databases necessary to support service life prediction of organic coatings, specifically assessing data quality, development of formats for computerizing these data, and the role of formal and informal standards in accepting these formats. Emphasis is placed on sharing and exchanging data across scientific disciplines, a major demand of service life prediction.

Each year, the quality and quantity of new technology used in computing continues to increase at an incredible rate. The impact of this growth on the gathering, manipulating and delivery of information is obvious to everyone, yet the best way to harness this new power is often not clear. In many areas of science and technology, the development of databases of properties of materials and systems is one facet of the information revolution. Many facts about the characteristics and performance of materials and systems that support every conceivable scientific and engineering activity, from research to product design to composition analysis to production planning and to failure analysis, can be aggregated into such databases and application software.

One engineering activity that will rely heavily on computerized data in the future is the prediction of the service life of complicated products. As computer power continues to increase and our ability to generate models of complex behavior based on greater

U.S. government work. Published 1999 American Chemical Society

understanding of nano- and micro-scale phenomena expands, models of service life prediction will grow in importance and acceptance. This volume is intended to address service life prediction of organic coatings, an exceedingly complicated activity. Because access to reliable data from many disciplines will be needed to perform accurate modeling of this service life, several issues need to be resolved to ensure data quality and ease of access. This paper addresses many of those issues, including the assessment of the quality of scientific and technical data and the databases containing these data, development of formats for capturing data in their full complexity, the role of standards in acceptance of those formats, and exchanging and sharing data on an interdisciplinary basis. Each of these subjects is considered in turn, with emphasis of the data needs related to service life prediction of organic coatings.

Overarching Considerations

Some additional background is needed for the discussion that follows. Prediction of the service life of organic coatings depends on a wide range of data, including meteorological, solar incidence and spectral characteristics, bulk materials properties, surface properties and material degradation. Each data type is generated by a separate and complicated discipline. The sensitivity of the prediction models to the input values demands that all data be of the highest possible quality. Experience has shown that quality must be built in, not added, i.e., it is unlikely that a prediction model can be designed to compensate for poor data. Furthermore, the needed data are generally not generated for the sole purpose of supporting service life predictions for organic coatings. Usually, the data are generated without regard to service life prediction and are intended for characterizating some phenomena or other purpose. Computer databases of these complex data should provide the following goals:
  • Support of interdisciplinary use
  • Support of interdisciplinary data generation and collection
  • Support of long term archiving
  • Capture of the full richness of the data and related metadata
  • Assessment of data quality
These goals naturally lead to the conclusion that building in quality demands an understanding of data needs and use, identification of the interfaces where data from one discipline are used in another discipline, and establishment of the time frames for data collection and use. With these general considerations in mind, let us consider more detailed issues.

Data Quality

The quality of scientific and technical (S&T) data is assessed from three points of view: documentation of the data generation; adherence to known physical laws; and comparisons among measurements. Each of these viewpoints is discussed briefly.

Documentation of the data generation. Any scientific and engineering measurement involves setting up an experiment, allowing one or more independent variables to change, and then measuring the response of the system. The same process applies whether the measurement involves an experiment or a calculation. To perform a measurement accurately requires that all independent variables have been identified and that all non-varying variables have been controlled adequately. Shoshoni *et al.* have named the variables that are fixed as "configuration" variables and those that vary as "instrumentation" variables (*1*). When assessing the quality of a result, the evaluator looks for documentation of the relevant independent variables and proof that the configuration variables have in fact been controlled adequately. For example, if a measurement is made at a given temperature, the evaluation determines whether the temperature was checked often enough and whether it was reset suitably if a deviation occurred.

Adherence to physical laws. The second viewpoint for determining data quality is based on how the data follow established physical laws. For materials data, examples include the laws of thermodynamics for phase equilibrium data and Hooke's law for elastic behavior. This viewpoint is not often useful for a single measurement. Usually physical laws are not known until a scientific field has matured, which in turn occurs after the generation and measurement of significant amounts of data. In these cases, new data are examined to see how they fit in with other data that adhere to the laws; for example, how a new measurement of the heat of formation of a compound fits with similar data determined by other reaction pathways. An extension of this viewpoint is to empirical behavior laws, common in materials science, which provide an excellent model for materials behavior over a wide range of material compositions and test conditions. Failure to follow the empirical laws raises many red flags to an evaluator.

Comparison to other measurements. Ideally, the evaluator would like to have demonstration that all independent variables have been identified and controlled, with the relevant physical laws well articulated. Such a situation, of course, never occurs, and a third evaluation viewpoint comes into play, namely, how well data compare to other measurements that purport to describe the same phenomena. When a sufficient number of tests are performed so that the body of comparable data justifies statistical analysis, rather interesting evaluations can take place. One major result might be the identification of hidden independent variables, not previously noticed. Such statistical analysis can play a major role in evaluating materials data, but rarely is enough information available to generate meaningful statistics. We are already aware of this lack of information through our experience, and in many areas of materials data, little or no attempt is made to intercompare data, except on a gross scale. For example, corrosion and tribological data, extremely important in product design, almost never have comparable or duplicate measurements.

Putting it all together. The process of data evaluation is a painstaking effort that takes time and expertise. A good recent discussion of the intricacies of the evaluation process has been given by Munro and Chen in a paper on evaluating data for high temperature superconductors (*2*). One result of the critical evaluation of data in a fairly new field is that few if any data have been generated with enough care. This is not surprising because most fields develop, first by having measurements to identify a phenomenon, second, by understanding the phenomenon in a gross manner, and finally by undertaking detailed measurements. Consequently, it is important that data evaluation efforts are fed back into the measurement community, so that experimental, test, and calculational methods are refined to improve data quality. To this end, publications such as the *Journal of Testing and Evaluation* and the *Journal of Physical and Chemical Reference Data* are very important in completing this feedback loop. The process of developing a body of reliable data is a long-term effort that requires countless improvements in experimental and testing methods. Measurements result from a single experiment or test; properties result from decades of measurements.

Database Quality

Computerized databases are often considered to be definitive sources of reliable information, yet if proactive steps are not taken to ensure database quality, such confidence is misplaced. Over the years, the Standard Reference Data Program at the National Institute of Standards and Technology has developed a comprehensive and highly effective approach to determining the quality of scientific factual databases. This review is done from three perspectives. The first is to ensure that the data in the database accurately represent the data as produced by the data evaluator or data generator. The second is that the database search and retrieval software find data that are in fact the correct data for the property, substance, systems and conditions selected. The third is on database reliability, namely, that the database works every time for every function without abnormal stopping.

The combination of a program for assessing data quality and ensuring database quality will result in reliable data that represent the best scientific and technical knowledge to date. While this in no way guarantees correct modeling results, because of potential problems in the model or because the data themselves are not good enough to support the model, at least the best possible data will be used. At the same time it must be recognized that data quality evolves and grows over time, often over decades. Data quality does not just happen; it is the result of a focused and proactive effort. The benefit of a strong data quality program is hard to overestimate, and as modeling becomes more important in the future, the community of scientists and engineers involved in service life prediction of organic coatings will require data quality programs to achieve the best possible engineering and resulting products.

Data Recording Formats

The process of computerizing scientific and technical data requires additional effort beyond data quality. Designing robust databases also requires considerable planning. One of the most important aspects of the planning process is the definition of all necessary and desired data for a database. For scientific and technical databases, there are two primary search paths: (1) What properties does a specified material or system have? and (2) What materials or systems have properties within a specified range? To support these search paths, every database designer must answer three questions: (1) How best to describe a material or system? (2) How to specify a measurement or calculation of a property of that material or system? and (3) How to specify the ancillary data associated with that measurement or calculation?

Because all database builders must answer these questions, individual disciplines have worked to define collective answers to these questions, and the resulting answers are commonly called *data recording formats*. The data recording formats serve many purposes. Their primary purpose is to facilitate the collection and exchange of all necessary and desired data. A common nomenclature is developed so that data from separate databases can be combined and compared. The data recording formats can form the basis for data dictionaries, thesauri, glossaries, shared vocabularies, and integrated data systems.

What is in a data recording format? A data recording format consists of a set of data elements with a name, definition, allowed values, representation (text, integer, etc.), examples, references, and possibly synonyms, antonyms, and relationships to broader or narrower terms. For complex materials and systems, a data recording format may easily have 100+ data elements; for measurements, 200+ data elements are not uncommon. Because very few data sets actually have that much information reported, the data recording format must differentiate between data elements that are *mandatory* and those that are merely *desirable*. Often, only 10% of the elements are necessary to make a data set worthwhile. If a data recording format is in place, the database building process can move ahead more smoothly because the data dictionary can be easily derived from the data recording format.

Materials and systems. When storing information on a material or system in a computerized database, three primary capabilities must be preserved - uniqueness, equivalency, and data exchange. Uniqueness means the ability to identify a specific material or system to whatever degree of completeness desired -- data exists for *stainless steel*, for *stainless steel 316*, for *stainless steel 316 with heat treatment xyz*. Equivalency means the ability to determine if two materials or systems are the *same* to some specified degree of detail. Exchanging and sharing of data means the ability to compare and combine data from different databases while maintaining completeness and consistency. *Stainless steel 316* in database A is the same as *S31600* in database B. Materials and system data recording formats include at least eleven major types of information:
1. Names and classes
2. Specifications
3. Source
4. Processing/assembly history
5. Shape and size
6. Composition and structure
7. Nature of constituents and components
8. Amounts of constituents and components
9. Shape and sizes of constituents and components
10. Relative location of constituents and components
11. Associativity of constituents and components

The reader is referred to (*3*) for more details on each information type.

Measurement, calculation and property results, and metadata. The same level of complexity of information types exists for these data. An important point to note is that a measurement result is virtually meaningless if enough ancillary data are not given. Today, the term *metadata* is commonly used to refer to these ancillary data, with the most common definition of *metadata* being *data about data*. For measurements, the types of information in a data recording format include:

1. Test or calculation method description
2. Measurement or calculation results
3. Analysis methods and results
4. Changes to standard test or calculational methods
5. Full description of special procedures

Standards

While data recording formats can be developed by any interested group, the economic value of some types of scientific and technical databases makes it useful for the formats to be placed on a more formal standards basis. This is especially true for engineering materials and systems, in contrast to scientific data. For engineering databases, traditional standards development organizations have become involved in these efforts. The American Society for Testing and Materials, the American Welding Society, the American Concrete Institute, and the Society for Protective Coatings are examples of organizations that have developed standard data recording formats. In the 1980s when this standardization work began, many participants were hopeful that as standard test methods were newly developed or revised, the responsible committee would include an appropriate data recording format. ASTM Committee E49 on Materials and Chemical Property Data wrote several generic guidelines (*3*) to foster such activity, but data recording formats are still more the exception than the rule.

In the scientific data world, where an economic motivation for standards is lacking, data recording formats have still been advanced, though usually on a less formal basis. Regardless of the nature of the developing group, if the development has involved representatives from all user communities, such formats have proved to be robust, cost-saving, and quite useful. For all types of S&T data, development of data recording formats has provided important insights about the ambiguity of present test methods, materials descriptions, and systems specifications. While the human eye and mind can easily interpolate or resolve ambiguities, the computer demands precision. Because virtually all test data are now captured by computers, data recording formats will only grow in importance and acceptance.

Interdisciplinary data exchange and sharing

On a broader scale, some areas of science and engineering have recognized that their activities are multidisciplinary, that the same information is used over and over again by subdisciplines, often with small but significant variations, and the need to combine, compare, and manipulate data from many data sources and disciplines is a way of life. These areas, such as product manufacturing, global climate change, earth observations, and chemical process modeling have started to develop rich and complex international standards. The International Standards Organization (ISO) has a very active committee, Technical Committee 184, Subcommittee 4 on Industrial Data (See for example http://www.nist.gov/sc4/ or contact ISO TC184/SC4 Secretariat, Manufacturing Engineering Laboratory, National Institute of Standards and Technology, Gaithersburg, MD 20899, USA.), preparing very comprehensive standards, especially for exchange protocols for data on the design and manufacture of a product. An exchange protocol defines the complete content structure of a data set that is exchanged or shared with another computer or application so that the exchanging parties can use the data set without ambiguity.

Some important features are well established for data recording formats and exchange protocols. These include the necessity for written documents that are well archived, review by all interested parties, the need to be facilitating, not regulating, and the need for supporting translation from the native database into the standard by means of clear definitions. In the long run, as is now well recognized, the definition of content is much more important than the structure of bits and bytes.

While some standards are already being developed for large interdisciplinary applications and uses, as described above, present efforts are not robust enough to be easily adopted by smaller efforts. The data exchange methodology is often not accessible because of the steep learning curve necessary to master. Service life prediction of organic coatings is clearly an interdisciplinary field. The diversity of data and information discussed in other papers in this volume gives clear voice to that reality. The data that will support the modeling efforts include data types also of interest to other interdisciplinary efforts - industrial product design and manufacture, and global climate change. How does the service life prediction community actually proceed to take advantage of what is already out there, yet move ahead in addressing its specific needs? Two specific types of actions are needed: (1) definition of the specific data sharing and exchange needs of the service life prediction community and (2) development of appropriate standards.

In terms of defining the special features for the service life prediction community, the first step is to define the *full range* of data needed. However, additional special issues facing the community include long-term archiving, long-term trends, messy data, multifaceted materials, and indeterminate data quality. Long-term archiving means that coated products are being designed for service lives of many decades; in many instances, much longer than the work-span of a single engineer or even companies. Not only is information needed on specific details, say 50 years after they were predicted, but enough detail must be present to support identification of long-term service trends and correlations with product features and service. The messiness of the data is a reflection of our present lack of knowledge of important independent variables in determining accurate service life prediction, especially for organic coatings. For example, simply characterizing the application of a coating well enough to model it accurately for 50 years of exposure to the elements provides an interesting challenge!

Regardless of this and other challenges, the service life prediction community must move ahead, because modeling efforts are not going to await solutions to all problems. To facilitate progress, we suggest several actions in the next few years. First, workshops should be held that define, as completely as possible, what data and metadata are needed to support service life prediction modeling over the long run. Disciplines concerned with these data, but outside the service life prediction community, should then be involved in building needed data-recording formats. Appropriate federal agencies and professional societies should be involved, with leadership coming from mainstream standards development organizations because of their experience in developing engineering standards. Finally, contact should be made with the ISO activity (See for example http://www.nist.gov/sc4/ or contact ISO TC184/SC4 Secretariat, Manufacturing Engineering Laboratory, National Institute of Standards and Technology, Gaithersburg, MD 20899, USA.) to determine the degree of overlap between its data exchange protocol and the needs of the service life prediction community, and how the service life prediction community can take advantage of existing technology.

Summary

Service life prediction of organic coatings is a modeling activity that depends critically on complex data for complex materials systems in complex environments. The long-term success of such modeling depends on defeating the GIGO syndrome in modeling - garbage in, garbage out. Quality must be built into data efforts to support this new field, in terms both of data and database quality assessment and of quality data recording formats. The interdisciplinary nature of service life prediction requires close interactions with many other scientific and engineering disciplines, among which data sharing will be mandatory. Finally, the service life prediction community must recognize that its special needs will be met only if it takes proactive steps to address its data needs.

Literature Cited

1. Shoshoni, A.; Olken, F.; Wong, H. In *The Role of Data in Scientific Progress*; Glaeser, P. S., Ed.; Data Management Perspective of Scientific Data; Elsevier Science Publishers: Amsterdam, 1985.
2. Munro, R.G.; Chen, H.; In *Computerization and Networking of Materials Databases*; Nishijima, S.; Iwata, S., Eds.; Data Evaluation Methodology for High Temperature Superconductors, ASTM STP 1311; American Society for Testing and Materials, 1997, Vol. 5, pp. 198-210.
3. *ASTM Standards on the Building of Materials Databases (American Society for Testing and Materials, Philadelphia, 1993);* Newton, C. H., Ed.; ASTM Manual on Building of Materials Databases; American Society for Testing and Materials: Philadelphia, PA, 1993.

Chapter 26

Uses and Delivery of Materials Information

J. G. Kaufman

The Aluminum Association, 900 19th Street, N. W., Washington, DC 20006

This paper addresses materials data needs from the users' perspective, and focuses on (a) the types and characteristics of information most often needed, notably numeric data, (b) the uses of the information, and (c) the delivery mechanisms available.

Experience has shown that most end-user searchers for material data want very specific numeric data, which are relatively complex in requiring units, may involve many orders of magnitude, and have many delimiting parameters that define their range of usefulness and applicability. Useful databases must address these characteristics of the data while providing search software that provides the flexibility and versatility to enable the searcher to get specific, well-documented answers.

The breadth of types of delivery systems for material property data are described, and the advantages and limitations of each type are provided.

Introduction

In this paper, we will address materials data needs from the users' perspective. This will not deal with the particular properties and information needed to evaluate organic coatings, but rather will focus on (a) the types of information most often needed, notably numeric data, (b) the uses of the information, and (c) the delivery mechanisms available.

In the sections that follow, we will first address the users themselves, noting the types of users, their characteristics, and their tasks. Then we will focus on the types of data needed, including the characteristics of those data. Next we will look at the options of where and how to access and manage the data, and finally we will look at representative sources for such data.

© 1999 American Chemical Society

This will be done with recognition that we face increasing needs for higher quality data and for easier access to the data. Historically, we find that though there are often many valuable sources of useful data, most individuals do know where they are or how to access them. Furthermore, what data are known about are of little use because they are likely to be very difficult to access and limited in documentation of their applicability and accuracy. The result is often inefficient or ineffective searches and, as a consequence, poor material decisions or inefficient designs.

We hope that broader understanding of users' needs and of the nature of the data will result in more useful data sources in the future.

The User Base

The user community for materials information may be characterized in a number of ways, including type and size of organization, and type of job responsibility. And it is quite likely that users of any specific database may vary greatly in both respects.

From the organization standpoint users may range from very large companies with considerable internal data resources who are looking simply for ways to keep those sources current, to small companies and private consultants who are heavily dependent upon external sources for most of their data needs. From the job function perspective, users may range from R&D scientists needing highly scientific characteristics to aid in the development of new or greatly improved materials, to designers who need statistically reliable performance data to assure that sound structures with desired performance result.
Two additional types of professionals worth noting are the data searchers themselves and the educators who are providing guidance to the next generation of users of all types.

The net effect of these two spectrums of users is that data sources must meet a wide range of needs, some requiring great breadth but not too much depth, and others requiring great depth of scientific and engineering knowledge in very specific areas. There is a tendency to try to build databases to meet both types of needs, but historically that has not been very successful. It is usually more efficient to build smaller individual databases tailored to specific needs.

Nature of the Data User Community

Professional searchers are a community among themselves. They spend hours each day searching, and so have the patience and ability to learn relatively complex search command-based systems. However, the end-user community of searchers (i.e., those who actually will use the data and, generally, without the assistance of professional searchers) have a notably different set of characteristics:

- Want numeric data, often rather complex in nature
- Have many other tasks in addition to searching for data, so must be able to do the search quickly and efficiently
- Do not have time to learn complex search command systems
- Have little patience with manuals to guide their searching
- Will need guidance in locating the data within a system
- Have expectations on how data should be presented (in logical scientific form)
- Will want to acquire the resultant data (own the database or download the needed data)

Thus databases developed to support the end-user audience must supply numeric data and provide access modes that allow searchers to intuitively interact in forming queries and specifying needed parameters. Further they need to provide the data in a useful and logical technical format and provide a capability for the user to obtain all of the needed background information.

An added aspect of this variability that must be accommodated in database search structure is the breadth of types of question. There are four basic types of question, with many variations of each:

- Need properties of a specific material or materials (material identification known)
- Need comparable properties of competing materials or current and potential candidate
- Need all materials having a specific property or combination of properties
- Need the material having the best combination of a specific set of properties

To handle all these three types of question, the ability to build queries in the different modes are required, inputting specific materials, specific properties, or combinations of the two.

Nature of Numeric Data

It is useful at this point to briefly consider the characteristics of numeric material property data, as they represent the toughest (most complex) pieces of information to incorporate in a database. Five critical elements might be noted.

First, numeric data must be handled as numbers in the database, searchable as values and ranges of values, not simply as strings as in the conventional bibliographic database. The absence of numerical search capability greatly limits the usefulness of properties databases to most end users.

Second, numeric values always have units, and in fact they have no meaning without those units. The units must always be displayed along with the numeric values.

Third, the numeric data values may within a single record vary by many orders of magnitude (example: modulus of elasticity in 10^7 psi or 10^8 MPa, and thermal conductivity in 10^{-6} BTU/hr/ft^2, often appear in the same record). As this example illustrates, there will usually be multiple units and unit classes in a single record. These factors present a significant component of their complexity.

Fourth, most numeric property data are dependent upon many other parameters, many of them also numeric information with all of the above complexity themselves. There are at least two classes of these dependent parameters: (a) those related to how the material was produced, and (b) those related to the conditions under which the property values were determined. In the former group are:

- Product form, size and shape
- Thermal processes by which it was produced
- Location and orientation from which test specimens were taken from the material.

The latter group may include the following parameters under which the property or performance might be measured:

- Temperature
- Time at temperature or under load
- Pressure
- Atmosphere
- External forces present

Fifth, and finally, there is the data quality issue. While this important element is the subject of another paper in this series, it is appropriate to note at this time that the statistical quality of material property data represents a quality factor that must be incorporated into the database itself. It is vital that the users know if the numeric values in the database are raw data (untreated test results), averages of several test results, statistically significant average or minimum values, or refereed data, i.e., those approved as standard by some applicable expert authority.

These five factors mean that, for every property data presented in the database, many other important parameters must also be presented to ensure that the users understand their applicability and limitations of the data. The complexity of the process is apparent, but it is essential as <u>the data points themselves have no value without all of the prescribing parameters that define their applicability</u>. Care must be taken to include them in the database, and to make them searchable as well.

Access to Numeric Data - The Tools

The nature of numeric data as outlined above, predetermines many of the requirements for the search software needed to access such data. Because the property data are numeric, the most useful answers will be obtained when the search software addresses the characteristics of such data, as follows:

- Provide capability for range (2000-2500) and tolerance (2000+/-100) searching as well as single point searching (2000)
- Provide for units conversion (e.g., English to International Standard systems)
- Provide for consistent rounding and significant figures algorithms
- Provide for logical table display of results.

It is key that material property database searching not be limited to string-recognition searching as is commonly used for textual or bibliographic databases.

One additional feature is desirable, though difficult to achieve. Because of the great variety and distribution of material property databases, it is highly desirable for the search software to permit multiple database access and search. At minimum this should be for all databases at any one location, and ultimately it should include multiple locations. This was the goal of the Material Property Data (MPD) Network, established and operated for about five years in conjunction with the Scientific and Technical Information Network (STN International), (see below) but the complexity of dealing with multiple systems proved too expensive to maintain and expand. Perhaps this will be possible to a greater extent in the future with Internet-based search systems.

Access to Numeric Data - The Sources

Experience has shown that most numeric data users will want to have their own self-contained database(s) under their own control. They may have them on their own PC, on searchable disks like CD-ROMs, on a workstation, or on a network host system. Their databases will focus on the specific materials and properties they use regularly.

Periodically, they will want to update their databases. They may do so by buying more disks, transferring data from associates, or gaining them through online access to available databases. There are advantages and limitations to all three types of access sources. To illustrate that, we can take a brief look at three of the most valuable sources of materials data today:

- Online - STN International
- Workstation/internal networking - MVisions from McNeal Schwendler
- Individual data disks - ASM International

Online Sources such as STN International - STN International is the worldwide scientific and technical information network operated in the USA by the American Chemical Society (ACS) and jointly with FIZ-Karlsruhe in Germany and the Japanese Information Center for Science and Technology (JISCT) in Tokyo. STN International has about 10 numeric databases covering the properties of materials and another fifteen covering chemicals and chemical compounds.

Systems such as STN International are accessed by computer/modem telephone link, and searchers are charged fees based upon time and the amount of information obtained from the system. The principal advantages of systems like STN International are: (1) the large number of databases and therefore type of numeric data accessible in a single connection, (2) the ability to use the same software in searching all of these databases, (3) the ability to search the databases in groups rather than individually, (4) the sophisticated numeric search capability on STN, and (5) the ability to access and retrieve only the specific data one needs without purchasing a large package possibly including much uninteresting data.

The principal limitations of such systems are closely tied to their advantages. The sophisticated search software is difficult for all but professional searchers to learn and use to maximum advantage. Occasional searchers like many of us do not know the search commands and usually will/can not take the time to learn them when we do want data.

The other perceived disadvantages are the fact that the databases are remote, and must be accessed and the desired data paid for each time. There is the mentality that the "cash register" is running and so one must hurry, and also that one small search mistake may cost a great deal. Such concerns usually fade with experience on such a system.

In summary such online systems are of value because of the wide range of sources provided and the ability to search those sources with a single software.

Workstation Data Systems such as MSC/MVISION - Workstation-based databases fill a major role in servicing design software, providing direct plug-in for data needed to solve specific engineering problems. MVISION is a premier example of that type of design and database system, originally designed by PDA Engineering, now a part of the McNeal-Schwendler Corporation. The system provides the tools for building and accessing, evaluating and incorporating the data in computer-aided engineering, design, and manufacturing software. The present version of MSC/MVISION contains about a dozen different sources of numeric data on materials.

The advantages of an MSC/MVISION system are (1) the ability to access and search all databases in the system with a single software, (2) the ability to transport those data directly into the engineering, design or manufacturing software needed to solve problems, and (3) the user's total control of the databases (in contrast to the remote databases in online systems).

The limitations are (1) the requirement to use the base communication system upon which such software is based, UNIX in the case of most workstation systems, including MSC/MVISION, (2) the ability to access only those databases incorporated into the particular workstation package, and (3) the cost of such systems, including the necessity of more powerful computers than might otherwise be necessary.

For engineering organizations with limited scope of design and data problems, the workstation package would appear to be outstanding because of the ability to integrate data with problem-solving software.

Disk-Based Databases such as ASM's Mat.DB - Floppy and CR-ROM disk-based systems are still among the most widely used except where professional searchers are involved. Since many material databases cover a rather specific group of materials or a particular type or group of properties or performance data, the databases are small or at least finite size and fit well onto such disks. The disks in turn can be used themselves or downloaded onto hard drive for permanent storage.

Mat.DB, produced by ASM International, is a DOS-based computerized materials library consisting of a database manager (the Mat.DB program) and fourteen data sets covering steels, aluminum alloys, titanium alloys, composites and other structural material classes. The Mat.DB product provides good illustrations of both the advantages and limitations of such data sources.

The advantages of Mat.DB are (1) it is completely self-contained and the purchaser has full control of it, (2) the several databases may all be accessed and searched with the same software, and (3) it is quite portable, being readily moved wherever it is needed.

The disadvantages of disk-based data systems are (1) each one tends to have its own unique data format design and search software, so the users must be familiar with al variety of such systems, (2) if one tries to combine data from one disk system with those from another, it is usually not practical because of the differences on data format design, and (3) such databases are usually not very versatile for handling a wide range of queries, and may not have very sophisticated numeric search software.

New disk-based databases are being developed all the time, and developers such as ASM International and William Andrew, Inc. (Plastics Design Library) are making every effort to better standardize data formats and search software, often being guided by ASTM Committee E-49 in this effort.

Combinations of Data Systems - Actually, for most users, it is not a case of one system or the other but of integrating different types of data systems to meet various needs. Some disk-based systems, as an example, include software for accessing online data sources, so one can add to their disk based database if useful data can be located elsewhere. The same can readily be done with workstation software, though no such systems are presently advertised to my knowledge.

Summary

In this overview, we have focused upon the nature of numeric property data for materials, and all of the major points are applicable in dealing with data for organic coating materials. Key to the development of valuable databases is recognition of the numeric nature of material property data and of the unique characteristics of numeric search software needed to adequately access and retrieve the desired data along with sufficient metadata so that we understand the limits of applicability of the data.

References

[1] Westbrook, J.H., Kaufman, J.G., and Cverna, F., "Electronic Access to Factual Materials Information: The State of the Art," MRS Bulletin, August, 1995, pp. 40-48. (Author's note: this reference contains a very complete list of references on this subject material, not repeated here)

[2] Westbrook, J.H., "Some Considerations in the Design of Properties Files for a Computerized Materials Information System," in The Role of Data in Scientific Progress, P. Glaeser, Editor, Elsevier, 1985.

[3] Kaufman, J.G., "Quality and Reliability in Materials Databases", in Computerization and Networking of Materials Databases, Vol. 3, T.I. Barry and K. Reynard, Editors, ASTM STP 1140, ASTM, Philadelphia, 1992.

Chapter 27

Managing and Tracking Automotive Paint Etch Studies Conducted in Jacksonville, Florida

Raymond Brockhaus[1] and Karlis Adamsons[2]

[1]Research Associate, Troy R&D Laboratory, DuPont Company, Troy, MI 48007
[2]Staff Chemist, Marshall R&D Laboratory, DuPont Company, Philadelphia, PA 19146

Database tools have been developed for managing and tracking of automotive OEM paint system resistance to environmental acid etch. This report documents the DuPont Automotive R&D effort requested by US OEM automotive customers during the last decade (1987/97), including the utility of using a flat file database on the efforts of managing the entire project. The "data-engine" was Filemaker Pro®, available on both Apple and PC (Intel® based) desktop computing platforms. The flat file environment includes use of line drawings (e.g., site and panel location maps), a variety of data manipulation and reduction tools, sample and reference material tracking features, as well as a user-friendly interface to communicate information to the end user.

The automobiles we currently produce and use globally are coated with protective and decorative layers. Due to market forces, government regulations, quality enhancements, need for cost containment, and end-user (car owner) increasing expectations, the coatings industry has constantly been striving for product improvements. Today's customer has expectations for long-term appearance performance, which entails design of systems with environmentally resistant chemistry and physical properties, in addition to long-term mechanical performance. Environmental acid rain etching became a significant marketing issue during the mid-80's to the mid-90's. This resulted from the evolution of car painting technology going from monocoat to a multi-coat (CC/BC) application strategy to provide desired appearance characteristics. Pigmentation of paints (for decorative purposes) was now incorporated into a basecoat (BC) layer, and covered by a protective clearcoat (CC) layer. The organic composition of each layer had also changed, going

© 1999 American Chemical Society

from primarily lacquer-based to enamel-based technology, which was (unfortunately) susceptible to acid etch damage from environmental rain, fog, or dew.

Introduction / Historical

The primary goal for developing database tools for managing and tracking of automotive OEM paint system resistance to environmental acid etch was to help designers in product development and service life prediction [1-6]. A typical automotive paint system over steel or plastic substrate is shown in Figure 1. Damage resulting from acid etch normally occurs at/near the surface of the uppermost layer, or clearcoat. Although the depth of the damage is often less than a few microns, the total surface area damaged can be quite substantial. A representative example of such acid etch damage is shown in Figure 2. The irregularly shaped regions, their size, and the extent of damage result in significant, and potentially dramatic, changes in appearance. This damage is permanent, and usually found to be unacceptable by the vehicle buyer or owner. Extensive research by the automotive paint manufacturers, specifically those developing and marketing topcoats, has been conducted in order to identify coatings more resistant to environmental acid etch. Our organization over the last decade has routinely been testing topcoats for acid etch resistance using both accelerated testing methodology and in outdoor locations that are prone to extensive damage of this nature [7-9]. This report will focus on and detail the development of a database to manage and track our outdoor-based acid etch studies.

A brief history into the evolution of automotive paint systems over the last two decades underscores the importance of acid etch resistance. In the mid to late 1980's the US automotive market was making a transition from the older paint systems containing lacquers, primers and black dip metal sealer to that illustrated in Figure 1. The newer paint systems contained the following series of layers: The topcoat, or "clearcoat", was composed of transparent and colorless binder polymers, cross-linker(s), and additives such as the UVA/HALS UV-screener package. The "basecoat" was composed of binder polymers, cross-linker(s), latex particles, and a high loading of pigment to provide color, and occasionally other pearlescent/metallic-flake appearance effects. The "primer", containing resins, filler and pigment particles, was designed to provide leveling and good adhesion between the basecoat and electrocoat layers, as thus good chip resistance, when steel substrate was used. And the "electrocoat" layer (i.e., E-Coat) was designed to give necessary corrosion protection and strong substrate (steel) bonding. Note that when plastic substrates (e.g., TPO) are used a primer interfaces with the substrate, and corrosion protection is not relevant. Since most of the early clearcoats were acrylic/melamine-based and had a high melamine cross-linker content, they proved to be highly susceptible to acid rain attack (i.e., acid etch). However, favorable driving forces toward this type of chemistry included regulatory, consumer quality, paint system cost/performance, and environmental

414

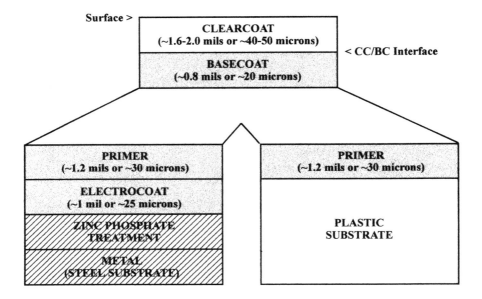

Figure 1. Multi-Layered Automotive Paint Systems over Steel
and Plastic Substrates; Referred to CC/BC Type Systems.

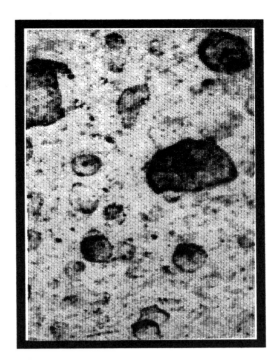

Figure 2. Clearcoat Acid Etch Type Degradation; Damage Shown on Various Acrylic/Melamine Formulations.

issues [10]. As a result, it became essential to explore a wide range of acrylic/melamine-based network compositions, as well as additives, in order to maximize utility of this technology and to achieve long-term service life. Because long-term maintenance of appearance is critical to customer satisfaction, designing and providing acid etch resistant paint systems was necessary for our continued business success.

Acid etch resistance testing of OEM topcoats is now done regularly to determine the hydrolytic stability at/near the surface. It is well known [10-16] that certain types of atmospheric, or airborne particulate, pollution (i.e., sulfur oxides, nitrogen oxides, engine exhaust soot/cinders) in combination with water (i.e., high humidity, dew, rain) can result in the formation of acidic solutions (or environments) capable of chemical bond hydrolysis. An exploded view of a chemical network typical of many styrenated-acrylic/melamine-crosslinked clearcoat systems is shown in Figure 3. Most commercial OEM clearcoats today use similar acrylic-based chemistry. These clearcoat networks contain both ester and ether type linkages that are susceptible to hydrolysis under these conditions. The acid etch damage can quickly result in what appears to be random patterns of microscopic and macroscopic pitting or erosion, often altering the appearance soon after exposure. The acid etch resistance testing provides an important tool for rating and ranking of coatings according to their hydrolytic stability.

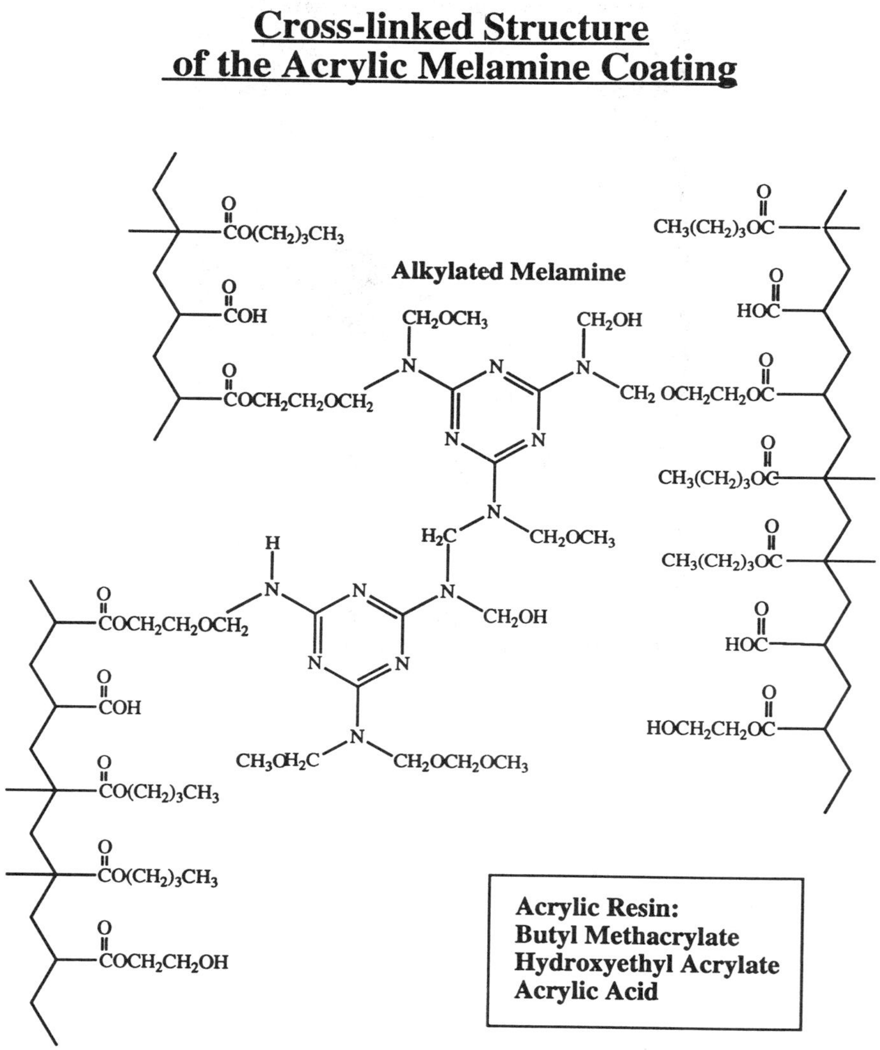

Cross-linked Structure of the Acrylic Melamine Coating

Figure 3. Exploded View of a Styrenated-Acrylic/Melamine-Based Automotive Clearcoat Network; Note Ether and Ester Type Chemical Bonds Susceptible to Acid Induced Hydrolysis.

The shipping area of Blount Island in Jacksonville, FL, was identified, by both OEM customers and various automotive paint suppliers, as an outdoor environment that typically resulted in acid etch type damage for topcoats. The shipping traffic and local industry are believed to be the primary sources of pollution creating an environment capable of producing acid etch. This environment has been consistently observed in this area. In the latter eighties

(~87-90) an exposure period was agreed upon, June through August, where automobile hoods, fenders, and bumpers could be located outside at a designated site and monitored on a periodic basis. These studies were thereafter done on a yearly basis. As more and more samples were introduced in the on-going annual studies, it became increasingly obvious that managing and tracking the data would be a considerable challenge.

The weathering and durability studies dealing with acid etch resistance of our topcoats are only a sub-set of the overall chemical and physical property, as well as mechanical performance, testing that is done on automotive paint systems. Databases have been developed to manage and track many other kinds of information. Examples include monitoring of gloss, distinctness of image (DOI), surface hardness, and color change data as a function of exposure time and conditions. In addition, coatings and multi-layered paint systems are potentially subject to many other causes of failure as is illustrated in Figure 4. This chart shows a failure-tree, highlighting environmental (e.g., acid etch under the pollution category), materials, processing, design, and application factors [4]. This failure-tree was created by scientists (J. Martin et. al.) at and associated with the National Institute of Standards and Technology (NIST; Gaithersburg, MD) following many years of study of automotive paint systems and architectural coatings. This is intended to provide a "Big Picture" perspective on the broad need for effective databases in this industry.

Original Database Design

Until the early nineties (~90-92) it was possible to do paper-based managing and tracking of the field samples and logging of data. As the number and type of samples kept increasing, the necessity for faster and more efficient means of doing this became very clear. Also, generating reports in a timely fashion and digestion of final results (i.e., trend analysis, degradation kinetics, comparison to historical data, obtaining feedback for improved future experiment design) was becoming increasingly more difficult.

Experience raised questions on the reproducibility of the exposure environment with regard to acid etch damage. The need for some degree of sample replication was determined. Not all areas at the designated Jacksonville, FL, site were identical in their ability to cause damage. Researchers started to concern themselves with the following issues: How many replicates of a system are required to give statistically meaningful results? Where within the site should the replicates be located? What controls are necessary? What is the minimum number of controls (replicate and type) that are required? Where should the controls be located? In a relatively short period of time the number of samples grew from a few hundred to many thousands. Answering questions such as this were critical, but well beyond a paper-based system.

In early 1992, the first attempt to create a computer-based management and tracking system was undertaken. Development was done using Claris's

418

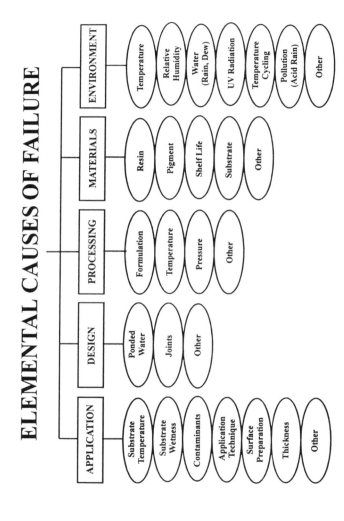

Figure 4. Causes of Failure for Automotive Paint Systems; Primary (Root) Causes are shown in Rectangles, Secondary (Specific) Causes are shown in Ovals. Chart Derived from Service Lifetime Prediction Manuscript, NIST Building Science Series **172**, "Methodologies for Predicting the Service Lives of Coating Systems, Building & Fire Research Lab, National Institute of Standards and Technology, 1994.

Filemaker Pro ® in flat file format resident on a Apple Macintosh ®. It was initially built as two files, which were merged into a single file at summer's end. The two sections were targeted at: a) collecting data on the paint system components, submitter(s), associated project/study, end use of material or piece, and types and quantities of sample prepared for testing (including controls and all needed replicates); and b) a set of fields to identify, locate (including rotations through various site locations), and log rating dates and values. A crude map of the testing facility grounds was included in the file as one of the display options available to the data file user/browser. The 1994 and 1997 (current) map layouts of the site racks/tables is shown in Figure **5-A** and Figure **5-B**, respectively.

Sample location tracking is done by assignment of rack/table, and column and row positions on each. As the map appears on the computer workstation screen, the rack/table rows start on the right hand side of the site and increases to the left. The "missing rows" were counted, even though the actual rack/table additions were not installed until later. The column numbering starts at the bottom and increases upward. Table positioning depends on the type and size of sample located there. For example, painted flat panels (~12"x18" are typical dimensions) are tracked according to row and column positions. Row position 1 is in the upper left of the rack/table, and goes across (1 through 4) to the right. Column position 1 is in the upper left of the rack/table, and goes down (1 through 7). The somewhat unusual layout of the site is a function of the site entrance/exit, roadways, physical obstructions (i.e., water well, storage shed, etc.), and undeveloped areas (available for future use).

Database Design Evolution

Although the database and data-reduction features continued to evolve, the basic system remained largely the same. Over the years the site has been expanded, filled in with new rack/table positions, and improved (i.e., removal of certain site obstructions) to promote more uniform exposure, and thus assessment of coating performance. The database did not require modification to handle the changes, except to add new records for new sample positions and updating the map.

Over the years the total number of submitted samples has steadily increased. Also, the relative percentage of each type and size of sample submitted has changed. When the site was first populated with samples, they were primarily hoods, hood sections, fenders and bumpers. Later, an increasing percentage of flat painted panels, many associated with experimental design studies, were submitted.

The database was designed to easily track missing data or data which had been identified as bad. This gives a researcher a means of filtering good from suspect data, permitting better trend analysis and assessment of degradation kinetics. If missing data becomes available, then it can be readily incorporated. An audit trail is maintained to document when and where samples were located, who logged new data, and dates of data entry.

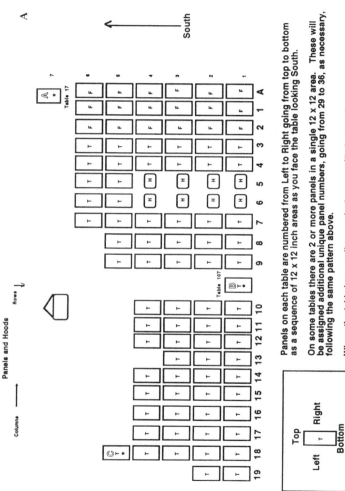

Figure 5. Map of Rack/Table Layout at Jacksonville, FL, Exposure Site; Rectangles Represent Rack/Table Locations.

A. Map from 7/1/94.

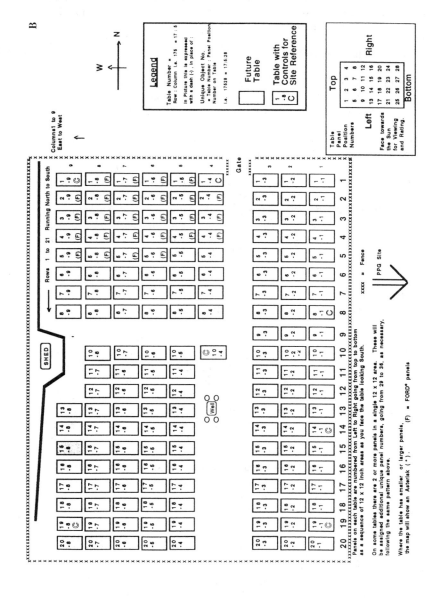

B. Map from 9/1/97. Key: T = Table, H = Hood Locations,
A/B/C = Tables containing smaller tables.

The current scheme of providing a unique ID for each sample was devised using the following formula: There can be a maximum of 9 rows of racks/tables across the field, 22 columns the length of the field, and 28 samples (of ~12"x 18" size) per table. A unique ID = a code of row, column, location on table. Thus, ID 1223 would be equivalent to row #1, column #2, and position #23 on the rack/table. Further, an ID 14706 would be equivalent to row #14, column #7, and position #06 on the rack/table.

Review of historical sample data is possible. Site locations are referenced to the map that was current at the time of exposure. This eliminated confusion due to site expansion, rearrangement, and distribution of samples by type and size.

Remote access to the Jacksonville, FL, exposure site database is done through the company intranet. Copies of the database files were placed on file servers at various remote R&D lab sites in a public access area in a read-only file environment. Files were established as multi-user access. This was a practical consideration since submitters in our OEM automotive coatings businesses are found globally, throughout the US and from Canada, Mexico, South America, Europe and Asia. The files were password protected for full access by two file maintenance personnel. Read-only mode of all displays and fields, but no record creation, modification, or deletion, is available to general users. No password is required for very limited, read-only access to select displays and fields.

Results & Discussion:
Database for Managing & Tracking Automotive Acid Etch Data

A variety of display layouts are available to users. New layouts can be easily developed after discussion with a user, usually requiring 30 minutes to several hours. One display layout was a "rack/table" with 28 positions, sample paint code or project, and ratings displayed on the user's display screen. Here the viewer of the database can see how all panel samples on that rack/table performed. As an example, an exploded view of rack/table positions 21 and 22 are shown in Figure **6**, with associated submitter and sample identification information, and acid etch rating data (including a running average) as a function of time. In each study/project record, the ratings were stored as 5 separate fields, initially, and from year 2 onward as a repeating field of various lengths from 5 to 12 values, depending on if special samples would have exposure extended into the Fall and Winter months. With repeating fields, built in functions were created to aid in providing running averages and count of readings, which help in finding, sorting, and viewing sets of results.

The rating file, shown in Figure 7, was used to generate blank paper recording sheets so that ratings data could be conveniently entered while in the field. Copies of these sheets would be sent via overnight courier to our base of operations (i.e., Troy R&D Lab, Troy, MI.) for final data entry. The approximate time from field observation, or rating, to secure data entry is about two days. At this point the researchers could access the database themselves and download the

ACID ETCH MONITORING		JACKSONVILLE, FLORIDA	
Location: Sample Table [73]			
Position #21	Tech: GEN-X	Position #22	Tech: GEN-X
Read=1 5 Read=2 5 Read=3 5 Read=4 6 Read=5 12 Read=6 12 Read=7 12	R-Avg [8.1] Notebook ID: 14292-1 Submitter: Omura, H. Date: 1994	Read=1 5 Read=2 5 Read=3 5 Read=4 5 Read=5 12 Read=6 12 Read=7 12	R-AVG [8.0] Notebook ID: 14293-1 Submitter: Omura, H. Date: 1994
System Chemistry Description: Acrylic/Melamine		System Chemistry Description: Acrylic/Melamine	

Figure 6. User Selected Database Output Screen showing Table Number, Submitter/Client, Description of System Chemistry, Notebook ID, Individual Acid Etch Ratings, Average of Acid Etch Ratings.

JACKSONVILLE, FLORIDA ACID ETCH MONITORING STUDY				DUPONT AUTOMOTIVE 1995									
System Code	System Replicate	Submitter or Client	Table & Loc	Ratings (DuPont Scale 0-12)									Av Rt
14292	1	Omura,H.	T73 21	5	5	5	6	7	.	.	.	.	5.6
14293	4	Omura,H.	T73 22	5	5	5	5	5	.	.	.	.	5.0
14294	1	Omura,H.	T73 23	5	5	5	6	9	.	.	.	.	6.0
14295	2	Omura,H.	T73 24	5	5	5	5	5	.	.	.	.	5.0
14296	4	Omura,H.	T73 25	5	5	5	5	6	.	.	.	.	5.2

Figure 7. User Selected Database Output Screen showing the Acid Etch Performance Ratings for a Given Sample; A Blank Sheet can be Printed for Purposes of Logging Rating Results.

ratings. They could also view other researcher's data, even from other remote sites. They could review how each of the control replicates were performing compared to the entire field, and in adjacent rows or columns of racks/tables.

Over the years of operation at the Jacksonville, FL, site our work has moved toward doing more designed experiments with formulations. In certain studies, the entire design matrix is duplicated and placed in separated locations around the test field to guard against random events having significant, and thus misleading, impact on expected coating performance. The database environment allows random sample placement with the power to quickly and easily find and display related sets of data. The search, find, and display functions can be programmed into macro script files for aid in data entry, data retrieval, and generation of special management reports.

Based on extensive studies performed in 1994 and 1995 it was determined that a minimum of four replicates is required, and a maximum of nine replicates is often desirable. More replicates have been determined to be unnecessary, and less has proven to be statistically unsound for effective decision making. The database played a key role in establishing these conclusions.

In the beginning of each year a template file is created, filled with records for each sample location, but lacking data in any of the fields except the unique sample ID number field. The records are populated as samples are handed over from the paint system developers to the acid etch exposure coordinator(s) for each research laboratory.

Old files can be used to retrieve and display a previous year's rating data on the control samples, to compare by visual inspection of the numeric data, or to compile a file for input to a statistics program for correlation or trend analysis. Data for prior years were merged into a single file where year-to-year performances could be easily examined. This master file has been found useful as a training and education tool for newer researchers entering this area of product development.

Summary

The database is currently in use. It has evolved into a user-friendly and powerful tool for researchers involved in coatings and paint system development. It is available to all remote DuPont research laboratories conducting acid etch performance studies at the Jacksonville, FL, site. The primary data obtained and logged is based on acid etch evaluations, however other kinds of chemical and physical property, as well as mechanical performance, data are now being collected for specific exposure and materials characterization studies.

The database supports the management and tracking of a (currently) large specimen population. It is our continuing expectation that the sample population will grow into the future, as will the need for additional tools to extract key information from the accumulated data. Other weathering studies are being introduced and conducted on many of the samples or sample sets. The database also aids our efforts in efficiently tracking notebooks for previously recorded experimental objectives, detail and analyses, associated principle investigators, research/memorandum reports, meeting summaries, etc..

Acknowledgement

The authors gratefully acknowledge contributions from the many submitters of automobile panels, hoods, fenders, and bumpers over the years. The database and data-reduction capabilities have continually evolved with close interaction with both periodic and regular users. It is expected that fruitful interactions with users will continue into the future. Special acknowledgement is extended to Stan Horvath (DuPont Marketing), Bernadette Colonna (DuPont R&D), Wyatt Mills (DuPont R&D), and Robert Matheson (DuPont R&D).

References

1. R.A. Dickie, Journal of Coatings Technology, **66**, No. 834, 29 (1994).

2. D.R. Bauer, Journal of Coatings Technology, **66**, No. 835, 57 (1994).

3. D.R. Bauer, Progress in Organic Coatings, 14, 193 (1986).

4. J.W. Martin, S.C. Saunders, F.L. Floyd, and J.P. Wineburg, NIST Building Science Series 172, Methodologies for Predicting the Service Lives of Coatings Systems, Building & Fire Research Lab, National Institute of Standards and Technology, 1994.

5. R.A. Dickie, Journal of Coatings Technology, **64**, No. 809, 61 (1992).

6. J.L. Gerlock, C.A. Smith, E.M. Nunez, V.A. Cooper, P. Liscombe, and D.R. Cummings, "Advances in Coating Technology: Predicting the Durability of Coatings", Proceedings of the 36th Annual Technical Symposium, Cleveland, OH., May 18, 1993.

7. P.H. Lamers, B.K. Johnson, W.H. Tyger; "Ultraviolet Irradiation of Melamine Containing Clearcoats for Improved Acid Etch Resistance"; Polym. Degrad. Stab., **55** (3), 309-322 (1997).

8. W.J. Blank; "Acid Etch Resistant Automotive Topcoat"; Patent (English), Application WO 95-US5541 950510, Priority US 94-250558 940527.

9. K. Shibato, S. Beseche, S. Sato; Adv. Coat. Inks Adhes., Asia-Pac. Conf., 4th (1994), 11 pp, Paper 42, Publisher: Paint Research Association, Teddington, UK.

10. J. Kamimura, G. Hyomen; "Countermeasure for Environmental Pollution In Automotive Industry: Coating Damage and Countermeasure for Acid Rain"; **46** (6), pp 481-6 (1995).

11. N. Rungsimuntakul, D. White, R. Fornes, R. Gilbert, J. Spence; "Study of the Effects of Acidic Pollutants on Automotive Finishes"; Report from North Carolina State Univ., Raleigh, NC, Gov. Rep. Announce, Index US, **93** (3), 8 pp (1993).

12. D. White, R. Fornes, R. Gilbert, A. Speer, J. Spence; "Physical Damage Formation on Automotive Finishes Due to Acidic Reagent Exposure"; Report from North Carolina State Univ., Raleigh, NC, Gov. Rep. Announce, Index US, **93** (3), pp 8 (1993).

13. R. Bradow, F. Bradow, P. Vandiver; "Estimating Damage to Automotive Finishes Associated with Acid Smut"; Proc. Annu. Mtg. Air Waste Manage. Assoc., 84th, Vol. **15B**, Paper 91/143.5, 13 pp (1991).

14. N. W. Paik, R. Keller; "Fly Ash Emissions from a Power Plant and Damage To Automobile Finishes"; J. Air Pollut. Control Assoc., **36** (7), pp. 821-3 (1986).

15. G.G. Campbell, G.G. Schurr, D.E. Slawikowski; "Techniques for Assessing Air Pollution Damage to Paints"; US Nat. Tech. Inform. Serv., PB Rep. No. 222377/4, 99 pp. (1972).

16. J.W. Spence, F.H. Haynie; "Paint Technology and Air Pollution: Survey And Economic Assessment"; US Nat. Tech. Inform. Serv., PB Rep. No. 210736, 49 pp. (1972).

Chapter 28

Methods for Representing and Accessing Material Property Data and Its Use With Decision Support Systems

Lawrence J. Kaetzel

National Institute of Standards and Technology, Building 226, B350, Gaithersburg, MD 20899

Improved methods are needed to access and understand coating knowledge that is stored in computers. Today, a proliferation of the Internet allows organizations to distribute information efficiently. However, shortcomings exist in the ability to assess its validity, quality, and completeness. This paper presents activities of the coating industry and government organizations that will improve decision-making for coating systems through improved understanding of the material properties and predicted performance.

Introduction

The coating industry is represented by a diverse collection of entities including raw material suppliers, coating manufacturers, facility owners, researchers, and trade associations. This diversity creates difficult problems for managing and sharing knowledge, in part, due to the sheer volume of information, the capability to provide convenient access and timely delivery, the ability to interpret content and formats (compatibility), and data quality issues. The need to address these problems and develop new strategies for representing and sharing knowledge is never more apparent then when seeking and retrieving information from the Internet. The Internet's capabilities and content are often overstated and in many instances provide the user with little to show for the time invested in searching and interpreting knowledge. Several factors are responsible for this dilemma:

- search capabilities are inadequate due to inconsistencies in the way knowledge is stored and displayed;
- computer hardware and software are often incongruous;
- users lack confidence in the developer or content;
- results are out-of-context or the terminology used is inconsistent; and

- user interfaces can be cumbersome to use (not logically organized or performance is lacking).

The National Institute of Standards and Technology (NIST) has initiated an activity to address the construction industry's information needs. The Computer-Integrated Knowledge System (CIKS) Network, along with industry and government partners, is developing solutions for construction industry knowledge users. CIKS is an activity of the NIST, Building and Fire Research Laboratory, High-Performance Construction Materials and Systems Program. Planning for the development of CIKS was initiated at a workshop held in June of 1996. Participants at the workshop identified pilot projects and opportunities for collaboration. The results of the workshop were published in a NIST report [1]. An Internet World Wide Web [W^3] is operational where the CIKS workshop report can be viewed on-line, as well as other CIKS information. The W^3 address is: http://www.ciks.nist.gov During the CIKS workshop, a coating material working group was established and this working group is currently involved in the development of several pilot projects that will be described in detail later in this article.

CIKS collaborative efforts include interactions with construction industry groups such as the Civil Engineering Research Foundation's (CERF), CONMAT (CONstruction MATerials) Council. In 1994, a CERF Coating Industry Working Group identified major research projects that address coating industry knowledge. These projects were described in a CERF report [2]. The Coating Industry Working Group identified Data and Cost Analysis and Technology Transfer as knowledge related areas that are important to the coating industry. These projects involve the development of standard data formats, engineering management systems, and the development of standard guides and procedures for successful use of high-performance coating systems. The report also identifies specific government agencies, private industry associations, and companies that have the expertise to address the research projects. Accomplishing the goals will require partnerships among the organizations. Figure 1 shows the interactive framework that has been established for CIKS and is currently being used as a framework for construction industry interactions.

An Overview of CIKS

The term "computer-integrated knowledge system" indicates that the scope of CIKS is broader than just compiling data and developing application programs. The CIKS system involves establishing a sound information infrastructure for consistent terminology, information standards, and interoperable systems that apply across construction industry disciplines and a variety of construction industry activities. Figure 2 shows a hierarchy and CIKS components. A meta model has been developed for the CIKS activity. For the purpose of this article, a "meta model" is a high-level representation of processes, knowledge formats and sources, and interfaces that describe a logical framework for CIKS. Describing the meta model is a complex exercise and would extend beyond the scope of this article. For readers

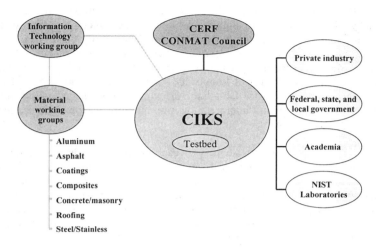

Figure 1: CIKS Partners and working groups.

who wish to gain a more complete understanding of the CIKS meta model, a detailed description can be found in a NIST Internal Report [3].

Implementation of CIKS includes making available prototype knowledge-based systems, developed jointly by NIST and industry, and testing existing production systems for interoperability. Construction industry practitioners, such as building designers, contractors, materials engineers, and specifiers are the intended users of CIKS.

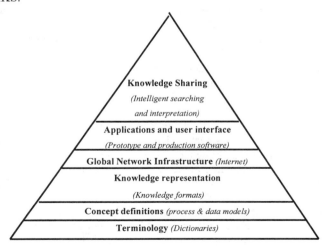

Figure 2: CIKS components and their hierarchy.

CIKS knowledge-based prototype systems incorporate virtually all forms of knowledge, from raw data to high-level human expert facts and rules-of-thumb. Knowledge sources involve both the public and private sector. Forms of generic and project knowledge to be represented in CIKS include computer-based models, databases, decision-support systems, standards, images, catalogues, handbooks, manuals, and integrated project knowledge bases. When complete, CIKS will incorporate features such as cataloguing, indexing, intelligent searching, retrieval, routing, browsing, query and interpretation, presentation systems, distance learning, help facilities, and collaborative authoring.

CIKS computerized systems use enabling information technologies that provide an open-systems framework that will ensure interoperability. Examples include intelligent search engines or agents that can find data in context on the Internet W^3 and multimedia capabilities that incorporate visual information that enhances knowledge interpretation. Standards developed by the International Standards Organization (ISO), the American National Standards Institute (ANSI), and the American Society for Testing and Materials (ASTM) will play a vital role in the implementation of CIKS.

The established goals for CIKS include developing:
- universal electronic access to distributed knowledge,
  information and data;
- application systems (prototype and production) that use
  the data, information, and knowledge to aid in (1) construction
  materials design, processing, selection and testing, and (2)
  facility design, construction (or installation and application),
  operation, maintenance, and repair;
- an open test bed at NIST for industry, academia, and government
  partners to build prototype systems , evaluate existing
  production systems, and to test enabling information technologies;
  and
- the implementation of commercial-scale systems, developed,
  deployed, and maintained by industry.

To meet these goals, new methods must be implemented that will bring about more coherent and reusable systems. This must start with the establishment of a consistent terminology for defining terms, such as those used by the coating industry when specifying products and equipment. Once a terminology base is established, process and data models must be developed. Process models or activity models, as they are often called, describe how knowledge is used. Data models show the data sources and data types for a process or activity model. These two types of models fall into a category described in the CIKS Meta Model as Data Management. Two activities and areas that involve the investigation and development of information models are Ontologies (process and data models), and the ISO, Standard for Product Data Exchange (STEP) [4,5]. Efforts to organize and communicate information using Ontologies are being addressed by the artificial intelligence community [6]. An example of this work can be found at the Stanford University, Knowledge Systems Laboratory in the development of the Ontolingua server [7]. The Ontolingua server allows collaborative authoring of ontologies via the Internet W^3 and may be useful in the development of CIKS.

The STEP effort has focused primarily on defining protocols for the exchange of computer aided design information (e.g., drawings and product specifications). However, to date, no significant work has involved construction materials knowledge. The use of Ontologies and STEP are being investigated to assess their application within the CIKS meta model.

For CIKS to achieve its goals within the projected 5-year timeframe, new collaborative agreements and partnerships with industry will be required. The test bed that has been established at NIST will play an important role. Opportunities will be provided for industry and organizations to test existing and prototype systems. This will involve on-site guest researcher appointments as well as remote access to the test bed via the Internet. A variety of computing platforms and information technologies will be involved in the testing. The capabilities currently operational in the test bed include multi-platform computer systems such Unix© and Microsoft NT^1 operating systems. Database management systems for designing, storing, and retrieving database information via the W^3 is operational. Information standards, such as Remote Data Access (RDA) [8] and ANSI SQL database standard [9] were used in the development of the computerized systems. Specific examples include the implementation of a paint proficiency sample database administered by the American Association of State Highway and Transportation Officials (AASHTO) Materials Reference Laboratory (AMRL). Members of the AMRL staff are currently extending the capabilities of the computerized system by developing new W^3 applications through the use the CIKS test bed. The new on-line capabilities will enhance the representation and communication of information involving proficiency sample tests for the AASHTO State Department of Transportation member agencies by providing more timely data acquisition and retrieval.

Another prototype CIKS has been developed for high-performance concrete and is described in a report published by NIST [10]. An example of an implementation of a CIKS high-performance concrete system was described in a recent article in Concrete International [11]. The system uses an Internet W^3 as an interface, and multiple forms of knowledge, such as computer-based models, material property databases, and bibliographic information. It is designed to predict the service life of chloride-exposed steel-reinforced concrete. Additional databases containing coating product data and asphalt materials are being considered for development at this time. Future applications will involve the use of intelligent

---

[1] The mention of commercial products and services in this paper does not constitute an endorsement of the National Institute of Standards and Technology. Their purpose is to show clarity and provide examples.

© is a registered trademark licensed exclusively by Xopen Company Limited.

agents, decision support systems, and a common user interface strategy for distributed systems. Many of these efforts will be carried out through collaboration with the CIKS Material Working Groups.

The CIKS network will potentially serve the whole range of stakeholders within the construction community, and the benefits to the construction industry should be considerable. The increased access to knowledge, coupled with improved materials and facility design, construction, and operation and maintenance should help reduce project delivery time, increase the service life of constructed facilities, and reduce maintenance and repair costs. Moreover, CIKS should help reduce cost increases/overruns and lost time caused by change orders, rework, and re-engineering. It will have a far-reaching impact on activities such as the development and issuance of standards and guides to the establishment of criteria for evaluating data. Perhaps most importantly, CIKS will provide an increased access to education and training and should raise the overall skill level of the construction industry's workforce and create a market-pull for innovative materials, technologies, and practices.

Knowledge Applications for the Coating Industry

Examples of Coating Industry Knowledge-Based Systems. Computerized Knowledge-Based Systems (KBS) are widespread among the coating industry. Examples of systems developed for construction industry application are identified in a selected bibliography later in this paper. Many of these systems are prototypes and were developed when information technologies (e.g. expert system shells, neural networks) were in their infancy. Features lacking in these systems include the ability to integrate with the business process and the ability to operate with computerized systems across different disciplines (e.g. designers, maintenance and operation staff) and processes (e.g., material and product selection, and cost analysis).

Many KBS are developed for in-house use. Examples include databases on coating products and services, research data, cost estimating, and decision-support systems. Many of these systems are rarely distributed to others, in formal or commercial formats. Reasons include their proprietary nature, lack of an incentive for distribution (marketability), and costs related to development and maintenance. In fact, many systems are developed and maintained by personnel responsible for developing the knowledge and their existence may be unrecognized by their colleagues.

Computerized systems designed for distribution, such as commercial products, require more effort to develop and maintain. These tend to be more formal systems where marketing strategies, distribution mechanisms, and customer (user) assistance are essential for a successful product and to return the development investment. Often knowledge content is the driving force for these systems. The Journal of Protective Coatings & Linings, "JPCL Archives II" is an example of such a product. Future versions of this product could be made available on-line via the

Internet, while still providing income to the publisher/distributor. Solutions currently being developed to allow electronic commerce and intellectual property distribution on the Internet will accelerate this method.

The Internet W^3 has altered thinking on how an organization's knowledge is kept. For example, through the use of an Intranet (an internal W^3), access to knowledge can be provided to personnel within an organization (e.g. marketing, testing, and research departments). Use of an Intranet results in benefits such as shorter development times, lower costs, and improved computer security. Although the Intranet is useful within organizations, use of the Internet will benefit organizations as new capabilities such as electronic commerce mature and increased network traffic capacity increases. To remain competitive, Internet-based information is no longer a convenience, but a necessity.

Perhaps the most significant constraint affecting the distribution of organizational and commercial knowledge-based systems is the lack of interoperability. Quite simply, interoperability means the ability to use knowledge and software among computerized systems. Incompatible data formats and computer hardware and software, incomplete or subjective data, inconsistent terminology, and the lack of electronic access are examples of specific factors that prohibit widespread use. Solving these problems will require collaboration by industry and government. Agreement must be reached on a common terminology, standard knowledge formats, criteria for establishing data quality, and common computer interfaces that provide seamless integration of knowledge to the user. The CIKS activity will address these issues.

The Society for Protective Coatings/Computer-Integrated Knowledge System (SSPC/CIKS) Joint Coating Working Group

One group that is currently addressing coating industry knowledge issues is the SSPC/CIKS Joint Coating Working Group. The Working Group comprises members from the SSPC Committee C.4.10 on "Knowledge-Base Systems for Coatings," and a CIKS Coating Industry Working Group formed during the CIKS June 1996 workshop. SSPC's Executive Director, Bernard Appleman chairs the group. Members of the group include public and private sector organizations representing coating formulators, consultants and engineers, facility owners, and researchers. Figure 3 shows a diagram of the group's interaction with the CIKS test bed and its role in the development of CIKS.

Collaboration with other industry and standards setting committees such as ASTM D-1 on "Paint and Related Coatings, Materials and Applications " and Committee E-49 on "Computerization of Material and Chemical Property Data" will be necessary. These collaborations will result in consistent terminology and standards for identifying, representing, and sharing coating material knowledge.

Specific focus areas of the SSPC/CIKS Joint Coating Working Group include:

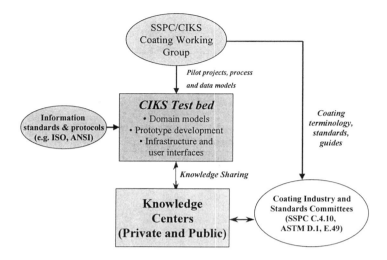

Figure 3: SSPC/CIKS Coating Working Group role in CIKS development.

- developing guides to assist coating industry knowledge users in using and integrating the Internet in business practice;
- improving the communication of computer-based information such as messages, computer-stored files, and access to application systems;
- developing standard formats for representing and exchanging coating product data;
- developing state-of-the-art reports describing current practice and enabling information technologies that have application within the coating industry.

Future projects requiring longer lead-times (2-years) to implement include the development of case-based reasoning (decision-making based on documented observations of coating performance), data dictionaries, and expert systems. Products from the Working Group will be disseminated in the form of SSPC Technology Updates, Guides, and the SSPC Coating Knowledge Center. The draft "Guide for the Identification and Use of Industrial Coating Material in Computerized Product Databases" exemplifies the Working Group's effort to develop standard guides and methods for representing coating material knowledge. Table 1 shows examples of the data elements proposed in the guide.

The draft document is being proposed as a SSPC Guide and would be used by coating manufactures, specifiers, and users (facility owners) for the communicating of coating product data. Figure 4 represents a diagram of the use of coating product data using the W^3. Product data sheets are now used to communicate this information. However, variations exist among manufacturers

Data segment	Example Data elements
Product description	Product name, product identification, generic type, system component
Intended use	Common application, substrate type, exposure environment, compatible undercoat and topcoat
Physical properties	Volume solids, solids by weight, mixed density, test methods
Mixing and application	Minimum and maximum dry film application thickness, theoretical coverage per volume, dew point, induction time, pot life
Key performance parameters	Corrosion resistance, weathering, abrasion resistance, test methods
Safety	NFPA health hazard, flammability
Manufacturer supplemental information	Manufacturer comments

Table 1: Example of product data segments and data elements.

when describing product data content, such as terms used and the type of data reported. As more companies use computerized systems such as the W^3 to disseminate information, standards such as the proposed coating product data guide will provide improved understanding (through consistent terminology and data elements). The ability to integrate product data among diverse computerized systems, such as company and facility owner project databases will be realized. Another benefit of the guide will be an increased understanding of coating material performance and data quality through the reporting of data elements such as those described in the "Key Performance Parameters" data segment. These parameters will be substantiated through the identification of test methods used to develop parameter values.

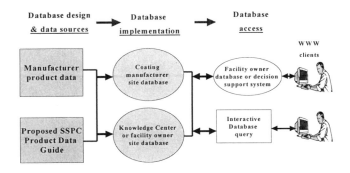

Figure 4: Diagram showing the generation and use of coating product data.

Current NIST Effort to Establish the CIKS Test Bed

NIST as an organization, is seeking to improve the delivery of its research results. The CIKS test bed will be a useful tool in providing construction material knowledge. The previously mentioned AMRL Paint Proficiency Sample Program is an example of the use of the test bed to provide access to technical data. Technical databases and decision-support systems are methods that have received the greatest attention thus far. Material property databases for cement and coatings can now be accessed through the W^3. The Uniform Resource Locator (URL) address for accessing the NIST technical databases is: http://ciks.nist.gov

Two decision-support systems have been developed by NIST during the past several years. The first of the two systems is the Highway Concrete Expert System (HWYCON). This system is designed to assist highway engineers in the diagnosis, selection, and repair and rehabilitation of highway concrete structures. It includes knowledge related to concrete pavements, bridges, and support structures. Several universities are also using the system as part of their material science and civil engineering curriculum. The Transportation Research Board sells HWYCON. The computerized system requires the Microsoft operating system, Windows version 3 or Windows '95. The distribution package contains a set of floppy diskettes and a report describing the design, installation, and operation of the system [12,13]. The second system called the Coating Expert Advisory System-I (COEX-I) [14] contains coating material knowledge and is designed to assist in the analysis of coating failures and selection of coating systems for stationary military structures. An overview of the COEX-I is described later.

Technical Databases

Technical databases contain many different types of data (see Figure 5). Examples include; product databases that describe material properties and manufacturer data, laboratory performance measurements, and outdoor exposure test results. It was stated earlier that significant differences occur among databases due to designer/developer preferences. These differences take the form of; inconsistent field names and contents; choice of computer hardware; and software that does not allow interoperability. The proposed SSPC Guide on coating product data formats is only the first step in providing compatible databases that can be used among computerized systems. To realize the full benefit of distributed database exchange, standard methods must be used to implement and disseminate the data. The steps in developing interoperable distributed databases include:

- establishing consistent database formats and terminology;
- establishing a logical schema to represent the physical data in the database;
- implementing of the database (acquire, computerize) using a database management system;
- providing electronic access through media distribution or electronically, via the Internet W^3.

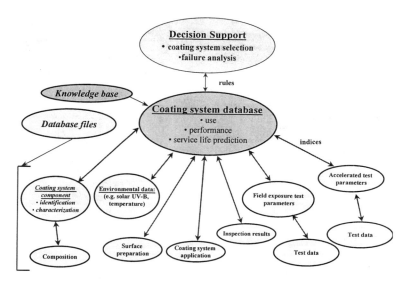

Figure 5: Examples of technical coating databases.

Significant operational enhancements can be achieved through electronic publication of databases using the W³. Interfacing database standards such as SQL (Structured Query Language) and the W³ client interface (e.g. WEB browser) reduces the need to develop multiple user interfaces for different computer platforms and can significantly reduce software development and maintenance costs. Access to databases can be provided in a more timely, and convenient manner. Figure 6 contains a diagram of the components of a distributed technical database system designed for W³ access. This model can be applied to virtually any type of database. The client (user) is provided the functionality of submitting queries (questions) to the database in an interactive mode. Typically, this is done using a W³ browser program. The process of converting input from the client involves converting the query into a SQL statement. In the instance shown in Figure 7, this is accomplished using the ISO Standard 9579 [8], "Remote Data Access" (RDA). This standard is a generic model providing database access and has been implemented at NIST for the purpose of interfacing W³ clients to SQL databases. The RDA standard was implemented using the C Programming Language. After receiving the SQL statement, the database management system retrieves the data from the physical database and produces a table containing the data elements (fields) and their values. This information is returned to the RDA component that formats the information for output in the Hyptertext Markup Language (HTML) and displays the information using the W³ client's browser. Added capabilities have been developed to also produce graphical plots. An interface has been developed to the NIST Dataplot Statistical and Graphical Analysis system [15]. This permits the graphical display of database information, interactively. Since the construction industry is comprised of companies with varying degrees of personnel and funding resources. It is necessary

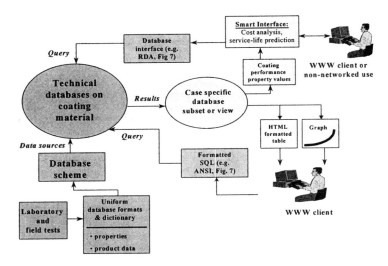

Figure 6: W³ implementation for technical databases.

to test knowledge system development using multiple platforms. Figures 7 and 8 show two methods that are used in the CIKS test bed to implement distributed technical databases using ISO, ANSI and de-facto standards for database storage, query, and retrieval. The use of newer, more flexible de-facto standards in the form of Microsoft Internet Information Server and the "Access" database management system, created opportunity for less costly hardware and software resources. For example, hardware and software costs for the resources shown in Figure 7 range from $50K to $75K dollars. Hardware and software costs for the resources shown in Figure 8 range from $5K to $10K dollars.

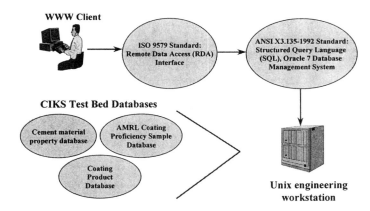

Figure 7: W³ database implementation, using an engineering workstation platform, ISO, and ANSI standards.

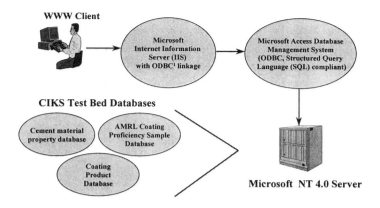

WWW Client

Microsoft Internet Information Server (IIS) with ODBC[1] linkage

Microsoft Access Database Management System (ODBC, Structured Query Language (SQL) compliant)

CIKS Test Bed Databases

Cement material property database

AMRL Coating Proficiency Sample Database

Coating Product Database

Microsoft NT 4.0 Server

[1] Object Database Connectivity

Figure 8: W³ database implementation, using microcomputer platform and de-facto standards.

Decision-Support Systems

For the purpose of this article, decision-support systems or as they are sometimes called, expert systems, are defined as follows:

"Computerized systems that can contain virtually any type of coating data and knowledge, such as technical databases, photographs, expert guidance, video and sound, computer-based models, and a logic module to operate (direct the logical instruction sequence) on the knowledge and provide an interface to the user."

Many attempts to develop decision-support systems have occurred in technical areas during the past 15 to 20 years. Examples of decision-support systems developed for construction materials users can be found in the bibliography at the end of this article. There are relatively few commercial decision-support systems available today. The most successful are operational within organizations that are committed to develop, maintain, and operate them within the organization. Historically, complex systems that cover a wide-interest area and involve high-level expert knowledge are costly to develop and maintain. However, advances in decision-support system development tools during the mid-1980's provided more cost effective development tools. These tools use the object-oriented programming [16] architecture and advanced techniques for the representation and use of knowledge, such as video, sound, and hypertext links. The result is a significant reduction in development time. Additional benefits in using object-oriented development tools include:

- ability to reuse data, knowledge, and procedures;
- relationships can be established among data and knowledge (inheritance);

- efficient graphical user interfaces are included in the tool;
- improved interfaces to external knowledge and programming modules.

COEX-I: An Object-Oriented Decision-Support System. One benefit of implementing decision-support systems is their ability to provide a systematic approach to problem-solving and knowledge dissemination. By incorporating the knowledge of expert(s) or specialists in coating materials and practices, improved levels of decision-making can result. For example, experts residing in a central location can extend or replicate their knowledge to field staff who need to evaluate the condition of coated structures, and perform repair and maintenance duties. Guidance for these individuals is typically found in printed form represented in manuals, guides, and standards. Computerizing the knowledge to include photographs, sound and video, and the guidance on the use of the guide significantly enhances knowledge understanding, resulting in cost savings and improved facility performance.

The COEX-I expert system was developed by a team of coating experts who had previously written the Military Handbook, "Handbook for Paints and Protective Coatings for Facilities" [17]. The group includes representatives from Department of Defense facilities who are involved in maintaining coated facilities. The group decided to develop a prototype decision-support system to assist military staff in analyzing coating failures that occur on stationary military structures such as water towers, buildings, and bridges. Section 11 of the guide covers the Analysis of Paint Failures and includes a decision tree that is designed to assist the user of the handbook. From this decision tree, rules (logic) were computerized in the form of question-and-answers. The rules provide a hierarchical structure to the knowledge and guide the user in problem solving, from the identification of visual observations found on the structure to recommendations given by the system. Recommendations include the identification of the coating failure, its cause, and guidance on remedial action(s). An additional capability was added to allow the user to specify criteria for coating system selection where total replacement is necessary for structural steel that shows blistering to the substrate. Figure 9 shows a diagram of the COEX-I system.

COEX-I is a prototype decision-support system that is being distributed for review and comment. Although the system is based on military structures, parts of the knowledge base apply to structural steel that is present in highway bridges. Examples include corrosion failures and videos contained in the system that provide guidance and inspection procedures for blistering and corrosion causes. Development tools used in the development of COEX-I are being applied to a new system designed to assist Federal Highway Administration (FHWA) and State Department of Transportation Engineers in the selection of coating systems for highway steel bridges. Knowledge contained in the system was developed through

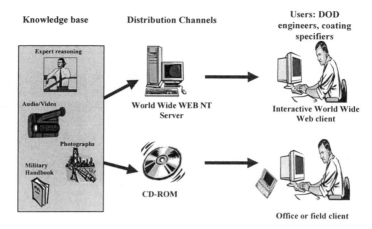

Knowledge base   Distribution Channels   Users: DOD engineers, coating specifiers

Expert reasoning

Audio/Video

World Wide WEB NT Server   Interactive World Wide Web client

Photographs

Military Handbook

CD-ROM

Office or field client

Figure 9: COEX-I system diagram.

various FHWA projects during the past decade. FHWA and NIST are developing the system, jointly, under FHWA, Turner-Fairbank Research Center sponsorship. It will be operational in 1998.

Summary

This article has presented an overview of an approach to solving industry-wide information needs for the coating industry, and through examples, has described several knowledge-based activities and applicable standards. The extreme diversity in user needs and variations in problem solving using computerized systems dictate the need for collaboration in the development of CIKS. These collaborations must occur between private sector companies, academia, and government agencies, and should result in a sound architecture that will enable a much greater degree of information sharing (interoperability), improved decision-making, and improved data quality. The CIKS test bed will provide opportunities to test new paradigms and prototype systems, information standards, and existing knowledge-based systems. A set of common terms and formats will permit seamless user interfaces that will result in shorter development times and cost savings for coating material users. Increased understanding of the performance of coating material will be realized through the identification of coating material properties and increased data quality. A realization of the SSPC Coating Knowledge Center will be accelerated.

References

[1] Clifton, J.R. and Sunder, S.S., "A Partnership for a National Computer-Integrated Knowledge Systems Network for High-Performance Construction Materials and Systems: Workshop Report," NISTIR 6003, National Institute of Standards and Technology, Gaithersburg, MD 1997.

[2] "Materials for Tomorrow's Infrastructure: A Ten-Year Plan for Deploying High-Performance Construction Materials and Systems," Civil Engineering Research Foundation, Washington, DC, 1994, Report #94-5011.

[3] Kurihara, T.Y. and Kaetzel, L.J., "Computer Integrated Knowledge System (CIKS) for Construction Materials, Components, and Systems: Proposed Framework, National Institute of Standards and Technology, Gaithersburg, MD, September 1997, NIST Internal Report 6071.

[4] W. Danner, "Standard for the Exchange of Product Model Data Development Methods: Specification of Semantics for Information Sharing," , National Institute of Standards and Technology, Gaithersburg, MD, September 1992, NIST Internal Report 4915.

[5] Carpenter, J. and Rumble, J., "STEP for Materials," in ASTM Standardization News, American Society for Testing and Materials, Conshohocken, PA, April 1997, Volume 25, Number 4, pp. 26-30.

[6] Falasconi, S., Lanzola, G. and Stefanelli, M., "Using Ontologies in Multi-Agent Systems," in Tenth Knowledge Acquisition for Knowledge-Based Systems Workshop (KAW '96), University of Pavia, Pavia, Italy, November 1996.

[7] Farquhar, A., Fikes, R. and Rice, J., "The Ontolingua Server: A Tool for Collaborative Ontology Construction Knowledge Systems Laboratory, "Stanford University, Berkeley, CA September 1996.

[8] Brady, K. and Sullivan, J., "User's Guide for RDA/SQL Validation Tests," , National Institute of Standards and Technology, Gaithersburg, MD, December 1996, NIST Internal Report 5725.

[9] "Database Language SQL," Federal Information Processing Standard (FIPS) 127-2, National Institute of Standards and Technology, Gaithersburg, MD, June 1993.

[10] Clifton, J.R., Bentz, D.P., and Kaetzel, L.J., "Computerized Integrated Knowledge Based System for High-Performance Concrete: An Overview,", National Institute of Standards and Technology, Gaithersburg, MD, February 1997, NIST Internal Report 5947.

[11] Bentz, D.P., Clifton, J.R., and Snyder, K.A., "Predicting Service Life of Chloride-Exposed Steel-Reinforced Concrete," in Concrete International, American Concrete Institute, Farmington Hills, MI, December 1996,Volume 18, No. 12, pp. 42-47.

[12] Kaetzel, L.J., Clifton, J.R., and Snyder, K.A., "Users Guide to the Highway Concrete (HWYCON) Expert System," Strategic Highway Research Program, National Research Council, Washington, DC, 1994, SHRP-C-406.

[13] Kaetzel, L.J., Clifton, J.R., Kleiger, P., and Snyder, K.A., "Highway Concrete (HWYCON) Expert System User Reference and Enhancement Guide," National Institute of Standards and Technology, Gaithersburg, MD, May 1993, NIST Internal Report 5184.

[14] "COEX-I Expert System for Steel Protection," in Building and Fire Research Laboratory, Research Update, National Institute of Standards and Technology, Gaithersburg, MD, 1996, Volume 2, Number 3.

[15] Filliben, J., "Dataplot Introduction and Overview", National Institute of Standards and Technology, Gaithersburg, MD, June 1984, NBS Special Publication 667.

[16] Henderson, P., "Object-Oriented Specification and Design with C++," McGraw-Hill Book Company, New York, NY, 1993.

[17] "Military Handbook: Handbook for Paints and Protective Coatings for Facilities," U.S. Department of Defense, Naval Facilities Engineering Command, Lester, PA, January 1995, MIL-HDBK-1110/1.

Selected Bibliography of Knowledge-Based Systems

H. Adeli, (Ed.), "Expert Systems in Construction and Structural Engineering," Chapman and Hall, New York, NY, 1988.

"Interactive Education: Transitioning CD-ROMS to the Web," in Computer Networks and ISDN Systems, Elsevier Science, 1994, No. 27, pp. 267-272.

"Chemistry Sites Proliferate on the Internet's World Wide Web," in Construction and Engineering News, November 1995, pp. 35-46.

"Developing Expert Systems," Federal Highway Administration, U.S. Department of Transportation, McLean, VA, December 1988, Publication No. FHWA-TS-88-022.

Cohn, L.F. and Harris, R.A., "Knowledge Based Expert Systems in Transportation: A Synthesis of Highway Practice," Transportation Research Board, Washington, DC, September 1992, National Cooperative Highway Research Program Synthesis 183.

Clifton, J.R. and Kaetzel, L.J., "Expert Systems in Concrete Construction," in Concrete International, American Concrete Institute, Farmington Hills, MI, November 1988.

Kuo, S.S., Clark, D.A., and Kerr, R., "Complete Package for Computer Automated Bridge Inspection Process," Transportation Research Board, Washington, DC, 1988.

"Knowledge Acquisition for Expert Systems in Construction," U.S. Army Corp's of Engineers, Construction Engineering Research Laboratory, Champaign, IL, September 1988, USACERL Special Report M-89/10.

Kaetzel, L.J. and McKnight, M.E., "Enhancing Coatings Diagnostics, Selection, and Use Through Computer Based Knowledge Systems," Proceedings of SSPC '95, The Society for Protective Coatings, Pittsburgh, PA, November 1995, SSPC 95-09, pp. 287-295.

Boocock, S.K. and Kaetzel, L.J., "Coating Industry Knowledge Base Systems: An Introduction to the SSPC Knowledge Center and the NIST Computer Integrated Knowledge Systems Network," The Society for Protective Coatings, Pittsburgh, PA, November 1996, Proceedings of SSPC '96, pp. 144-149.

Marshall, Jr., O.S., "Development of the PAINTER Engineered Management System," U.S. Army Corp's of Engineers, Construction Engineering Research Laboratory, Construction Engineering Research Laboratory, Champaign, IL, January 1994, USACERL Special Report FM-94.

Kaetzel, L.J., Clifton, J.R., Kleiger, P., and Snyder, K.A., "Highway Concrete (HWYCON) Expert System User Reference and Enhancement Guide," National Institute of Standards and Technology, Gaithersburg, MD, May 1993, NIST Internal Report 5194.

Sharpe,R., Marksjo, B.S., Ho, F., and Holmes, J.D., "WINDLOADER: Wind Loads on Structures Advisor," CSIRO Division, Australia, July 1989, Building Construction Engineering, SP-012.

Kaetzel, L.J. and Clifton, J.R., "Expert/Knowledge-Based Systems for Cement and Concrete: State-of-the-Art Report," Strategic Highway Research Program, Transportation Research Board, Washington, DC, 1991, SHRP-C/UWP-91-527.

Pielert, J.H. and Kaetzel, L.J. Kaetzel, "Cement and Concrete Materials Databases and the Need for Quality Testing," in Materials Science of Concrete III, American Ceramic Society, Westerville, OH, 1992, pp. 337-358.

Kaetzel, L.J. and Clifton, J.R., "Expert Systems for Building Materials and Structures: Expert/Knowledge Based Systems for Materials in the Construction Industry," RILEM Journal of Materials and Structures, RILEM, 1995. Volume 28, pp. 160-174.

Arnold, C., Drisko, R., Griffith, J., Neal, J., Nguyen, T., and Yanez, J., "An Expert System Application for Paints and Coatings: Painting Advisor," Proceedings of SSPC '87, The Society for Protective Coatings, Pittsburgh, PA, 1987, SSPC 87-07, pp. 1-6.

Author Index

Subject Index

A

Accelerated laboratory
features, 3
source of service life data, 2
Accelerated life tests (ALT)
major limitations, 171
modeling relationship between service
life and environmental conditions,
166–167
modeling unit-to-unit or spatial
variability in environmental
conditions and use-rates, 167–169
models, 152
potential pitfalls, 164
relating results with field performance,
165–166
strategy for analyzing ALT data, 153
suggestions for planning and executing
experiments, 164–165
using prior information, 159–164
See also Microelectronics; Ultra-
accelerated natural sunlight exposure
testing
Accelerated testing
different methods of accelerating
reliability test, 150–151
example of product development
protocol, 27
using prior information, 159–164
See also Microelectronics; Ultra-
accelerated natural sunlight exposure
testing
Accelerated weathering
accelerated testing wars, 145–146
brief summary of testing methods (1906
to present), 133–134, 137
company with cutting edge methodology,
140
confidence level with QUV-A
Weathering Tester, 140
degradation by UV-A and UV-B lamps,
141, 142*f*, 143*f*
dissatisfaction with broad use of devices
with UV-B bulbs, 140–141
exploring UV-A testing device, 144–145
key terms, 131, 133
lack of correlation of UV-B leading to
UV-A testing devices, 141

misuse of device as marketing tools,
145–146
no single infallible predictor of outdoor
durability, 137, 140
percent gloss retention (Florida versus
Arizona retention), 132*f*
prediction of durability, 130–131
proven value of devices, 145
QUV-A and QUV-B fluorescent-
UV/condensation testers, 140
"real world" myth, 131
tools of coatings scientists, 146–147
Acceleration models
empirical acceleration, 151
physical acceleration, 151
Acceptable quality level (AQL),
maximum level of failure, 186
Accessing material property data. *See*
Material property data systems;
Materials information
Accidental damage, failure mode, 24
Accumulated damage, failure mode, 24
Acid etch defects
approach, 217
automotive top coat, 217*t*
comparison to conventional
measurement technique, 226, 227*f*
image acquisition condition, 217
objectives, 216–217
test procedure by VIEEW (Video Image
Enhanced Evaluation of Weathering),
216
VIEEW results, 222, 223*f*
See also Video Image Enhanced
Evaluation of Weathering (VIEEW)
system
Acid rain etching of coatings. *See*
Automotive paint etch studies
AC impedance spectroscopy
advances in corrosion protection
measurements, 14
See also Organic coatings
Action spectra
based on responses to monochromatic
exposure, 45
See also Spectral weighting functions
(SWF)
Aerosol containers

468

Tl